KV-254-294

NEURONAL AND COGNITIVE EFFECTS OF OESTROGENS

WITHDRAWN
FROM
UNIVERSITY OF PLYMOUTH
LIBRARY SERVICES

The Novartis Foundation is an international scientific and educational charity (UK Registered Charity No. 313574). Known until September 1997 as the Ciba Foundation, it was established in 1947 by the CIBA company of Basle, which merged with Sandoz in 1996, to form Novartis. The Foundation operates independently in London under English trust law. It was formally opened on 22 June 1949.

The Foundation promotes the study and general knowledge of science and in particular encourages international co-operation in scientific research. To this end, it organizes internationally acclaimed meetings (typically eight symposia and allied open meetings and 15–20 discussion meetings each year) and publishes eight books per year featuring the presented papers and discussions from the symposia. Although primarily an operational rather than a grant-making foundation, it awards bursaries to young scientists to attend the symposia and afterwards work with one of the other participants.

The Foundation's headquarters at 41 Portland Place, London W1N 4BN, provide library facilities, open to graduates in science and allied disciplines. Media relations are fostered by regular press conferences and by articles prepared by the Foundation's Science Writer in Residence. The Foundation offers accommodation and meeting facilities to visiting scientists and their societies.

Information on all Foundation activities can be found at http://www.novartisfound.org.uk

900431812 7

Novartis Foundation Symposium 230

NEURONAL AND COGNITIVE EFFECTS OF OESTROGENS

2000

JOHN WILEY & SONS, LTD

Chichester · New York · Weinheim · Brisbane · Singapore · Toronto

Copyright © Novartis Foundation 2000

Published in 2000 by John Wiley & Sons Ltd,
Baffins Lane, Chichester,
West Sussex PO19 1UD, England

National   01243 779777
International (+44) 1243 779777
e-mail (for orders and customer service enquiries): cs-books@wiley.co.uk
Visit our Home Page on http://www.wiley.co.uk
or http://www.wiley.com

All Rights Reserved. No part of this book may be reproduced, stored in a retrieval
system, or transmitted, in any form or by any means, electronic, mechanical, photocopying,
recording, scanning or otherwise, except under the terms of the Copyright, Designs and
Patents Act 1988 or under the terms of a licence issued by the Copyright Licensing Agency,
90 Tottenham Court Road, London, W1P 9HE, UK, without the permission in writing
of the publisher.

Other Wiley Editorial Offices

John Wiley & Sons, Inc., 605 Third Avenue,
New York, NY 10158-0012, USA

WILEY-VCH Verlag GmbH, Pappelallee 3,
D-69469 Weinheim, Germany

Jacaranda Wiley Ltd, 33 Park Road, Milton,
Queensland 4064, Australia

John Wiley & Sons (Asia) Pte Ltd, 2 Clementi Loop #02-01,
Jin Xing Distripark, Singapore 129809

John Wiley & Sons (Canada) Ltd, 22 Worcester Road,
Rexdale, Ontario M9W 1L1, Canada

Novartis Foundation Symposium 230
ix+282 pages, 47 figures, 12 tables

British Library Cataloguing in Publication Data

A catalogue record for this book is available from the British Library

ISBN 0 471 49203 5 ✓

Typeset in $10\frac{1}{2}$ on $12\frac{1}{2}$ pt Garamond by Dobbie Typesetting Limited, Tavistock, Devon.
Printed and bound in Great Britain by Biddles Ltd, Guildford and King's Lynn.
This book is printed on acid-free paper responsibly manufactured from sustainable forestry,
in which at least two trees are planted for each one used for paper production.

Contents

UNIVERSITY OF PLYMOUTH

Item No. 900 4318127

Date 10 AUG 2000 Z

Class No. 612·405 NEU

Contl. No. ✓

LIBRARY SERVICES

Participants

E.-E. Baulieu INSERM U488 and College de France, 80 Rue de General Leclerc, Blg Gregory Pincus, 94276 Le Kremlin-Bicetre, France

J. B. Becker Psychology Department, Reproductive Sciences Program and Neuroscience Program, The University of Michigan, 525 East University Avenue, Ann Arbor, MI 48109, USA

C. Behl Max-Planck-Institute of Psychiatry, Kraepelinstrasse 2-16, D-80804 Munich, Germany

C. L. Bethea Division of Reproductive Sciences, Oregon Regional Primate Research Center, Beaverton, OR 97006, USA

H. Fillit The Institute for the Study of Aging, 767 Fifth Avenue, Suite 4200, New York, NY 10153, USA

S. E. Gandy Department of Psychiatry, New York University, The Nathan S. Kline Institute for Psychiatric Research, 140 Old Orangeburg Road, Orangeburg, NY 10962-2210, USA

R. B. Gibbs University of Pittsburgh School of Pharmacy, 1004 Salk Hall, Pittsburgh, PA 15261, USA

J.-Å. Gustafsson Karolinska Institutet, Department of Medical Nutrition, Novum, S-141 86 Huddinge, Sweden

V. W. Henderson Department of Neurology, University of Southern California School of Medicine, 1420 San Pablo St (PMB-B105), Los Angeles, CA 90089, USA

A. E. Herbison Laboratory of Neuroendocrinology, The Babraham Institute, Cambridge CB2 4AT, UK

Y. Hurd   Department of Clinical Neuroscience, Psychiatry Section, Karolinska Hospital, S-171 76 Stockholm, Sweden

C. Kirschbaum*   University of Trier — FPP, Dietrichstrasse 10-11, 54290 Trier, Germany

P. J. Kushner   Metabolic Research Unit, Box 0540, University of California San Francisco, San Francisco, CA 94143-0540, USA

E. R. Levin   Division of Endocrinology, Veterans Affairs Medical Center, Long Beach, Long Beach, CA 90822 and Departments of Medicine and Pharmacology, University of California Irvine, Irvine, CA 92717, USA

V. N. Luine   Department of Psychology, Hunter College of City University of New York, New York, NY 10021, USA

B. McEwen (*chair*)   Howard and Margaret Milliken Hatch Laboratory of Neuroendocrinology, The Rockefeller University, Box 165, 1230 York Avenue, New York, NY 10021-6399, USA

D. D. Murphy   Department of Pathology, University of Pennsylvania, 3rd Floor Maloney, 3600 Spruce Street, Philadelphia, PA 19104, USA

D. Pfaff   Department of Neurobiology and Behavior, Box 275, The Rockefeller University, 1230 York Avenue, New York, NY 10021-6307, USA

S. M. Resnick   Gerontology Research Center, Box 03, National Institute on Aging, National Institutes of Health, 5600 Nathan Shock Drive, Baltimore, MD 21224-6825, USA

B. B. Sherwin   Department of Psychology & Department of Obstetrics and Gynecology, McGill University, 1205 Dr. Penfield Avenue, Montreal, Quebec, Canada H3A 1B1

J. W. Simpkins   Center for the Neurobiology of Aging and Department of Pharmacodynamics, University of Florida, Box 100487, Gainesville, FL 32610, USA

---

*Current address: Department of Psychology, University of Dusseldorf, Universitaetstr. 1, D-40225 Dusseldorf, Germany

S. S. Smith   Department of Neurobiology and Anatomy, MCP-Hahnemann University, EPPI, 3200 Henry Ave, Philadelphia, PA 19129, USA

C. D. Toran-Allerand   Departments of Anatomy & Cell Biology, and Neurology, and Centers for Neurobiology & Behavior, and Reproductive Sciences, Columbia University College of Physicians and Surgeons, 630 West 168th Street, New York, NY 10032, USA

C. S. Watson   Department of Human Biological Chemistry and Genetics, University of Texas Medical Branch, Galveston, TX 77555, USA

O. Wolf (*Novartis Foundation bursar*)   Neuroimaging Laboratory, Department of Psychiatry, NYU School of Medicine, Room THN 314, 550 First Avenue, New York, NY 10016, USA

C. S. Woolley   Department of Neurobiology and Physiology, Northwestern University, Evanston, IL 60208, USA

Chairman's introduction

Bruce McEwen

Howard and Margaret Milliken Hatch Laboratory of Neuroendocrinology, The Rockefeller University, Box 165, 1230 York Avenue, New York, NY 10021-6399, USA

The study of reproductive events — sexual behaviour and neuroendocrine function — has been the main topic in the study of the action of gonadal hormones in the brain. However, we have known for a long time that these hormones, particularly oestrogens, have other effects, including effects on motor coordination, cognitive function, the occurrence of epilepsy, pain mechanisms, dementia, depressive illness and the modulation of premenstrual syndrome (e.g. McEwen 1999). Nevertheless, the fields of hormone action and behavioural neuroscience have only recently come to focus on these non-reproductive effects of ovarian hormones, and this has been due to the enormous progress in neuroscience and molecular endocrinology.

The origins of this meeting go back several years to a meeting that Professor Etienne Baulieu organized and chaired in 1995, which summarized evidence that gonadal hormones have important effects outside of the traditional reproductive axis (Ciba Foundation 1995). To set the stage for this Novartis Foundation Symposium, I would like to summarize my own interpretation of the history of this field.

Initial evidence at the cellular level for oestrogen actions in the brain was based upon the uptake and retention of [³H]oestradiol, using the technique of steroid autoradiography. Figure 1 shows a map by Pfaff & Keiner (1973) of [³H]oestradiol localization in the female rat hypothalamus and amygdala, where many of the oestrogen-concentrating neurons are located. But in these maps based upon autoradiographs, one sees that in the hippocampus there are occasional dots that represent oestrogen-concentrating cells. These cells were largely ignored at the time, because there was much work to be done on the hypothalamus and the amygdala, but the oestrogen-concentrating cells in the hippocampus have since been mapped by steroid autoradiography (Loy et al 1988) and by immunocytochemistry with antibodies against the alpha form of the oestrogen receptor (DonCarlos et al 1991, Weiland et al 1997). A recent study by Orikasa and colleagues using non-isotopic *in situ* hybridization for α oestrogen receptor mRNA shows the pattern of labelling of interneurons in the rat hippocampus (Fig. 2).

1

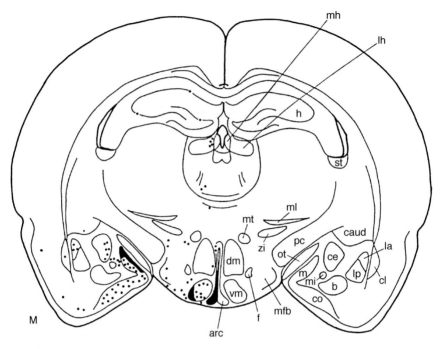

FIG. 1. Map of the female rat brain at the level of the dorsal hippocampus and ventromedial hypothalamus, showing presence of cells that retain [³H]oestradiol in hypothalamic area (arcuate nucleus and ventromedial nucleus) and amygdala. Note also that dots depict scattered [³H]oestradiol concentrating neurons in the dorsal hippocampus. From Pfaff & Keiner (1973) by permission.

Similarly, scattered oestrogen-concentrating cells in other parts of the nervous system, such as the midbrain raphe and brainstem, have turned out to be equally interesting in their own right, as will become evident in this volume. For example, an early study documented, by steroid autoradiography, the presence of [³H]oestradiol uptake in catecholaminergic neurons of the brainstem (Heritage et al 1980).

From this and subsequent work, it is possible to make the generalization that in major, widely projecting neural systems (noradrenergic, serotonergic, cholaminergic and dopaminergic) there are either demonstrated oestrogen binding sites or oestrogen effects that strongly indicate these systems are subject to the influence of ovarian hormones. But how do these effects come about, particularly if known oestrogen binding sites are scarce or even not detectable? We recognize now that there is more than one type of oestrogen receptor (ER). Besides the ERα, we know of the ERβ receptor in at least two forms, β1 and β2, and there is a suggestion that there may be other ER subtypes. There are also

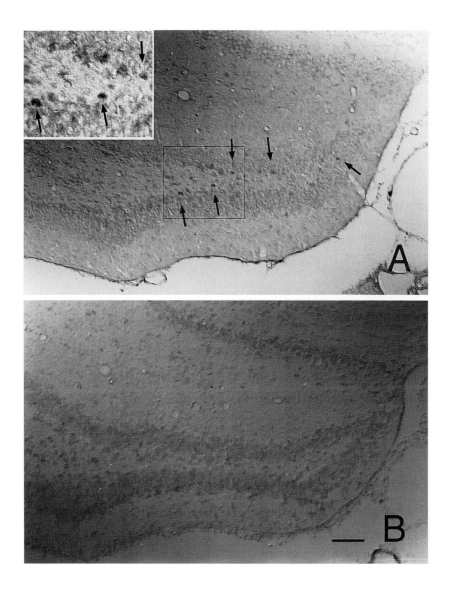

FIG. 2. *In situ* hybridization for ERα and β in the hippocampus of 14 day-old female rat. Positive signals for ERα are detectable in the hilus region of the dentate gyrus (indicated by arrows, in A) and stratum radiatum of Cornus Ammonis (not shown). In contrast, no signals for ERβ message (B). Insert in A is enlargement of the region indicated by a square.

multiple oestrogen response elements in genes, which allow the intracellular oestrogen receptors to regulate different pathways. The original steroid autoradiography using [³H]oestradiol was not able to distinguish between ERα and ERβ, as long as they were high affinity receptors. The use of selective antibodies and cDNA or cRNA probes has thus become very important in the discrimination between these two receptor types (for example, see Fig. 2).

Besides the intracellular oestrogen receptors, there is also evidence that there are other kinds of oestrogen receptors involved in neuroprotection in which the steroid specificity is different from the traditional receptor, being equally favourable to 17α-oestradiol as to 17β-oestradiol. It has been recognized for a long time that if oestrogen is applied either systemically or locally, some nerve cells respond very rapidly — so rapidly that it is hard to explain these effects as traditional genomic actions. In a 1977 paper, Kelly, Moss and Dudley made iontophoretic application of 17α-oestradiol hemisuccinate in the preoptic area of the thalamus, recording rapid responses of nerve cells that indicated that there are putative membrane receptors (Kelly et al 1977). In a more recent paper, the late Bob Moss and colleagues demonstrated the presence of oestrogen effects in ERα knock-out (ERKO) mice on the activity of ionic currents in hippocampal neurons triggered by the excitatory amino acid, kainic acid (Gu et al 1999). These rapid effects of oestrogens on kainic acid induced currents in the hippocampus were not blocked or mimicked by the oestrogen receptor antagonist, ICI 182 780 (Fig. 3). Because ICI 182 780 blocks or mimics oestrogen actions mediated by both ERα and ERβ, this set of findings reinforces the idea that there is a totally different mechanistic pathway involving a membrane oestrogen receptor.

The other reason for mentioning these papers is that many of you are aware that Bob Moss sadly died several months ago, awaiting a heart transplant. He was invited to this meeting and I would hope that we could dedicate the symposium to his memory. He was a pioneer and a leader in studying the non-genomic actions of oestrogens.

Other work on rapid oestrogen effects further emphasizes the existence of a unique oestrogen receptor mechanism associated with membrane events. The work on oestrogen effects on calcium ion currents in neurons of the neostriatum (Mermelstein et al 1996) will be discussed later in this volume by Jill Becker.

As to the mechanisms of events associated with non-genomic and cell surface actions of ovarian steroids, one of the first papers on this subject was from Etienne Baulieu's laboratory and demonstrated the direct inhibition by progestins of adenylate cyclase activity in oocytes (Finidori-Lepicard et al 1981). This study draws attention to the fact that second messenger systems are also subject to activation and inhibition by steroid hormones. More recently, Migliaccio et al (1993) showed that the MAP kinase system is a target of oestrogen action. In one particular case (Migliaccio et al 1996), the transfection of

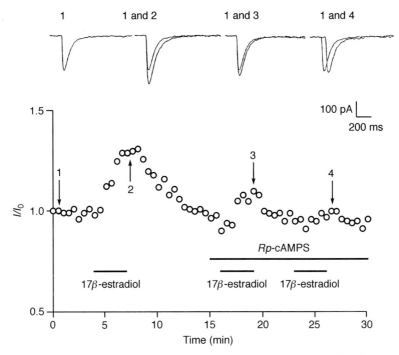

FIG. 3. Effect of Rp-cAMPs on 17β-oestradiol potentiation of kainate-induced currents in a neuron dissociated from the ERKO mouse. In the neuron responsive to 17β-oestradiol (50 nM, 3 min), application of Rp-cAMPs (50 μM) gradually blocked the action of 17β-oestradiol on kainate-induced currents. Representative current traces selected at specific time points (*arrows*) are displayed at the top of the figure. The current traces in 1 and 4 are artificially separated to allow visualization of the identical current amplitudes.

ERα into COS-7 cells resulted in the activation by oestradiol of the phosphorylation of a protein substrate, and this effect could be blocked with ICI 182 780, indicating that an intracellular ERα is involved in some form of intracellular signalling. This topic is discussed in this volume by Ellis Levin and Dominique Toran-Allerand.

Finally, oestrogen effects on cognition and protection from Alzheimer's disease are among the most striking and intriguing aspects of this emerging field of non-reproductive actions of ovarian hormones. As will be summarized by Barbara Sherwin in this volume, oestrogen treatment of women with low oestrogen levels improves verbal memory. In one striking example, oestrogen treatment reversed the impairment of verbal memory due to the suppression of ovarian hormones by gonadotropin-releasing hormone agonists (Sherwin & Tulandi 1996). In addition to these actions, oestrogens have been shown to have protective effects on the ageing nervous system, but the exact mechanism behind this is not clear. The maintenance of normal synaptic function is one part of this protection, but we also recognize that there are other potential mechanisms, such as the inhibition of the formation of the toxic amyloid β protein and the interaction of oestrogens with the generation or destruction of free radicals. These represent other potential pathways by which oestrogens will be neuroprotective, and much remains to be learned about the underlying cellular and molecular mechanisms that are involved. We will hear about this in this volume in the context of data from Alzheimer's disease.

This brief history of this field highlights what I think are some of the main issues in the study of the neural and cognitive effects of oestrogens. These are dealt with in much greater breadth and depth in the chapters and discussion which follow.

References

Ciba Foundation 1995 Non-reproductive actions of sex steroids. Wiley, Chichester (Ciba Found Symp 191)

DonCarlos LL, Monroy E, Morrell JI 1991 Distribution of estrogen receptor-immunoreactive cells in the forebrain of the female guinea pig. J Comp Neurol 305:591–612

Finidori-Lepicard J, Schorderet-Slatkine S, Hanoune J, Baulieu E 1981 Progesterone inhibits membrane bound adenylate cyclase in *Xenopus laevis* oocytes. Nature 292:255–257

Gu Q, Korach KS, Moss RL 1999 Rapid action of 17β-estradiol on kainate-induced currents in hippocampal neurons lacking intracellular estrogen receptors. Endocrinology 140:660–666

Heritage AS, Stumpt WE, Sar M, Grant LD 1980 Brainstem catecholamine neurons are target sites for sex steroid hormones. Science 207:1377–1380

Kelly M, Moss R, Dudley C 1977 The effects of microelectrophoretically applied estrogen, cortisol and acetylcholine on medial preoptic–septal unit activity throughout the estrous cycle of the female rat. Exp Brain Res 30:53–64

Loy R, Gerlach J, McEwen BS 1988 Autoradiographic localization of estradiol-binding neurons in rat hippocampal formation and entorhinal cortex. Brain Res 467:245–251

McEwen BS 1999 The molecular and neuroanatomical basis for estrogen effects in the central nervous system. J Clin Endocrinol Metab 84:1790–1797

Mermelstein PG, Becker JB, Sunneier DJ 1996 Estradiol reduces calcium currents in rat neostriatal neurons via a membrane receptor. J Neurosci 16:595–604

Migliaccio A, Pagano M, Auricchio F 1993 Immediate and transient stimulation of protein tyrosine phosphorylation by estradiol in MCF-7 cells. Oncogene 8:2183–2191

Migliaccio A, Di Domenico M, Castoria G et al 1996 Tyrosine kinase/p21[ras]/MAP-kinase pathway activation by estradiol–receptor complex in MCF-7 cells. EMBO J 15:1292–1300

Pfaff DW, Keiner M 1973 Atlas of estradiol-concentrating cells in the central nervous system of the female rat. J Comp Neurol 151:121–158

Sherwin BB, Tulandi T 1996 'Add-back' estrogen reverses cognitive deficits induced by a gonadotropin-releasing hormone agonist in women with leiomyomata uteri. J Clin Endocrinol Metab 81:2545–2549

Weiland NG, Orikasa C, Hayashi S, McEwen BS 1997 Distribution and hormone regulation of estrogen receptor immunoreactive cells in the hippocampus of male and female rats. J Comp Neurol 388:603–612

Mechanism of oestrogen signalling with particular reference to the role of ERβ in the central nervous system

Eckardt Treuter, Margaret Warner* and Jan-Åke Gustafsson*[1]

*Karolinska Institutet, Department of Biosciences at Novum, S-141 57 Huddinge, and *Karolinska Institutet, Department of Medical Nutrition, Novum, S-141 86 Huddinge, Sweden*

Abstract. The discovery of a second oestrogen receptor (ER), ERβ, has drastically changed our view of oestrogen action. Since the two ERs, ERα and ERβ, have somewhat different tissue and cellular distribution as well as ligand binding specificity the possibility exists that they have different biological roles. Indeed, several observations seem to indicate that they may even have opposite effects so that ERβ diminishes the activity of ERα. The CNS contains both ERα and ERβ and it is conceivable that they may have specific and individual roles in oestrogen signalling in the brain.

2000 Neuronal and cognitive effects of oestrogens. Wiley, Chichester (Novartis Foundation Symposium 230) p 7–19

Oestrogen signalling and oestrogen receptors

Oestrogens, including the natural endogenous oestrogen 17β-oestradiol, play important roles in mammalian reproduction, and they are responsible for the proper function of a variety of physiological processes. Oestrogens are able to promote the development of breast cancer and contribute to tumorigenesis in reproductive tissues. They can affect cardiovascular, skeletal, immune and nervous systems and are being implicated in a growing number of human disease states including heart disease, osteoporosis and Alzheimer's disease (for review see Kuiper et al 1998, Enmark & Gustafsson 1998). Oestrogen signalling is mediated by oestrogen receptors (ERs), ligand-inducible transcription factors belonging to the nuclear receptor (NR) superfamily (Mangelsdorf et al 1995). ERs are modular transcription factors containing an N-terminal constitutive transcriptional

---

[1]This chapter was presented at the symposium by Jan-Åke Gustafsson, to whom correspondence should be addressed.

activation domain (AF-1), a central DNA-binding domain, and a multifunctional C-terminal domain responsible for ligand binding, dimerization and ligand-regulated transcriptional activation (LBD/AF-2). ERs usually modulate transcription of their target genes by binding to promoter response elements.

In addition, cross-talk mechanisms may allow ERs to influence the activity of other transcription factors (see Paech et al 1997). ERs are unique steroid receptors because they exist as two related ER subtypes (paralogues=encoded from different genes), ERα and ERβ (Kuiper et al 1996). Both ERs display quite similar ligand-binding characteristics (Kuiper et al 1997), thus expanding the regulatory potential of oestrogens and anti-oestrogens in the sense that different cells may respond differently to the same stimulus, depending on the ER subtype composition. This is of concern considering that anti-oestrogens (e.g. tamoxifen) that modulate ER activity are currently being used to treat breast cancer and may account for negative side effects such as increased incidence of endometrial cancer. It is possible that the two ER subtypes differentially contribute to these effects, since differences have been reported with regard to their antagonist-dependent activation properties at AP-1 sites. These functional differences, together with their specific expression patterns in tissues, suggest that ERα and ERβ may play distinct roles in gene regulation.

ER activation via the ligand-inducible AF-2 domain

Similarly to other nuclear receptors, ERs have to communicate with chromatin and components of the transcriptional machinery (Mangelsdorf et al 1995). ERs possess two independent activation domains, termed AF-1 and AF-2. AF-1 is a ligand-independent activation region in the N-terminus of ERs and AF-2 is the ligand-dependent activation region in the C-terminus and constitutes an integral part of the multifunctional LBD. AF-1 shares structural and functional features with other eukaryotic activation domains, and although no specific cofactors have been identified yet, it may mediate phosphorylation-dependent activation (Tremblay et al 1999, and references therein). The majority of the cofactors function through the AF-2 domain which therefore plays a central role in transforming the ligand signal.

So how do ligands trigger the signal via AF-2? Although ER ligands bind exclusively to the LBD, it is important to note that only the binding of agonists triggers AF-2 activity, while the binding of antagonists does not. Recent structural analyses on ERα complexed with 17β-oestradiol or raloxifene/tamoxifen show that each ligand class induces a different LBD conformation (Brzozowski et al 1997, Shiau et al 1998). These structural rearrangements have been recognized to be critical for ligand-induced transcriptional activation by promoting the recruitment of coactivators. In particular, the precise positional rearrangement of helix 12 upon binding of agonistic ligands seems to be a prerequisite for coactivator binding. Conversely, antagonistic ligands are thought to induce a different 'autoinhibitory' conformation which prevents the association of coactivators. Indeed, recent functional studies and structures of various LBDs demonstrate that the nuclear receptor AF-2 constitutes the NR-box binding surface (Shiau et al 1998, Darimont et al 1998). AF-2 consists of a hydrophobic peptide-binding groove on the surface of the LBD formed by residues from helices 3 to 5 (static region, signature region) and 12 (flexible region, AF-2 core). Consistent with the functional conservation of ligand activation mediated by specified coactivators, these residues are highly conserved between all ligand-activatable receptors including the two ER subtypes.

Transcriptional cofactors for the AF-2 domain

During the past five years, protein–protein interaction screenings and biochemical approaches have led to the identification of a large number of putative AF-2 cofactors which may act as corepressors, coactivators or diverse coregulators at different functional levels (Freedman 1999).

Transcription intermediary factor TIF2 (P1P1)

One important class of AF-2 coactivators includes three related members of the p160 coactivator family (SRC-1, TIF2/GRIP1, AIB1/ACTR) and the two related coactivators CBP/p300 (for references see Leers et al 1998). The detection of histone acetyltransferase (HAT) activity in these coactivators functionally connects them to chromatin, since acetylation of lysine residues of histones is associated with chromatin derepression and is required for transcriptional activation *in vivo*. Since multiple interactions have been detected between different HAT coactivators, larger ER–HAT complexes may exist (Freedman 1999). Indication for a function of p160 coactivators in ER-mediated gene expression comes from the discovery of frequent AIB1 gene-amplification in ER-positive breast and ovarian cancer cells (Anzick et al 1997) as well as from recent SRC-1 knockout studies.

We have identified and characterized the p160 coactivator TIF2 (Leers et al 1998). Specifically, we have investigated mechanistic aspects of the interaction with the AF-2 domain and could identify small peptide interaction motifs within TIF2. These motifs with the consensus sequence LXXLL, also called NR-box or LCD/LXD, exist in most of the AF-2 binding proteins (for references see Leers et al 1998). The leucine core of the motif constitutes the primary docking site for cofactors to conserved AF-2 residues, providing the explanation why AF-2 constitutes the coactivator interaction surface (Shiau et al 1998).

Nuclear orphan receptor SHP and receptor-interacting protein RIP140

Our studies on two additional ER cofactors illustrate that not all AF-2 binding proteins act simply as coactivators. We have identified two AF-2 interacting proteins, RIP140 and SHP, which exhibit rather negative coregulatory functions, because they can antagonize p160 coactivators such as TIF2 *in vivo* and compete for binding to the AF-2 *in vitro* (Treuter et al 1998, Johansson et al 1999). We suggest that these cofactors form a unique category of coregulators that are distinct from conventional corepressors. SHP is itself an unusual orphan receptor with homology to the LBD but lacking a DNA-binding domain. Although originally suggested to dimerize with certain other receptors, our studies on the interactions of SHP with ERs (and other receptors) reveal that the original name ('short heterodimer partner') could be a misnonym: instead of interacting as a dimerization partner, we found that SHP binds directly to the AF-2 of ligand-activated ERs and thus functions in the same way as AF-2 cofactors (Johansson et al 1999). Furthermore, since SHP contains repression domains, it could potentially connect liganded ERs to the corepressor complex. This is of particular interest because the role of corepressors in ER signalling is yet unclear.

ERβ and the CNS

Women are at greater risk for Alzheimer's disease than men and oestrogen replacement can help to reduce the risk and severity of Alzheimer's-related dementia in postmenopausal women (Kawas et al 1997). Loss of ovarian function has long been known to result in reversible changes in mental function and behaviour (Gibbs 1998). At present there is a great need for understanding of the mechanisms through which ovarian hormones modulate neuronal function within the CNS. This is an extremely important clinical issue since oestrogen replacement therapy is associated with some severe health risks. The major unwanted side effects of hormone replacement therapy include endometrial cancer, gall bladder disease and breast cancer.

Before the discovery of the second oestrogen receptor, ERβ (Kuiper et al 1996), the idea of an oestrogen which would exert positive effects on the brain and bone but spare the uterus and breast was almost unimaginable. Now that we know that ERβ is highly expressed in the brain but not in the uterus (Laflamme et al 1998, Alves et al 1998), it is possible to think of ERβ-selective ligands which would have the desired tissue specificity. With the development of ERα (Cooke et al 1998) and ERβ (Krege et al 1998) knockout mice, called ERKO and BERKO, respectively, the tools are now available to determine the relative importance of ERα and ERβ in the CNS. Both oestrogen receptors appear to be involved in a multitude of regulatory events, the details of which will be worked out within the next few years.

Several mechanisms have been proposed to explain the beneficial actions of oestrogen in the CNS (Birge 1998, Rodriquez & Grossberg 1998). These include improvement in cerebral blood flow, reduction in oxidative stress, regulation of neurotransmitter synthesis, enhancement of neuronal sprouting and regulation of apolipoprotein E (ApoE). Some of these functions are already known to be regulated by ERβ, but the task of assigning functions to ERα and ERβ in regulation of the CNS is only just beginning.

Oestrogen receptors in the cardiovascular system

The cardiovascular effect of oestrogen is currently under intense investigation. Oestrogen decreases expression of adhesion molecules, lowers systemic blood pressure, promotes vasodilatation, decreases platelet aggregability, inhibits vascular smooth muscle cell proliferation, possesses potent antioxidant and calcium antagonist activities, inhibits adrenergic responses and down-regulates platelet and monocyte reactivity (Lüscher & Barton 1997).

ERβ is expressed in the vessel wall (Lindner et al 1998). An important piece of information was gained from the study of ERKO mice. This is that oestrogen can exert its protective effect on blood vessels in the absence of ERα (Rubanyi et al 1997). From this observation we conclude that ERβ is of major importance in the protective role of oestrogen in the cardiovascular system. Upon damage of vessels by removal of the endothelium, there is a tremendous increase in ERβ expression in the smooth muscle cell layer of the vessel wall and treatment with oestrogen inhibits proliferation of the smooth muscle cells (Lindner et al 1998). Proliferation of smooth muscle in the vessel wall is an early/important event in vascular pathology.

Another extremely important function of oestrogen receptors in the cardiovascular system is regulation of endothelial nitric oxide synthase (NOS) (Kauser & Rubanyi 1997). Oestrogen regulates the level of NOS in the cardiovascular system and oestrogen-induced increases in coronary blood flow are antagonized by inhibitors of NO synthesis (Lang et al 1997). Very preliminary data show that in ERKO mice NO production in the endothelium is reduced (Rubanyi et al 1997) but the effects of oestradiol on NOS in ERKO mice have not been studied. The levels and regulation of NOS in BERKO mice is a priority but has not yet been examined. Interestingly, the ovarian phenotype in BERKO mice (Krege et al 1998) is similar to that observed in NOS$^{-/-}$ mice (Jablonka-Shariff & Olson 1998). This indicates that ERβ may be an important regulator of NOS.

Oestrogen receptors in neurotransmitter systems

Oestrogen has a positive effect on several neurotransmitter systems that are assumed to be involved in the regulation of emotions, behaviour and cognition

(Yaffe et al 1998). It enhances synaptic sprouting and also induces synaptophysin, a presynaptic protein involved in neurotransmitter release (Plunkett et al 1998).

In the area of behaviour, oestrogen regulates aggression, sexual drive, impulsivity and hostility. Studies with ERKO mice (Lindzey et al 1998) revealed the importance of ERα in these brain functions. Experiments in rats have revealed that choline acetyltransferase (ChAT) decreases in neurons in the medial septum and nucleus basalis magnocellularis after ovariectomy relative to age-matched, gonadally intact controls. Short-term oestrogen replacement can partially restore ChAT in these brain regions (Gibbs 1998). Basal forebrain cholinergic neurons project to the hippocampus and cortex and loss of their function may contribute to the risk and severity of cognitive decline associated with ageing and Alzheimer's disease in postmenopausal women. The localization of ERβ in neurons in the forebrain (Laflamme et al 1998, Alves et al 1998) indicates that ERβ may be important in cognition.

Oestrogen and ApoE

Human ApoE plays a prominent role in cholesterol transport, plasma lipoprotein metabolism and in recovery of the brain after noxious stimuli or neuronal injury (Mahley et al 1996). ApoE polymorphism has been related to significant modifications of lipoprotein profile, as well as to the incidence of different pathologies including cardiovascular disease, Alzheimer's disease, and vascular dementia. One of the three common alleles of apoE, apoE4, is over-represented in Alzheimer's subjects compared with age- and sex-matched controls (Stone et al 1998). ApoE is found within the two characteristic neuropathologic lesions of Alzheimer's disease — extracellular neuritic plaques representing deposits of amyloid β peptide (Aβ) and intracellular neurofibrillary tangles representing filaments of a microtubule-associated protein called tau. The apoE4 protein is thought to be less effective in neuronal repair processes and interacts strongly with Aβ to form deposits or plaques in the brain. Oestrogen plays a role in regulation of ApoE levels (Srivastava et al 1997), but regulation of ApoE in ERKO and BERKO mice has not yet been examined.

Oestrogen receptors in oxidative stress

Oxidative stress-induced neuronal cell death has been implicated in different neurological disorders and neurodegenerative diseases such as Alzheimer's disease (Gordge 1998). Anti-oestrogens are potent antioxidants, protecting cells from damage by radicals and other toxic by-products of metabolic oxidation. Transcriptional regulation of the quinone reductase gene, involved in protection against oxidative stress, is mediated by an electrophilic/antioxidant response element (EpRE/ARE). In MCF-7 cells, anti-oestrogens elicited an increase in expression of EpRE/ARE reporter constructs and oestradiol inhibited this anti-oestrogen-dependent gene activity (Montano & Katzenellenbogen 1997). ERβ is more efficacious than ERα in transactivation of EpRE/ARE-containing reporter gene constructs (Montano & Katzenellenbogen 1997). These findings suggest that the protective antioxidant effects of anti-oestrogens are mediated by ERβ.

References

Alves SE, Lopez V, McEwen BS, Weiland NG 1998 Differential colocalization of estrogen receptor β (ERβ) with oxytocin and vasopressin in the paraventricular and supraoptic nuclei of the female rat brain: an immunocytochemical study. Proc Natl Acad Sci USA 95:3281–3286

Anzick SL, Kononen J, Walker RL et al 1997 AIB1, a steroid receptor coactivator amplified in breast and ovarian cancer. Science 277:965–968

Birge SJ 1998 Hormones and the aging brain. Geriatrics (suppl) 53:S28–S30

Brzozowski AM, Pike AC, Dauter Z et al 1997 Insights into the molecular basis of agonism and antagonism in the oestrogen receptor as revealed by complexes with 17β-oestradiol and raloxifene. Nature 389:753–758

Cooke PS, Buchanan DL, Lubahn DB, Cunha GR 1998 Mechanism of estrogen action: lessons from the estrogen receptor-α knockout mouse. Biol Reprod 59:470–475

Darimont BD, Wagner RL, Apriletti JW et al 1998 Structure and specificity of nuclear receptor–coactivator interactions. Genes Dev 12:3343–3356

Enmark E, Gustafsson J-Å 1998 Nyupptäckt östrogenreceptor. Läkartidningen 95:1945–1949

Freedman LP 1999 Increasing the complexity of coactivation in nuclear receptor signaling. Cell 97:5–8

Gibbs RB 1998 Impairment of basal forebrain cholinergic neurons associated with aging and long-term loss of ovarian function. Exp Neurol 151:289–302

Gordge MP 1998 How cytotoxic is nitric oxide? Exp Nephrol 6:12–16

Johansson L, Thomsen JS, Damdimopoulos AE, Spyou G, Gustafsson J-Å, Treuter E 1999 The orphan nuclear receptor SHP inhibits agonist-dependent transcriptional activity of estrogen receptors ERα and ERβ. J Biol Chem 274:345–353

Jablonka-Shariff A, Olson LM 1998 The role of nitric oxide in oocyte meiotic maturation and ovulation: meiotic abnormalities of endothelial nitric oxide synthase knock-out mouse oocytes. Endocrinology 139:2944–2954

Kauser K, Rubanyi GM 1997 Potential cellular signaling mechanisms mediating upregulation of endothelial nitric oxide production by estrogen. J Vasc Res 34:229–236

Kawas C, Resnick S, Morrison A et al 1997 A prospective study of estrogen replacement therapy and the risk of developing Alzheimer's disease: the Baltimore Longitudinal study of Aging. Neurology 48:1517–1521 (erratum: 1998 Neurology 51:654)

Krege JH, Hodgin JB, Couse JF et al 1998 Generation and reproductive phenotypes of mice lacking estrogen receptor β. Proc Natl Acad Sci USA 95:15677–15682

Kuiper GG, Enmark E, Pelto-Huikko M, Nilsson S, Gustafsson J-Å 1996 Cloning of a novel estrogen receptor expressed in rat prostate and ovary. Proc Natl Acad Sci USA 93:5925–5930

Kuiper GG, Carlsson B, Grandien K et al 1997 Comparison of the ligand binding specificity and transcript tissue distribution of estrogen receptors α and β. Endocrinology 138:863–870

Kuiper GG, Carlquist M, Gustafsson J-Å 1998 Estrogen is a male and female hormone. Sci Med 5:36–45

Laflamme N, Nappi RE, Drolet G, Labrie C, Rivest S 1998 Expression and neuropeptidergic characterization of estrogen receptors (ERα and ERβ) throughout the rat brain: anatomical evidence of distinct roles of each subtype. J Neurobiol 36:357–378

Lang U, Baker RS, Clark KE 1997 Estrogen-induced increases in coronary blood flow are antagonized by inhibitors of nitric oxide synthesis. Eur J Obstet Gynecol Reprod Biol 74:229–235

Leers J, Treuter E, Gustafsson J-Å 1998 Mechanistic principles in NR box-dependent interaction between nuclear hormone receptors and the coactivator TIF2. Mol Cell Biol 18:6001–6013

Lindner V, Kim SK, Karas RH, Kuiper GG, Gustafsson J-Å, Mendelsohn ME 1998 Increased expression of estrogen receptor-β mRNA in male blood vessels after vascular injury. Circ Res 83:224–229

Lindzey J, Wetsel WC, Couse JF, Stoker T, Cooper R, Korach KS 1998 Effects of castration and chronic steroid treatments on hypothalamic GnRH content and pituitary gonadotropins in male wild-type and estrogen receptor-alpha knockout mice. Endocrinology 139: 4092–4101

Lüscher TF, Barton M 1997 Biology of the endothelium. Clin Cardiol (suppl) 20:II3–II10

Mahley RW, Nathan BP, Pitas RE 1996 Apolipoprotein E. Structure, function, and possible roles in Alzheimer's disease. Ann NY Acad Sci 777:139–145

Mangelsdorf DJ, Thummel C, Beato M 1995 The nuclear receptor family: the second decade. Cell 83:835–839

Montano MM, Katzenellenbogen BS 1997 The quinone reductase gene: a unique estrogen receptor-regulated gene that is activated by antiestrogens. Proc Natl Acad Sci USA 94:2581–2586

Paech K, Webb P, Kuiper GG et al 1997 Differential ligand activation of estrogen receptors ERα and ERβ at AP1 sites. Science 277:1508–1510

Plunkett JA, Baccus SA, Bixby JL 1998 Differential regulation of synaptic vesicle protein genes by target and synaptic activity. J Neurosci 18:5832–5838

Rodriguez MM, Grossberg GT 1998 Estrogen as a psychotherapeutic agent. Clin Geriat Med 14:177–189

Rubanyi GM, Freay AD, Kauser K et al 1997 Vascular estrogen receptors and endothelium-derived nitric oxide production in the mouse aorta. Gender difference and effect of estrogen receptor gene disruption. J Clin Invest 99:2429–2437

Shiau AK, Barstad D, Loria PM et al 1998 The structural basis of estrogen receptor/coactivator recognition and the antagonism of this interaction by tamoxifen. Cell 95: 927–937

Srivastava RA, Srivastava N, Averna M et al 1997 Estrogen up-regulates apolipoprotein E (ApoE) gene expression by increasing ApoE mRNA in the translating pool via the estrogen receptor α-mediated pathway. J Biol Chem 272:33360–33366

Stone DJ, Rozovsky I, Morgan TE, Anderson CP, Finch CE 1998 Increased synaptic sprouting in response to estrogen via an apolipoprotein E-dependent mechanism: implications for Alzheimer's disease. J Neurosci 18:3180–3185

Tremblay A, Tremblay GB, Labrie F, Giguère V 1999 Ligand-independent recruitment of SRC-1 to estrogen receptor β through phosphorylation of activation function AF-1. Mol Cell 3:513–519

Treuter E, Albrektsen T, Johansson L, Leers J, Gustafsson J-Å 1998 A regulatory role for RIP140 in nuclear receptor activation. Mol Endocrinol 12:864–881

Yaffe K, Sawaya G, Lieberburg I, Grady D 1998 Estrogen therapy in postmenopausal women: effects on cognitive function and dementia. JAMA 279:688–695

DISCUSSION

Pfaff: In your experiments with neuroblastoma cells in which you transfected in one of the other oestrogen receptors (ERs), gave oestrogens and then saw sprouting, do you have any reason to suppose that the sprouting phenomena are following on from primary changes in the nerve cell body, where so many growth-related reactions take place (Cohen & Pfaff 1981, 1992, Cohen et al 1984, Chung et al 1984, Jones et al 1985, 1986, 1990)? Or would you hypothesize that the ERs are located on the dendrites and are having a direct effect there?

Gustafsson: We have only just begun to look at this. It seems that some of the molecular events that occur are specific for ERα or ERβ, which would indicate that dendrite sprouting is secondary to what goes on in the body of the cell.

Baulieu: Do you find ERβ in glial cells?

Gustafsson: Yes, there are examples where ERβ occurs in glial cells. We are now looking in meningioma cells—although these are not glial cells, they are non-neuronal cells. Meningioma cells seem to contain ERβ, which is consistent with the well-known finding that these cells are oestrogen sensitive. Indeed, oestrogen has been used as a treatment for meningiomas.

Baulieu: This is interesting, because many meningioma cells have abundant progesterone receptors (PRs) but no or very little ERα.

This leads me on to my second question. From the correlation you showed for breast cancer, PR induction is more a function of ERα activity than ERβ. Could you comment on this specificity of induction? And at the molecular level, what might be the mechanism differentiating ERβ and ERα in terms of induction of PR?

Gustafsson: This phenomenon seems to be tissue specific. In the brain, both ERα and ERβ appear to be involved in PR induction, as shown by the studies by Merchenthaler and colleagues, who obtained induction of PR even in ERα$^{-/-}$ mice (Shughrue et al 1997). In certain tissues I suspect that PR will be induced by ERα, in others it will be induced by ERβ, and in others perhaps both.

As to the mechanism, the obvious possibility is that the promoter region of the PR gene will have the potential to respond to both ERα and ERβ, but in different tissues access to the respective response element may be enhanced or inhibited by tissue-specific DNA binding factors.

Toran-Allerand: You described the ER distribution during development, and you mentioned that ERβ might be important for imprinting prenatally. It is important to remember that although there is a lot of ERβ in the adult cerebral cortex, during development there is also the clear presence of ERα, particularly in the cerebral cortex. ERα is responsive to oestradiol, is developmentally regulated and then becomes much more reduced in its cortical distribution with age. However, following injury to the cerebral cortex as occurs with cerebrovascular disease or experimental stroke, Phyllis Wise (Dubal et al 1999) and others have shown that there is an up-regulation of ERα. Perhaps cortical ERα has a very specific role during development with respect to cerebral cortical differentiation and then the levels fall dramatically. ERβ then may have other functions, but when the cortex is injured in the adult there is a re-expression of ERα and oestrogen again may have a therapeutic role, for example in a return to quasi-normal cerebral cortical functions.

I would also add that the cellular distribution of immunoreactive ERα by immunohistochemistry does not necessarily give an accurate picture. This is because in order to obtain immunoreactivity, one needs a significantly higher level of antigen present than the levels of ERα needed for a physiological effect. Is it possible that in some of the regions where there is no apparent ERα by immunoreactivity, one could make specific radiolabelled ligands in order to do autoradiography? One could then show whether or not ERα is present at low levels in regions where it is currently not thought to exist.

Gustafsson: I'm sure that you are right. However, I'd rather concentrate on areas where there is more ER. One should never trust only one technique. A reasonable requirement is that you should try to get both mRNA and protein. The problem may often be the opposite. There are many studies out there showing lots of mRNAs, particularly in terms of ERα, where there were many splice products and so on which for a while obscured the field. But then it was found that at the protein level, nothing was seen of these splice products. As far as I can understand the only functionally important ERα demonstrated so far is the wild-type ERα, whereas the situation does seem to be different for ERβ. ERβ2 is expressed at the protein level. The data so far are, however, preliminary.

Herbison: Azcoitia et al (1999) have recently published work describing ERβ expression in astrocytes.

As you pointed out at the beginning of your paper, there are many nuclear hormone receptors, including a large number of orphan receptors with no clear function. I have a suspicion that there may be more than two ERs. No doubt you have had the opportunity to cross your ERα and ERβ knockout mice. Is there any residual oestrogen binding in these animals?

Gustafsson: That is obviously an issue of great interest to us. We have just killed many of these double knockouts which we have been breeding for various experiments. They are extremely costly to get: only 1 in 16 of the offspring are the double knockouts. This week we are examining specific binding in various tissues, so soon I will be able to answer that question. It is tantalizing that when you look in a database for ER sequences, the available sequences can be separated into three groups. In addition to the α and β groups, there is a third group which mainly consists of fish ER. We have used these fish ER sequences for a long time to try to fish out an ERγ, so far without success. We are continuing, and we are also using proteomics to try to get at this issue. So far there is no third ER, but I would definitely not exclude that possibility.

Bethea: I have a question regarding genistein as an antagonist. You are speculating that ERβ is antiproliferative, and yet it is having an antiproliferative effect when it is bound to genistein, which you said was an antagonist. Does this mean that if ERβ is bound to an agonist that it would be proliferative?

Gustafsson: It is a question of concentrations. At a certain level you can have an agonist function, and at another level you can have an antagonist function, and at certain other levels you start to activate ERα. The level of phytooestrogens is extremely important. It is completely conceivable that if you take genistein at low levels there will only be an agonist-type activation of ERβ, which might well be beneficial. In the UK, some health food representatives have recommended that as an alternative to breast implants women should take mega quantities of phytooestrogens. Indeed, some women have reported (happily) that when they did this they got bigger breasts, which poses a certain risk to them because at those concentrations the phytooestrogens would affect ERα. The fact that ERα and ERβ counteract each other is consistent with the vessel and prostate data, and also with the *in vitro* data. The heterodimer assumes the characteristics of ERβ, and on an oestrogen response element ERβ is less efficient than ERα. There is one further, even more exciting possibility. ERβ2, when heterodimerizing with ERα, allegedly completely shuts down ERα. There you have an excellent antiproliferative agent. ERβ2 is a lousy binder of oestradiol: it binds it 100-fold less efficiently than ERβ1. But there might be ligands so far undiscovered that fit in the ligand-binding pocket of ERβ2. This is a fascinating possibility.

Toran-Allerand: At very high doses, genistein acts as a receptor tyrosine kinase inhibitor, and if one were to take huge amounts of it, one might find oneself in a lot of trouble.

Simpkins: In hippocampus, cortex or cerebellum, what percentage of neurons show positive binding for either ER?

Gustafsson: ERβ is more widely distributed than ERα. As a rough figure, I'd say that 20% of the total neurons would be ERα or ERβ positive.

Pfaff: The safest thing to say is that the numbers are low but not zero. In complicated systems, small numbers of neurons can make a big difference.

Kirschbaum: Is there any evidence for significant sex differences in the distribution of ERβ in the brain?

Gustafsson: That is an interesting question, but we simply have not had time to address this.

Kirschbaum: I would be interested in looking at the potential differences in the effectiveness of such drugs with respect to cognitive processing.

McEwen: There is a difference between the presence of the receptor and whether the receptor is programmed to have a particular effect. There are now plenty of examples where males have ERs but they don't turn on the same gene products or don't turn them on to the same degree.

Gibbs: As a preface to the upcoming sessions on oestrogen in dementia and hippocampal function, could you comment on the distributions of ERα and ERβ specifically in the hippocampus?

Gustafsson: I cannot say any more than ERβ does exist in the hippocampus and the cortex.

Pfaff: I would add that in the hippocampus it is exclusively within interneurons, and not in pyramidal cells.

Bethea: In the monkey we have just finished *in situ* hybridizations, and we see a very high concentration of the ERβ in the dentate gyrus, and lower concentrations in CA1, 2 and 3.

Woolley: Is that mRNA or protein?

Bethea: mRNA.

Woolley: So one of the questions is how the distribution of the mRNA differs from the distribution of the protein.

McEwen: The most extreme discrepancy exists with some of the original reports of ERβ mRNA in cerebellum and all over the hippocampus, and the original steroid autoradiography which didn't show very much binding in the cerebellum and only scattered cells in the hippocampus.

Pfaff: We have unpublished data that the presence of ER protein in a nerve cell can have an effect on cognition independent of the ligand. We are pursuing this with Jan-Åke Gustafsson.

Watson: Which epitopes on the molecule do your new antibodies which are much better at detecting ERβ recognize? Are they different from the ones that are commercially available? Since you have done a lot of immunocytochemistry, have you ever seen anything that looks like membrane staining for ERα or ERβ?

Gustafsson: We have a bunch of good antibodies, both polyclonal and monoclonal, that we have produced and collected over the years. It was extremely difficult to produce them. The best one we have is 503, and is against a hybrid protein, namely the human ER with an 18-amino acid insert (rodent sequence, as in ERβ2) in the LBD, expressed in baculovirus. We have injected this hybrid protein into chicken and got excellent polyclonal antibodies. But they are not so good for Westerns. We have other antibodies that are better for this purpose.

Toran-Allerand: Although steroid autoradiography does not distinguish between ERα and ERβ, we have shown using an iodinated oestrogen (rather than the traditional tritiated steroid, which is much less radioactive), that there are lots of oestrogen binding sites in the cerebral cortex, for example, in the developing mouse. We have even found binding sites in the striatum. Perhaps we have to re-consider the distribution of oestrogen binding sites in the brain in view of the fact that in the past the specific activity of the ligands was really not high enough to show cells with low levels of oestrogen binding. When we carried out steroid autoradiography in the cerebral cortex, with the tritiated ligand we had to wait 742 days, whereas with the iodinated ligand we obtained very good results in less than four weeks.

Fillit: Most women so far on hormone replacement therapy have received the conjugated equine oestrogens. I know this is a complex mixture, but is anything known about their relative affinities for the different ERs?

Gustafsson: That is a very relevant question: it is something we would like to study, but we haven't done it yet.

References

Azcoitia I, Sierra A, Garcia-Segura LM 1999 Localization of estrogen receptor β immunoreactivity in astrocytes of the adult brain. Glia 26:260–267

Chung S, Cohen R et al 1984 Ultrastructure and enzyme digestion of nucleoli and associated structures in hypothalamic nerve cells viewed in resinless sections. Biol Cell 51:32–34

Cohen RS, Pfaff DW 1981 Ultrastructure of neurons in the ventromedial nucleus of the hypothalamus in ovariectomized rats with or without estrogen treatment. Cell Tissue Res 217:451–470

Cohen RS, Pfaff DW 1992 Ventromedial hypothalamic neurons in the mediation of long-lasting effects of estrogen on lordosis behavior. Prog Neurobiol 38:423–453

Cohen R, Chung S, Pfaff DW 1984 Alteration by estrogen of the nucleoli in nerve cells of the rat hypothalamus. Cell Tiss Res 235:485–489

Dubal DB, Shughrue PJ, Wilson ME, Merchenthaler I, Wise PM 1999 Estradiol modulates bcl-2 in cerebral ischemia: a potential role for estrogen receptors. J Neurosci 19:6385–6393

Jones K, Pfaff D, McEwen BS 1985 Early estrogen-induced nuclear changes in rat hypothalamic ventromedial neurons: an ultrastructural and morphometric analysis. J Comp Neurol 239:255–266

Jones K, Chikaraishi D, Harrington C, McEwen BS 1986 *In situ* hybridization detection of estradiol-induced changes in ribosomal RNA levels in rat brain. Mol Brain Res 1:145–152

Jones K, Harrington C, Chikaraishi D, Pfaff DW 1990 Steroid hormone regulation of ribosomal RNA in rat hypothalamus: early detection using *in situ* hybridization and precursor-product ribosomal DNA probes. J Neurosci 10:1513–1521

Shughrue PJ, Lubahn DB, Negro-Vilar A, Korach KS, Merchenthaler I 1997 Responses in the brain of estrogen receptor α-disrupted mice. Proc Natl Acad Sci USA 94:11008–11012

Oestrogen receptor function at classical and alternative response elements

Peter J. Kushner, David Agard, Wei-Jun Feng, Gabriela Lopez, Andrew Schiau, Rosalie Uht, Paul Webb and Geoffrey Greene*

Metabolic Research Unit, Box 0540, University of California San Francisco, CA 94143-0540 and * The Ben May Institute for Cancer Research and Department of Biochemistry and Molecular Biology, University of Chicago, Chicago, IL 60637, USA

Abstract. The oestrogen receptor (ER), bound to classical response elements (EREs) in the promoter of target genes, activates transcription by recruiting coactivator proteins. We will describe structural studies that show that oestrogens allow the formation of a hydrophobic cleft on the surface of the ER that serves as a docking site for coactivators. Anti-oestrogens displace part of the receptor, which then occludes the site, blocking coactivator access. In addition to activating at classical EREs, the ER activates transcription at alternative elements such as AP-1 sites. These bind the Jun/Fos proteins but not ER. Interestingly both oestrogen and tamoxifen activate transcription at AP-1 sites. We propose a mechanism whereby oestrogen and anti-oestrogen allow ER to activate transcription from alternative response elements. ER binds to the coactivators, CBP and GRIP1, that have been recruited by Jun/Fos and through this contact 'triggers' these coactivators into full activity. In this circumstance the ER is part of the coactivator complex for Jun/Fos.

2000 Neuronal and cognitive effects of oestrogens. Wiley, Chichester (Novartis Foundation Symposium 230) p 20–32

The oestrogen receptor (ER) is a ligand-regulated transcription factor that in its best known mode of action binds to oestrogen response elements (EREs) in the promoter region of target genes. ER can also activate transcription from alternative target genes that contain sites for other transcription factors such as AP-1, CREB and SP-1, but that do not contain a binding site for ER. In this review we describe recent studies that have revealed in molecular detail how oestrogen and anti-oestrogens regulates ER activities at EREs, and other studies that provide a more speculative model for how ER functions at AP-1 sites.

ER domains and their function

The ER is a modular protein with three functional domains (Fig. 1; Parker 1998). In the centre of the protein is the DNA-binding domain (DBD) that directs the

20

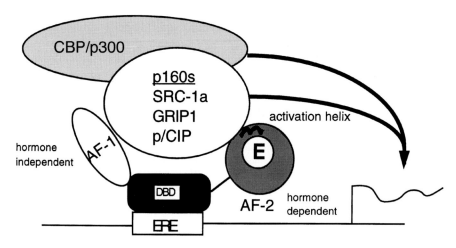

FIG. 1.   Model of ER action. The ER DBD binds the ERE and thus positioned serves as a tether for p160/CBP coactivators through the hormone independent AF-1 and the hormone dependent AF-2. The latter requires the function of the activation helix (helix 12).

protein to EREs in the promoter region of target genes. Replacement of the ER DBD with a heterologous DNA-binding domain (e.g. that from the yeast GAL4 protein) will direct the receptor to a different set of target genes (those with GALREs). The N-terminal domain contains a weak and constitutive activator function, AF-1. A second and stronger activation function known as AF-2 is in the C-terminal ligand-binding domain (LBD). AF-2 is active when bound to oestrogen, DES or other agonists and is inactive when bound to tamoxifen, raloxifene, ICI 182 780 or other anti-oestrogens.

Initially ERs are inactive and bound by their LBD to a high molecular weight complex containing the heat shock proteins HSP70 and HSP90. When oestrogen, which has worked its way to the nucleus binds the receptor, the ER dissociates from the heat shock protein complex, dimerizes and binds to EREs. Once bound to the ERE, the ER uses AF-1 and AF-2 to stimulate transcription from the promoter. Recent studies reveal that AF-1 (Webb et al 1998, Tremblay et al 1999) and AF-2 (Xu et al 1999) work by binding to coactivators of the p160 family, which has three members SRC-1, GRIP1 and p/CIP. These proteins bind the ER LBD very tightly in the presence of oestrogen. The p160s in turn are complexed with CBP (or p300) and p/CAF. CBP, p/CAF and p160s each contain histone acetyltransferase activity (HAT) (Bannister & Kouzarides 1996, Chen et al 1997, Ogryzko et al 1996, Spencer et al 1997, Yang et al 1996), and it has been hypothesized that the HAT activity of the coactivator complex acetylates key lysines in the N-terminus of histones H3 and H4 with the consequent remodelling of chromatin to a looser state. The remodelled chromatin allows access to the

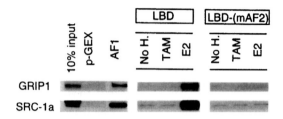

FIG. 2.   The AF-1 and ligand binding domain (LBD) of ER bind to the p160s SRC-1 and GRIP1. A GST pull down assay with labelled SRC-1a and GRIP1 binding either to p-GEX control beads or beads with the ER AF-1, LBD or LBD with an AF-2 mutation (LBD-[mAF-2]) as indicated and incubated with no hormone (No H.), tamoxifen (TAM) or oestradiol (E2).

transcriptional machinery and increased transcription. It is possible that the coactivator complex, in addition to remodelling chromatin, contacts components of the transcriptional apparatus and modulates their activities.

It should be noted that anti-oestrogens, such as tamoxifen, compete with oestrogen for binding to a single site on the ER. Once bound, however, anti-oestrogens have similar effects as oestrogen on the first steps of receptor activation. Thus tamoxifen frees ER from heat shock proteins, allows it to dimerize, and bind the ERE. Tamoxifen does not, however, allow the receptor once bound on DNA to stimulate transcription, because it does not allow binding of p160 proteins by the LBD (Oñate et al 1995).

The binding of ER through AF-1 and AF-2 are illustrated in Fig. 2. In this glutathione-S-transferase (GST) 'pull down' assay, the ER LBD, or AF-1 domain are attached to GST beads as indicated. The beads are reacted with labelled coactivators, either GRIP1 or SRC-1, and the captured proteins are displayed on an SDS gel. Notice that the coactivators bind to the AF-1 region rather weakly. The coactivators bind to the AF-2 region strongly, but only in the presence of oestrogen. Tamoxifen does not allow any binding to AF-2. Mutations in helix 12 also abolish all binding as will be discussed below.

Details of coactivator recruitment by DES-ER LBD and blockade of recruitment by tamoxifen LBD

The molecular mechanism whereby oestrogen or DES allow p160 binding has become clear from recent studies including some of our own (Feng et al 1998, Ding et al 1998, Darimont et al 1998, Shiau et al 1998, for review see Xu et al 1999). Feng et al (1998) made a library of mutants in surface residues of the thyroid hormone (TR) and ERs using the X-ray structure of the former as a guide. Amino acids on the surface of the LBD were mutated to arginine, which has the largest soluble side chain in the expectation that the bulk of the side chain would interfere with protein–protein interactions. This mapping revealed that a hydrophobic cleft on the LBD surface formed by helices 3, 4, 5 and 12 forms the coupling surface for coactivators. The entire floor and walls of this cleft are hydrophobic, but two charged residues are prominent on the rim. One of the charged residues is a glutamic acid 454 in helix 12, another is lysine 388 at the end of helix 3. Both residues are conserved in nuclear receptors and both were previously identified by Malcolm Parker's lab as crucial for AF-2 activity (Danielian et al 1992, Henttu et al 1997). X-ray crystallographic studies of the ER LBD bound to oestradiol solved by investigators at the University of York showed that helix 12 overlies the hormone (Brzozowski et al 1997). Indeed, the hormone is completely buried beneath helix 12, which forms the lower border of the hydrophobic cleft. In comparison, in the X-ray structure of the LBD with the AF-2 antagonist raloxifene, an extension of the ligand displaces helix 12, which is rotated 110° upward and now sites across the hydrophobic cleft. This suggests that raloxifene and other anti-oestrogens might somehow block coactivator access to the hydrophobic cleft.

To confirm this suggestion, it was necessary to identify how coactivators recognized the cleft. Several studies showed that GRIP1 and related p160s recognized the ER and TR through nuclear receptor boxes (NRBoxes), which have the sequence LXXLL (Heery et al 1997, Le Douarin et al 1996, Voegel et al 1998, Ding et al 1998). Thus an attractive idea is that the leucine side chains of this potential amphipathic helix project in the hydrophobic cleft. We were able to confirm this idea directly by obtaining the X-ray structure of the ER LBD with DES and an NRBox2 of GRIP1, ILHRLL (Shiau et al 1998). In this structure, which is illustrated below, the leucine side chains of the NRBox indeed project directly down into the hydrophobic cleft of the LBD each leucine making contacts with multiple side chains including those identified by the mutational studies. The two charged residues on the rim of the cleft form hydrogen bonds with nitrogen and carbonyl targets in the backbone of the NRBox helix. They thus act as a charge clamp determining that the NRBox will form a short helix when in contact with the LBD. Thus the coactivator LXXLL motifs recognize the agonist–ER LBD complex through an extensive hydrophobic interaction and a charge clamp with the hydrophobic cleft. In a companion structure of the LBD with tamoxifen, helix 12 is again rotated and sits in the hydrophobic cleft. Amazingly, helix 12 of ER has a mimic of the coactivator recognition motif. LXXML and the side chains of this mimic project directly downward into the hydrophobic cleft almost identically as they do in the complex with the coactivator. Thus antagonists such as tamoxifen or raloxifene completely block coactivator recruitment by occluding the hydrophobic cleft from the coactivator NRBox.

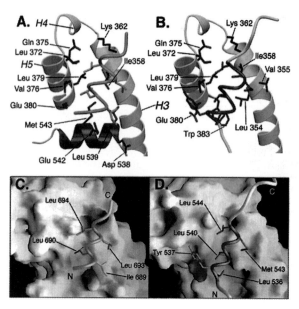

FIG. 3.   The oestrogen receptor hydrophobic cleft with an agonist (DES) and occupied by the coactivator GRIP1 (A, C), or with the antagonist tamoxifen (B, D) and occupied by helix 12 of the receptor itself. Notice that the coactivator recognizes the cleft through three leucine residues (690Leu-X-X-Leu-Leu) whose side chains plug into the cleft. Helix 12 recognizes the cleft by a mimic of the coactivator (536Leu-X-X-Met-Leu).

ER action at alternative response elements

If the ER works at an ERE by recruiting coactivators through the hydrophobic cleft, how does it activate at an AP-1 site? The AP-1 proteins Jun and Fos bind to these sites but ER does not. Indeed, Jun and Fos binding to the AP-1 site is needed for ER action and ER appears to increase the intrinsic transcriptional activity of Jun/Fos when bound to the site (Webb et al 1995). An additional twist to the paradox is that tamoxifen in some cell types is a potent activator of transcription at AP-1 target genes (Webb et al 1995). How is it possible for the ER to activate AP-1 target genes that bind Jun/Fos and not ER, and how can tamoxifen allow ER to activate AP-1 targets when tamoxifen blocks AF-2, the major activation function in ER? To explain these paradoxes we propose the model shown below (Webb et al 1999). The basis idea is that Jun/Fos are known to stimulate transcription by recruiting the coactivator CBP/p300 by direct contact. CBP/p300 in turn recruits a p160. This is the same coactivator complex that ER recruits at an ERE — but the surface through which the ER recruits the complex is not utilized by Jun/Fos. ER can thus 'join' the complex by binding through the unoccupied surfaces of the p160. In the presence of oestrogen the ER will contact the p160 primarily with

ER activates at EREs by recruiting
coactivators

ER activates at alternative
response elements by triggering
coactivators

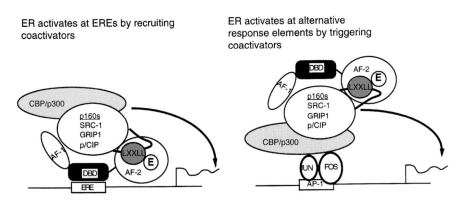

FIG. 4.   Model of ER action at AP-1 sites. Jun/Fos transcription factors bound to AP-1 sites recruit a complex of CBP/p300 and a p160 such as GRIP1 through contacts with the CBP/p300 component. ER can then bind the p160 through AF-1 and AF-2 and in so doing triggers the coactivators into full activity. note that ER action at AP-1 can be obtained with the 'flip horizontal' command applied to ER action at an ERE.

the hydrophobic cleft contacting the LXXLL motif of the p160. In the presence of tamoxifen this contact will be inactive, but the weaker AF-1 function will still serve to bring the ER to the promoter. Once ER is at the promoter we propose that it triggers the coactivator complex into full activity. That is to say, the contact between ER and the coactivator causes the coactivator to become fully active.

Although this notion of 'triggering' is highly abstract, the model does make some striking predictions. For example, since the ER is proposed to work at the level of the coactivators and not at the level of Jun/Fos, it should be possible to see ER working on coactivators tethered in some different manner to DNA. In unpublished work we show that this prediction is fulfilled (P. Webb & P. J. Kushner, unpublished results). When CBP is tethered to DNA via fusion to the DNA binding domain of the yeast GAL4 protein, it is only weakly competent to activate transcription from GAL4 response elements. However the addition of ER and oestrogen activates the tethered CBP. Thus the ER acts as a 'coactivator' when CBP is on the DNA! Furthermore, the parameters of ER coactivation of tethered CBP, such as the role of different ER and coactivator functions, closely parallel the parameters of ER activation at AP-1. Thus we think that triggering by protein–protein contacts may play a major, and previously unappreciated, role in ER reprogramming of cellular patterns of gene expression.

References

Bannister AJ, Kouzarides T 1996 The CBP co-activator is a histone acetyltransferase. Nature 384:641–643

Brzozowski AM, Pike AC, Dauter Z et al 1997 Insights into the molecular basis of agonism and antagonism in the oestrogen receptor as revealed by complexes with 17β-oestradiol and raloxifene. Nature 389:753–758

Chen H, Lin RJ, Schiltz RL et al 1997 Nuclear receptor coactivator ACTR is a novel histone acetyltransferase and forms a multimeric activation complex with P/CAF and CBP/p300. Cell 90:569–580

Danielian PS, White R, Lees JA, Parker MG 1992 Identification of a conserved region required for hormone dependent transcriptional activation by steroid hormone receptors. EMBO J 11:1025–1033 (erratum: 1992 EMBO J 11:2366)

Darimont BD, Wagner RL, Apriletti JW et al 1998 Structure and specificity of nuclear receptor–coactivator interactions. Genes Dev 12:3343–3356

Ding XF, Anderson CM, Ma H et al 1998 Nuclear receptor-binding sites of coactivators glucocorticoid receptor interacting protein 1 (GRIP1) and steroid receptor coactivator 1 (SRC-1): multiple motifs with different binding specificities. Mol Endocrinol 12:302–313

Feng W, Ribero RC, Wagner RL et al 1998 Hormone-dependent coactivator binding to a hydrophobic cleft on nuclear receptors. Science 280:1747–1749

Heery DM, Kalkhoven E, Hoare S, Parker MG 1997 A signature motif in transcriptional co-activators mediates binding to nuclear receptors. Nature 387:733–736

Henttu PM, Kalkhoven E, Parker MG 1997 AF-2 activity and recruitment of steroid receptor coactivator 1 to the estrogen receptor depend on a lysine residue conserved in nuclear receptors. Mol Cell Biol 17:1832–1839

Le Douarin B, Nielsen AL, Garnier JM et al 1996 A possible involvement of TIF1α and TIF1β in the epigenetic control of transcription by nuclear receptors. EMBO J 15:6701–6715

Ogryzko VV, Schiltz RL, Russanova V, Howard BH, Nakatani Y 1996 The transcriptional coactivators p300 and CBP are histone acetyltransferases. Cell 87:953–959

Oñate SA, Tsai SY, Tsai MJ, O'Malley BW 1995 Sequence and characterization of a coactivator for the steroid hormone receptor superfamily. Science 270:1354–1357

Parker MG 1998 Transcriptional activation by oestrogen receptors. Biochem Soc Symp 63:45–50

Shiau AK, Barstad D, Loria PM et al 1998 The structural basis of estrogen receptor/coactivator recognition and the antagonism of this interaction by tamoxifen. Cell 95:927–937

Spencer TE, Jenster G, Burcin MM et al 1997 Steroid receptor coactivator-1 is a histone acetyltransferase. Nature 389:194–198

Tremblay A, Tremblay GB, Labrie F, Giguère V 1999 Ligand-independent recruitment of SRC-1 to estrogen receptor β through phosphorylation of activation function AF-1. Mol Cell 3:513–519

Voegel JJ, Heine MJ, Tini M, Vivat V, Chambon P, Gronemeyer H 1998 The coactivator TIF2 contains three nuclear receptor-binding motifs and mediates transactivation through CBP binding-dependent and -independent pathways. EMBO J 17:507–519

Webb P, Lopez GN, Uht RM, Kushner PJ 1995 Tamoxifen activation of the estrogen receptor/AP-1 pathway: potential origin for the cell-specific estrogen-like effects of antiestrogens. Mol Endocrinol 9:443–456

Webb P, Nguyen P, Shinsako J et al 1998 Estrogen receptor activation function 1 works by binding p160 coactivator proteins. Mol Endocrinol 12:1605–1618

Webb P, Nguyen P, Valentine C et al 1999 The estrogen receptor enhances AP-1 activity by two distinct mechanisms with different requirements for receptor transactivation functions. Mol Endocrinol 13:1672–1685

Xu L, Glass CK, Rosenfeld MG 1999 Coactivator and corepressor complexes in nuclear receptor function. Curr Opin Genet Dev 9:140–147

Yang XJ, Ogryzko VV, Nishikawa J, Howard BH, Nakatani Y 1996 A p300/CBP-associated factor that competes with the adenoviral oncoprotein E1A. Nature 382:319–324

DISCUSSION

McEwen: You describe here oestradiol and tamoxifen having parallel agonist effects. However, in your previous publication (Paech et al 1997), I thought they were doing opposite things.

Kushner: Indeed, with ERβ tamoxifen and oestrogen have opposite effects at AP-1. With ERα, tamoxifen will mimic oestrogen, but only in certain tissues, for example, the uterus and not the breast. ERβ with tamoxifen will activate at these AP-1 sites even more strongly than ERα, and seems to do it universally. In fact, it seems to prefer tamoxifen to oestrogen in this pathway. You can actually block tamoxifen effects with oestrogen. It seems that the preferences are changed, but both of them will activate in some tissues bound to anti-oestrogens.

McEwen: In the brain, Vicky Luine did some work several years ago using a tamoxifen-like compound, CI-628, and oestradiol (Luine & McEwen 1977, Jellinck et al 1982). Both of these induced choline acetyltransferase in the basal forebrain. On the basis of what you're saying, with present knowledge would this most likely be an ERα-mediated event?

Kushner: Did oestrogen and CI-628 work equally well?

Luine: Oestrogen was much more effective.

Kushner: That sounds more like ERα to me. With ERβ it is the other way around: the antihormone would probably be a more potent activator transcriptionally, if this effect is due to activation of transcription.

McEwen: That is another, separate issue.

Simpkins: Did you tell us how the ER was mutated in the K206A mutant?

Kushner: No.

Simpkins: Is it possible that cells are expressing other kinds of ERs that are like this K206A mutant and whose normal function is to activate the AP-1 site?

Kushner: The K206A mutation is in the base of the first zinc finger in the DBD. We picked that site because previous studies with the glucocorticoid receptor had shown that the site which is conserved in all nuclear receptors changed the behaviour of receptors on AP-1. The glucocorticoid receptor normally inhibits AP-1 target chains. If you make the analogous mutation K461A, it becomes an activator. Since the ER was an activator, we decided to look to see what would happen.

This is puzzling because it is in the DBD. We know that the mutation does not increase the ability of the receptor to activate at the classical response pathway; it only has an effect at this AP-1 pathway. And we have some preliminary evidence that it disrupts some kind of corepressor interactions with the receptor: it disrupts the ability of the receptor to bind and titrate repressors. Perhaps there are corepressors bound to the receptors which then modulate the receptors' ability to interact with coactivator proteins when the receptors are not on DNA.

Levin: Another level of complexity would be to ask, either at the classical ERE or AP-1 site mechanisms, how these then change interactions with basic transcriptional machinery. This could be an overwhelming question, since there are at least 40 proteins potentially involved. How does the scenario that you paint change the protein–protein interactions?

Kushner: We don't know, but there are two strong possibilities. First, the oestrogen receptor in contact with the coactivators may change the assembly of coactivators. Second, it may change the enzymatic activities of the coactivator. Our hunch is that the coactivators change conformation and increase or modulate HAT activity by this contact, but that's just a guess.

Pfaff: Thinking about your competition experiments the other way around, is it proven that there are physiologically important competitions among nuclear receptors for coactivators?

Kushner: Yes, although it depends upon your standard of proof. One of the most dramatic cases is that the thyroid hormone receptor represses transcription of the genes involved in thyroid hormone synthesis. It represses transcription of the hypothalamic–pituitary axis in general. The α subunit of the glycoprotein hormones is turned off, and so forth. This appears to be through so-called NTREs (negative thyroid hormone response elements) which are activated by the receptor in the absence of ligand. When you add ligand they are repressed. There is evidence that this repression occurs through steroid coactivates. This is described in some recent papers from Larry Jameson's lab (Tagami et al 1999).

Toran-Allerand: One way to activate the AP-1 site is, if one were activating membrane ERs which share signalling pathways with growth factors, one could bypass the nucleus in its entirety.

Kushner: When we first started to explore this area, that was our working hypothesis. We spent a year and a half looking at the activation of MAP kinases under cell culture conditions in which the ER action at AP-1 sites was optimized: we couldn't see that activation in our cells under those conditions. Some other experiments eventually convinced us that it couldn't be that pathway, but this could certainly occur and then might synergize with the pathways that I've been describing.

Luine: With regard to the glucocorticoid receptor interaction with AP-1, both Bruce McEwen and myself have seen sex differences in the effects of stress on rats. Is the same kind of interaction occurring in males and could this be a level where one can sort out the sex differences?

Kushner: It's certainly possible. Rosalie Uht in my lab has been studying the interaction of glucocorticoid receptors and ERs at these AP-1 sites. The glucocorticoid receptor is known to repress, while the ER activates. We looked at what happens if both are present, in competition. It turns out that the most abundant receptor wins out. It is possible to block ER action with enough glucocorticoid receptor, and then this repression can be relieved by increasing ER. We have also seen this in a neural cell line, GT1.

Luine: We see the females being less sensitive to glucocorticoids at least on the behavioural level.

Kushner: That also depends on having receptors co-expressed in the same cells. This has not really been shown in many cases.

Luine: We're looking at cognitive function, so we have been focusing on the hippocampus where the glucocorticoid receptor is very abundant.

Herbison: Is there any evidence for the regulation of the expression of gonadal steroid receptor expression of coactivators? Could this be a physiological mechanism for the modulation of ER-dependent transcription?

Kushner: There's some. Some of the coactivators seem to be equally abundant in different cell types, while others appear to be regulated, especially p300. It is also possible that coactivator's activity is regulated: there is a lot of suggestive evidence that phosphorylation and other post-translational modifications regulate activity of the coactivator, so even if they're equally abundant they may not be fully active in different cells.

Pfaff: I was impressed by your beautiful pictures of helix 12 and the windshield wiper analogy. Could you comment on the temporal aspects of this? Are there data on this, or standard biophysical speculations? In the current issue of *Molecular Endocrinology* there's a paper about oestrogen-stimulated catabolism of ER, and many of us are concerned with temporal fluctuations in oestrogen presence in the brain.

Kushner: That's a tough question. The windshield wiper must be occurring quickly, and it is known that the oestrogen receptor is short-lived once bound to ligand, perhaps lasting only 3–5 min. This is obviously going to be an important issue, but I don't think anybody has looked at it in depth. The corepressor proteins are also known to be subject to regulatory catabolism and there is evidence that neuronal cells of certain sorts have a protein that specifically degrades corepressors. These are very simple systems, and we have tried to set up so that they work, to isolate one of the regulatory pathways.

Pfaff: In the temporal domain there's a tradition of work that started with Harris & Gorski (1978), which was extended in the neural and behavioural domain by people working with Bruce McEwen and myself (Parsons et al 1981, 1982a,b), which showed that pulsatile applications of oestrogens are sufficient to make up for a long application of oestrogens. Are there circumstances that you can imagine where repeated cycles of on/off would be more effective than a continuous presence of oestrogen?

Kushner: We could propose scenarios but I don't know of evidence for such things. All one has to imagine is that coactivators are induced and it takes some time, and the first application of oestrogen somehow helps the coactivator induction or corepressor degradation, so the second pulse is going to be even more potent and so forth. There are examples of 'secondary responses' with all the steroid receptors, where the first application of hormone activates certain genes and then the second application activates a whole group of other genes. The mechanism behind this is poorly understood.

Henderson: How many response element-like regions are there in addition to ERE, AP-1 and raloxifene response elements?

Kushner: I think there will be a large number, because many transcription factors utilize the same coactivators. It may well be possible for steroid receptors to modulate transcription indirectly at those other transcription factors.

Watson: Aren't coactivators and corepressors phosphoproteins? Phosphorylation may regulate their activity, rather than their synthesis or degradation.

Kushner: That could be the case. We have some evidence that the activities of the p160 coactivator are regulated through epidermal growth factor pathways and MAP kinases. There are going to be nodes of integration of information.

Bethea: I heard that the raloxifene response element data (Yang et al 1996) had been retracted.

Kushner: They partially retracted their data in *Science* (Yang et al 1997). The problem was with the mapping of the raloxifene response element. They did deletion mapping to show the existence of the response element, with about 10 different deletions, the last of which abolished the raloxifene response. Later on they found out that the last deletion went into the CAT reporter gene itself, and eliminated it. There are raloxifene-responsive genes, but whether this is the element or not is unclear.

Baulieu: Do you have any information as to whether ERα or ERβ is involved in the AP-1 pathway? Is there a difference between them?

Kushner: Yes, ERβ is especially potent at these AP-1 sites, particularly in the presence of anti-oestrogen.

Baulieu: Do all the anti-oestrogens work the same in terms of the helix 12 story?

Kushner: We would like to know this; there is so far no publication about the ICI–oestrogen receptor complex. Perhaps Jan-Åke Gustafsson knows about this: there were some rumours that he had a structure, and everyone would like to know whether it's going to be similar. But it looks like raloxifene and tamoxifen are very similar in their effects on the conformation of the receptor, and have the same mechanism of action.

Baulieu: You might have seen a paper that we published with M. G. Catelli and colleagues using some of your cells (Devin-Leclerc et al 1998). We studied an original anti-oestrogen, basically an oestradiol with a side chain branching at the C11-β position. The mechanism of action involves the oestrogen–receptor complex remaining mostly cytoplasmic and forming protein synthesis-dependent aggregates. I don't know whether the activity of other anti-oestrogens also involves these changes of receptor distribution and interactions.

Kushner: I don't know, that's another dimension. The hormones can affect the cellular distribution of the receptor.

Bethea: About five years ago, Donald McDonnell had a cover on *Molecular Endocrinology*, showing this sort of conformational slinky diagram of the oestrogen receptor (McDonnell et al 1995). He speculated that there was going to be this range of conformations, with type 1 to type 4 ligands. Does this fit your windshield-wiper model, or is there going to be a windshield wiper that can be stuck in the middle?

Kushner: It does not fit. Raloxifene and tamoxifen have some differences. For example, tamoxifen is very oestrogen-like on the uterus and raloxifene is much less so. When you look at the structures, they appear very similar. However, the ligands themselves protrude from the receptor, so they are exposed, and these are different. Consequently, the receptor–ligand complex has a different conformation in that sense.

References

Devin-Leclerc J, Meng X, Delahaye F, Leclerc P, Baulieu EE, Catelli MG 1998 Interaction and dissociation by ligands of estrogen receptor and Hsp90: the antiestrogen RU58668 induces a protein synthesis-dependent clustering of the receptor in the cytoplasm. Mol Endocrinol 12:842–854

Harris J, Gorski J 1978 Evidence for a discontinuous requirement for estrogen in stimulation of deoxyribonucleic acid synthesis in the immature rat uterus. Endocrinology 103:240–245

Jellinck PH, Luine VN, McEwen BS 1982 Differential effects of catecholestrogens, progestins and CI-628 administered by constant infusion on the central and peripheral action of estradiol. Neuroendocrinology 35:73–78

Luine V, McEwen BS 1977 Effects of an estrogen antagonist on enzyme activities and ^{3}H estradiol nuclear binding in uterus, pituitary and brain. Endocrinology 100:903–910

McDonnell DP, Clemm DL, Hermann T, Goldman ME, Pike JW 1995 Analysis of estrogen receptor function *in vitro* reveals three distinct classes of antiestrogens. Mol Endocrinol 9:659–669

Paech K, Webb P, Kuiper GGJM et al 1997 Differential ligand activation of estrogen receptors ERα and ERβ at AP1 sites. Science 277:1508–1510

Parsons B, Rainbow T, Pfaff DW, McEwen BS 1981 Oestradiol, sexual receptivity and cytosol progestin receptors in rat hypothalamus. Nature 292:58–59

Parsons B, McEwen B, Pfaff DW 1982a A discontinuous schedule of estradiol treatment is sufficient to activate progesterone-facilitated feminine sexual behavior and to increase cytosol receptors for progestins in the hypothalamus of the rat. Endocrinology 110:613–619

Parsons B, Rainbow T, Pfaff DW, McEwen BS 1982b Hypothalamic protein synthesis essential for the activation of the lordosis reflex in the female rat. Endocrinology 110:620–624

Tagami T, Park Y, Jameson JL 1999 Mechanisms that mediate negative regulation of the thyroid-stimulating hormone α gene by the thyroid hormone receptor. J Biol Chem 274:22345–22353

Yang NN, Venugopalan M, Hardikar S, Glasebrook A 1996 Identification of an estrogen response element activated by metabolites of 17β-estradiol and raloxifene. Science 273:1222–1225

Yang NN, Venugopalan M, Hardikar S, Glasebrook A 1997 Correction: raloxifene response needs more than an element. Science 275:1249

General discussion I

McEwen: Peter Kushner, could you say some more about the ligand-independent functions of the oestrogen receptors?

Kushner: For a while it was thought that the oestrogen receptor (ER) was completely inactivate in the absence of ligand. When it was first cloned and used for transfection studies, oestrogen needed to be added for there to be any effect in the simplified model systems that were used. In fact, this first receptor turned out to have a mutation that was a cloning artefact, at valine 400, which should have been a glycine. This mutation destabilized the receptor and also seemed to increase the interaction of the receptor with heat shock proteins, so that the receptor was fully bound in this heat shock protein complex. However, some of the wild-type receptor appears to be active even in the absence of hormone. It can bind to DNA and it is quite active in this AP-1 pathway. There may be activity by the unliganded receptor. This activity could be modulated by interactions with corepressor proteins, which are known to bind to the thyroid hormone receptor and the retinoic acid receptor, and keep them inactive in the absence of hormone. The situation with the ER is much more controversial, but there's some suggestion that corepressors may play a similar role in keeping the receptor inactive in the absence of hormone.

Toran-Allerand: There's a group in India who have suggested that the unliganded (i.e. unbound) ER in the goat uterus is a receptor tyrosine kinase which loses its activity in the presence of the hormone (Anuradha et al 1994).

McEwen: Donald Pfaff, you commented earlier that there is expression of ERβ message in places like the cerebellum and the hippocampus, where there is sparse evidence at this point for any kind of high affinity oestrogen binding, and the jury is still out on protein expression. You implied that there may even be a ligand-independent expression of a protein that can have some kind of regulatory function. Could you say more on this?

Pfaff: I said this because when we looked at the cognitive performance of ER knockout (ERKO) animals, in the absence of either ovaries or testes, there's a difference in the decline of incorrect responses between ERKO and wild-type (S. Ogawa et al, unpublished results). The direction of the difference, in turn, is different between genetic females and genetic males. This would be some indication that the protein is doing something in the absence of a ligand.

Cheryl Watson earlier asked a question about phosphorylation of the coactivators of corepressors. Peter Kushner, in your universe of thought about ERs, either on DNA or off the DNA, how large a role does phosphorylation of ER itself have to play to explain data that you have or you know about?

Kushner: For the data we know about, very little. Other labs have reported that the AF-1 function in the N-terminus of the receptor is regulated by phosphorylation through MAP kinases, and that epidermal growth factor (EGF) will partially activate the ER even in the absence of ligand by increasing the activity of that AF-1 function. This maps to a Ser118 in the N-terminus. We have found it hard to get the EGF activation ourselves.

Pfaff: Mutating is different from having it there unphosphorylated, isn't it?

Kushner: That's right. Vincent Giguère has recently looked at a similar site in ERβ, which has no conventional AF-1, and doesn't interact with coactivators through that region (Tremblay et al 1999). It doesn't seem to have the same sort of functions as ERα, but it has a MAP kinase-inducible AF-1 function. Vincent has shown that when two particular residues in ERβ are phosphorylated, the N-terminus binds more strongly to the coactivators. There is an inducible function there.

Watson: ERKOs have an interruption in the gene past the midway point of exon 2. Exon 1 is mainly UTR, but from exon 2 a partial protein is being produced, the portion responsible for interactions with many comodulators. If we are talking about all these interactions that may not necessarily depend on the DNA binding domain, in those mice some of the unexplained phenomena might have to do with a portion of the ER being produced, and its interactions with other proteins.

Kushner: Ken Korach's lab looked really hard at that. They were quite concerned because they saw some residual oestrogen binding protein in the uterus of the ERα knockouts. But now they think that was ERβ. They feel fairly confident that there is no ERα protein made in these mice.

Watson: But they were focusing on the possibility that alternate splicing may produce a smaller protein with the ligand binding domain, because they were worried about explaining ligand binding. I'm wondering about the other end — the N-terminal domain that has so much potential for interacting with a lot of other proteins, and has little to do with binding.

Pfaff: In the brain, if you use immunocytochemistry with antibodies towards the C-terminal end, there is no evidence that the translation has picked up again. There is a huge neomycin resistance sequence in place of exon 2. If you use antibodies against the N-terminus, in a very small number of neurons on the medial side of the preoptic area you can see faint immunoreactivity. This has been seen by two or three labs now. So although I think that technically you're right, the explanatory power of what you're talking about seems to be quite small, at least in the brain.

McEwen: The only way to get at the question as to whether there is a protein that is expressed that doesn't bind oestradiol, is to find some way of introducing an antisense to that particular message that could inactivate it, perhaps in an inducible way. Then one could look to see whether this blocks or alters cell function. We are not quite there yet, but this will clearly be possible soon.

Gibbs: If I understand you correctly, the AF-1 domain of ERα will bind coactivators in a ligand-independent way. Does this mean that ERα is always bound up with coactivators swimming around in the cell? In contrast, ERβ does not have the AF-1 domain and so it is not complexed with these coactivators unless it is bound by ligand (via AF-2). Is this how you are thinking?

Kushner: That is in large part true. The AF-1 interaction is weak, so if ERα is on an oestrogen response element the binding isn't strong enough for it to pull coactivators out of solution. It needs some help from AF-2.

Gibbs: Does that explain why people don't routinely see co-precipitation the coactivators with ERα?

Kushner: Yes, not unless you have hormone.

Levin: I had two general points that I wanted to address. First, one of the problems of genistein, particularly in the studies of cell proliferation, is that it is probably inhibiting tyrosine kinases that are activated by the growth factors liberated by the damage to the intimal lining. It may be that it is an inhibitor of proliferation completely unrelated to ER effects.

Second, I would be very worried about giving oestrogen antagonists to elderly men, who are very prone to developing cardiovascular disease.

McEwen: In that connection, I read recently that if raloxifene or tamoxifen are given to a perimenopausal woman who still has oestrogens, as opposed to a woman who has lost oestrogen function, there's an accelerated loss of bone mass, despite the fact that in postmenopausal women they cause some maintenance of bone mass. Then there is the additional problem that postmenopausal women differ in the degree to which they still produce some oestrogen endogenously, which is related to body fat and adrenal androgen aromatization and so on. These women, given these SERMs, may actually not benefit at all. Couple this to the fact that we are only beginning to discover all the different things that oestrogens do in the brain, possibly with different mechanisms, and we are left with a real can of worms in terms of trying to figure out any kind of therapeutic strategy at all.

Fillit: This is why we need to do clinical studies with these drugs to see what the global effects are in terms of specific outcomes, such as mortality, bone density, fracture rates and cognitive decline.

Pharmacogenomics may also be of great interest here: looking at individual responses and side effects. This may allow individualized therapy for people.

Kushner: Sometimes when we consider the possibilities it looks almost hopeless! There are so many different organ systems and even different parts of the brain itself that are affected by oestrogen. There are two receptors and these interact. What we would like to avoid is doing experiments on people, but if we don't have any good systems in cell culture and animal systems that's what we will end up doing. One wonders whether there is some hope in these systems. In a way, I think there is. There are only a few receptors — two, maybe three or even four — and only a few surfaces that these receptors use to interact with other proteins. Once we get all of those, we will be able to look at example ligands and get all the combinatorial possibilities, of which I suspect that there will only be a few hundred. That way we'll find out what we can do.

Resnick: I was just going to echo your comments about how little we actually know about some of the cognitive effects of these drugs. From what is published, we know nothing about the effects of tamoxifen on cognition. This is particularly worrisome, because these drugs are being administered to many healthy women for prevention of breast cancer. People may be familiar with the STAR (Study of Tamoxifen and Raloxifene) trial which is just beginning. This trial is a raloxifene/tamoxifen comparison for women at high risk for breast cancer and does not include a placebo group. As recent studies have demonstrated a 50% reduction in the risk for breast cancer in women taking tamoxifen, some oncologists argued that inclusion of a placebo group would be unethical. These drugs will be given to thousands of 'high risk' women, but to be high risk you only have to be over age 65. This has enormous potential implications.

Simpkins: Stan Birge (Washington University) describes referrals of women to his geriatric clinic from the breast cancer trials who are on tamoxifen and show unexplained and apparently fairly precipitous decline in cognitive function. There was a recent report about 100 women who were going through breast cancer therapy that included tamoxifen, and they described cognitive decline in those women also (Van Dam et al 1998). This raises a flag about tamoxifen, even in the absence of presumed ovarian oestrogen secretion.

Levin: I have treated patients who have had this experience. If you look at just the breast cancer data where tamoxifen/raloxifene is used, it is effective prophylaxis for just five years. Is that really the only organ in which this is going to happen? The bone data and cardiovascular data are all from short-term studies. We have no idea of what the long-term effects are, and the brain is completely unexplored. I went to the last endocrine meeting with the idea that I would come away with a firm set of recommendations for my patients, even knowing that each one is going to be individualized, and I was as confused as ever. I think the whole endocrine world is confused right now therapeutically.

Toran-Allerand: When I first started working with oestrogen over 20 years ago, tamoxifen application was a very good way of changing the pattern of sexual differentiation of the brain. One gave tamoxifen and obtained a blockage of oestrogen action. This is something that many people who don't study the brain but who have discovered potential value of anti-oestrogens are completely ignorant of. Potentially it is a very serious problem. We have a drug, tamoxifen, which is known to affect the brain as an oestrogen antagonist. Currently, physicians are using tamoxifen for treating young women at high risk for breast cancer. One simply does not know what this is going to do 20 years later in women who for one reason or another might have a propensity to develop Alzheimer's disease.

Sherwin: I agree with Susan Resnick: there is not one prospective controlled clinical trial of tamoxifen on cognitive function in women. There is now one controlled clinical trial of raloxifene, which was published in 1998. This was done by the Eli Lilly company. The treatment groups were raloxifene and placebo, but they left out the critical control group, which would have been oestrogen. They found that there was no difference between raloxifene and placebo after twelve months of treatment with respect to cognitive function.

Resnick: It was relatively short-term, and the tests were not the most sensitive measures, which was of some concern. Some of the tests were relatively non-standard.

Sherwin: All other information about SERMs and cognitive functioning in women is anecdotal.

Bethea: Aren't there huge trials taking place with tamoxifen in the prevention of breast cancer? Can't they give them some cognitive tests?

Resnick: That's exactly what we are trying to initiate. I work for the National Institute on Aging, and we have just issued a Research and Development Contract to Sally Shumaker of Wake Forest University School of Medicine to do an ancillary study to the Women's Health Initiative (WHI), adding cognitive testing to the WHI randomized clinical trial to look at the effects of active treatment (oestrogen or oestrogen plus progesterone) versus placebo. We would like to develop a similar project to add cognitive testing to the STAR trial. However, the problem we will have to address is the lack of a placebo control group in the STAR trial. If tamoxifen and raloxifene have the same directional effect, we are not going to see a difference. As I mentioned before, in view of the reduced risk for breast cancer in women treated with tamoxifen, many oncologists believe it would be unethical not to offer women a comparable compound.

Luine: As far as I know there has never been a test in rats or mice in terms of tamoxifen or any anti-oestrogen on cognitive function.

Resnick: I believe these rodent studies are being conducted at present in several labs, but the data are still very limited.

McEwen: I wonder whether it might be a better therapeutic strategy to design a SERM that doesn't get into the brain and then worry about how to deal with the brain separately. Jim Simpkins, a number of years ago you were involved with the technology from Nicholas Bodor of getting oestrogens into the brain through an oestrogen conjugate with a compound that was oxidized and 'locked' into the brain, and yet was metabolized peripherally. Do you want to say anything about that?

Simpkins: The strategy was to create an oestrogen that in the brain was locked in chemically and then cleared peripherally. Effectively, the strategy works. The problem that the technology came up against, which really was a business issue, was the prospect that you might be treating the brain with oestrogens, and depriving the periphery. But there are technologies out there that could be used to treat the brain quite separately from peripheral tissues.

McEwen: This is at least a rational strategy. I am surprised that the pharmaceutical industry hasn't fully investigated these possibilities.

Simpkins: A major question asked by the pharmaceutical companies concerned what was going to happen to bone or heart if we are indeed effectively treating the brain, and our answers were ineffective. We are now discovering that we may be antagonizing ERs in the brain with tamoxifen, and we don't know what the long-term consequences of that are. In effect we are doing those large clinical trials in reverse of the strategy we were proposing.

Levin: Perhaps we can begin to profile women individually. If they have no propensity to osteoporosis or cardiovascular disease, the issues is preventing hot flushes and dementia. With knowledge of how these compounds work in different systems, we can take a rational pharmacological approach. This is my hope.

Simpkins: That's a good point, but the pharmaceutical company perspective is that they want a product that they can use to treat everyone. They don't want to rely on doctors to say that their compound is good to treat one in 10 women, for example.

Baulieu: I am very much in favour of studying tamoxifen effects on the brain. I am a little surprised by the pessimistic attitude of this group in towards the potential for tamoxifen effects at the brain level, since not much is known of its action on neurons and glial cells, and currently no prediction can be made.

McEwen: Part of the pessimism in this room, if I understand it, is due to the over-optimism on the part of various drug companies that it's going to be the panacea. We are urging caution.

Sherwin: I wanted to respond to the idea that if a woman didn't have risk factors for cardiovascular disease and osteoporosis, it would be possible just to treat her brain with oestrogen-like compounds. The fact is that we're looking at protection of the quality of life for ageing women. This strategy might be fine in a 55 year old, but far more women die of cardiovascular disease and are disabled by osteoporosis than have dementia. Since the incidence of these diseases increases with increasing age, we need to think about the treatment of older women.

Murphy: Getting back to the idea about using oestrogens to treat conditions such as osteoporosis, this is how many pharmaceutical companies who are currently marketing oestrogen replacement therapy started out. Dr Ventimiglia at Pfizer has a huge bone unit, and in the past they have never looked at the effects of the drugs they are marketing for bone loss on cognition. They are now, however, and have set up a neurodegenerative unit. I think we are going to start to see more of the pharmaceutical companies looking at the cognitive effects of drugs they currently market for other indications.

Gibbs: You are right, and I think this has been driven by business issues. The companies perceive large markets in the treatment of bone loss and cardiovascular disease and I suspect they haven't wanted to threaten their potential profits in those markets by discovering some adverse effect in the brain. They are now beginning to do these studies because the FDA is specifically asking about the brain effects of these drugs. It is not that they haven't thought about it: quite understandably I suspect they have chosen not to do those studies until they have to.

Resnick: In talking to some people in the pharmaceutical industry, I found that part of the reason they're doing brain studies is because people have been getting information from the media about possible effects on dementia and memory. The marketing people in the pharmaceutical industry are saying they need more information to address these issues.

Luine: I wanted to comment on the dose and the regimen. When you talk about therapeutics you are talking about chronic treatment of people. Most of us don't do experiments like that. We're interested in sexual behaviour or changes in the receptor, for instance, so there is very little information in terms of what happens when you give oestrogen chronically to an animal. To be optimistic, maybe there is so much down-regulation of the central receptors that there is no longer an anti-oestrogen effect. Clearly, we are in a realm in which we need a lot more research just on oestrogen treatment regimes.

Bethea: I was able to get the brains from adult monkeys that had been maintained on tamoxifen for 30 months. I compared the tryptophan hydroxylase levels with monkeys that had been on primarin for 30 months or spayed. These data have not been published yet, but tamoxifen did have a very antagonistic effect compared with oestrogen after 30 months.

Henderson: I would guess that from an industry point of view something like Alzheimer's disease is a defined clinical entity, but something like age-associated cognitive decline is much more nebulous. It may be more difficult for a company to obtain an FDA indication for a drug that improves normal cognitive function than for one that diminishes Alzheimer symptoms. The same line of reasoning might hold true for improving mood in women who don't have clinically diagnosed depression.

Gandy: One may end up defining a new disease in the same way that AIDS led to the identification of HIV encephalopathy. There may be a 'tamoxifen encephalopathy' that is different from Alzheimer's disease or anything else.

References

Anuradha P, Khan SM, Karthikeyan N, Thampan RV 1994 The nonactivated estrogen receptor (naER) of the goat uterus is a tyrosine kinase. Arch Biochem Biophys 309:195–204

Tremblay A, Tremblay GB, Labrie F, Giguère V 1999 Ligand-independent recruitment of SRC-1 to estrogen receptor beta through phosphorylation of activation function AF-1. Mol Cell 3:513–519

Van Dam FSAM, Schagen SB, Muller M J et al 1998 Impairment of cognitive function in women receiving adjuvant treatment for high-risk breast cancer: high-dose versus standard-dose chemotherapy. J Natl Cancer Instit 90:210–218

Nuclear receptor versus plasma membrane oestrogen receptor

Ellis R. Levin

Division of Endocrinology, Veterans Affairs Medical Center, Long Beach, Long Beach, CA 90822 and Departments of Medicine and Pharmacology, University of California, Irvine, Irvine CA 92717, USA

Abstract. The co-existence of both plasma membrane and nuclear oestrogen receptors has changed our thinking about the mechanisms of the actions of this sex steroid. To date, however, the plasma membrane receptor has not been isolated. However, many emerging data implicate this receptor in the rapid, non-genomic effects of oestrogen, and this is seen when the membrane receptor effects a variety of signal transduction events. Although discrete actions of oestradiol could be mediated through the plasma membrane receptor, there is probably often a coordination of effects mediated through both receptors, perpetuating and magnifying the cell biological effects of the steroid.

2000 Neuronal and cognitive effects of oestrogens. Wiley, Chichester (Novartis Foundation Symposium 230) p 41–55

Nearly 25 years ago, Pietras and Szego provided evidence for a plasma membrane oestrogen binding protein which rapidly responded to oestradiol (E2) (Pietras & Szego 1977, 1980). The idea that a membrane receptor exists and may mediate some of the cell biological functions of E2 has been supported by evidence from several laboratories. At the same time, investigations of the rapid effects of other steroid receptor superfamily ligands has provided additional support for the existence of plasma membrane receptors (Blackmore et al 1991, Gametchu 1987, Nemere et al 1994, Wehling 1995). Ultimately, however the isolation of the membrane oestrogen receptor (ER) must be accomplished to understand the structure–function relationships. Nevertheless, evidence has begun to emerge that the membrane ERs importantly contribute to the actions of this steroid in both traditional and non-traditional target cells. In addition, there may be other membrane-related mechanisms by which the ER effects cellular actions. This paper will compare the plasma membrane-mediated events to those of the nuclear receptor, and their relative impacts for cell biology.

Cellular actions of the membrane ER

Many functional studies of the rapid, non-genomic effects of E2 provide support for the existence of the membrane ER. E2 can trigger an intracellular Ca^{2+} spike (Tesarik & Mendoza 1995), stimulate adenylate cyclase (Aronica et al 1994) or activate phospholipase C in a few seconds to minutes (Le Mellay et al 1997). The membrane ER might also be involved in the positive or negative regulation of transcription, considered to be the primary function of the nuclear ER (Halachmi et al 1994). This action appears to require modification of cytosolic signal transduction pathways, such as the ERK MAP kinase pathway. In MCF-7 breast cancer cells, E2 can stimulate the activity of signalling cascades that activate ERK (Migliaccio et al 1996). This action may contribute to the proliferative effects of E2 in breast cancer since many growth factors depend on ERK activation for communication to the nuclear growth programme (Pagès et al 1993). Subsequently, it has been shown that E2 can activate c-Fos transcription, probably through ERK (Watters et al 1997). It is likely that this kinase phosphorylates proteins which bind the serum response element on the Fos promoter (Janknecht et al 1993). E2 is also capable of negatively modulating transcription via the membrane ER, by inhibiting activating signal transduction (Morey et al 1998). The actions of the two receptors to modulate transcription are shown in Fig. 1.

It is established that E2 can modulate the production of nitric oxide (NO) in endothelial or other cells (Yallampalli et al 1994, Hayashi et al 1995). E2 can alter NO synthase mRNA levels in various cell types, but there may be additional actions that result in the rapid up-regulation of this enzyme's activity. ERα importantly contributes to the basal NO production in mouse aorta (Rubanyi et al 1997). More recently, NO synthase activation was reported to be rapidly up-regulated through an ERK-related action in cultured endothelial cells, and was attributed to the plasma membrane ER (Chen et al 1999). Neuroendocrine effects of the membrane ER have been recently reported. In cultured median eminence explants, the membrane ER stimulates endothelial cell-derived release of NO and secretion of gonadotropin-releasing hormone (GnRH) (Prevot et al 1999). E2 also causes the rapid release of prolactin from pituitary cells (Pappas et al 1994). Possible signal transduction events responsible for these actions were not defined.

Recently, expression of ERα or ERβ in Chinese hamster ovary (CHO) cells that do not normally express ER, has led to several proposals about the nature and derivation of the membrane ER. Expression of a single cDNA for either ER receptor gave rise to both membrane and nuclear binding proteins (Razandi et al 1999). Although the affinities of ER for E2 in both compartments was nearly identical, a much greater abundance of ER was detected in the nucleus. This ratio was as much as 40–50:1 and is consistent with data from cultured human cells which

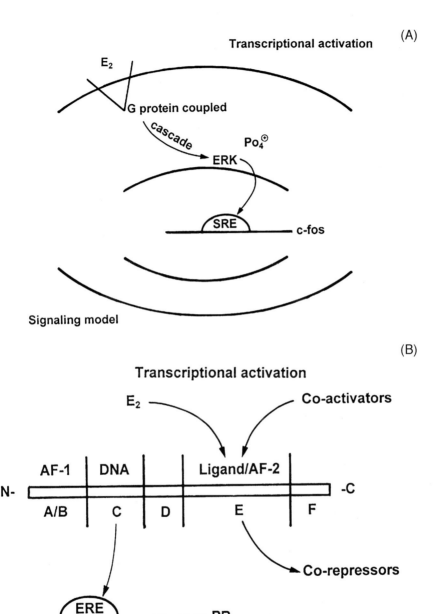

FIG. 1. Schema of transcriptional regulation originating from (A) membrane ER or (B) nuclear ER. ERE, oestrogen response element; ERK, extracellular signal regulated protein kinase; PO_4, phosphorylation; PR, progesterone receptor; SRE, serum response element.

normally express ER. Both ERα or ERβ membrane receptors can signal to ERK in this model, which is necessary for cell proliferation in this setting. The membrane receptors activate $G_{\alpha}q$ and G_{α}, consistent with studies in the literature that E2 can activate phospholipase C, Ca^{2+} flux and cAMP generation. Membrane ER can also modulate c-Jun MAP kinase activity, thereby potentially contributing to a variety of cell proliferation, differentiation and metabolic functions. In fact, basal JNK activity is suppressed by ERα while this kinase's activity is stimulated in ERβ-expressing CHO cells.

Structure of the membrane ER

Little is known about the structure of the membrane ER. If cells which normally express ER utilize the same strategies for production as transfected CHO cells, certain conclusions can be drawn. First, the membrane and nuclear ER must have a similar amino acid composition. This is based upon the fact that antibodies raised against the ligand binding or other domains of the classical nuclear ERα can identify the membrane ER in several cell types (Pappas et al 1995, Morey et al 1997). At least in transfected CHO, when E2 is cross-linked to either membrane or nuclear ER, the ER–E2 complexes migrate comparably on gel, and the receptor affinities for E2 are nearly identical (Razandi et al 1999). We speculate that a post-translational modification of a small number of ER must occur in the endoplasmic reticulum, facilitating the movement to and insertion into the plasma membrane bilayer. This may be accomplished through lipid modification of the membrane ER, perhaps in conjunction with a transporter protein.

Oestrogen and the cardiovascular system

The development of cardiovascular diseases such as atherosclerosis in post-menopausal women can be prevented substantially by oestrogen (Stampfer et al 1991). The mechanisms are varied and often interactive and are mediated by both the membrane and nuclear receptor (Table 1). First, E2 modulates the lipid status of the postmenopausal woman. Oestrogen inhibits low density lipoprotein (LDL) and lipoprotein a synthesis, and stimulates high density lipoprotein (HDL) synthesis. E2 inhibits the inactivation of the fibrinolytic system, by inhibiting the formation of plasminogen activator 1. Importantly, E2 inhibits the oxidation of LDL, perhaps mediated through interactions with matrix proteins and free oxygen radicals. These latter actions are most likely mediated through complex signalling interactions between membrane ER and the arterial wall matrix, including the integrins and basement membrane proteins.

E2 also appears to have substantial rapid effects to preserve coronary blood flow. These include stimulating NO synthase activity, as already mentioned. Stimulation

TABLE 1   Effects of oestrogen to prevent cardiovascular disease

Rapid (membrane mediated)
   Vasodilation
   NO synthase activation
   Gating Ca^{2+} and K^+ channels
   Vasoactive peptide synthesis, action

Chronic (nuclear mediated)
   LDL decrease
   HDL increase
   Lipoprotein a decrease
   Vascular endothelial growth factor synthesis increased
   Plasminogen activator inhibitor 1 decreased

Unclear or mixed effect
   Antioxidant
   Cell–matrix interaction
   Reduces platelet adhesiveness
   Inhibits vascular smooth muscle proliferation and migration

of NO formation leads to generation of cGMP, and could account for the observation that E2 rapidly stimulates the gating of large conductance, Ca^{2+} and voltage-activated K^+ (BKca) channels on coronary arteries (White et al 1995). E2 also rapidly inhibits L-type Ca^{2+} channels on vascular smooth muscle cells (Nakajima et al 1995), thereby reducing the actions of vasoconstrictors which act through this mechanism. In fact, many vasodilators demonstrate this important action. This results in vasorelaxation and improved perfusion of the heart. Oestrogen rapidly stimulates adenylate cyclase and generates cGMP, acting through a Ca^{2+}-sensitive mechanism in pulmonary vascular smooth muscle; this contributes to the vasodilatory effects of the steroid (Farhat et al 1996). These actions are likely to result from G protein-coupled receptor signalling in vascular cells, resulting in the activation of many downstream molecules. Compartmentalization of these signalling molecules occurs in combinations which are unique to a particular cell type. In this way, E2 can either stimulate or inhibit effects on a wide variety of cells, including vascular cells. Finally, E2 inhibits the production and actions of vasoactive peptides, such as ET-1, which importantly contribute to the pathogenesis of cardiovascular disease (Morey et al 1997, 1998).

Oestrogen and bone

E2 appears to influence physiology via the membrane receptor in other organs. This sex steroid initiates aspects of bone modulation through signalling. In osteoblasts, E2 rapidly activates influx of Ca^{2+}, increases cytosolic Ca^{2+}, inositol trisphosphate and diacylglycerol formation through a pertussis toxin-sensitive G protein, (probably G_i) (Lieberherr et al 1993). E2 also stimulates an increase in cAMP, cGMP and intracellular Ca^{2+} in human pre-osteoclastic cells (Fiorelli et al 1996). In the latter study, cell surface ERs were demonstrated by immuno-flourescence. In chondrocytes, E2 stimulates alkaline phosphatase activity via a protein kinase C-related mechanism (Sylvia et al 1998). When isolated chondrocyte membranes are incubated with E2, membrane fluidity is rapidly dissipated and lipid metabolism altered (Schwartz et al 1996). Oestrogen can also inhibit the ability of parathyroid hormone (PTH) to stimulate osteoclast formation, by blocking PTH-induced cAMP formation. This is likely to originate from a membrane ER interaction with adenylate cyclase (Kaji et al 1996).

Oestrogen and the CNS

A variety of CNS actions of E2 have been attributed to effects occurring at the plasma membrane. Both *in vitro* or *in vivo*, in models of experimental cerebral ischaemia, E2 can preserve neurovascular endothelial cells and neurons (Shi et al 1997, Behl et al 1995) and limited the extent of cerebral damage and mortality (Simpkins et al 1997a). E2 also preserved hippocampal and other neurons following experimental seizure in rats (Panickar et al 1998). β-amyloid-induced toxicity in neuroblastoma cells is decreased by the sex steroid (Gridley et al 1998), and several parameters of neuronal function as well as spatial and other forms of memory are preserved in rats by the administration of oestrogen (Simpkins et al 1997b). In women, replacement oestrogen results in improved cognitive function after menopause (Schmidt et al 1996) and improve motor speed and non-verbal processing in girls with Turner's syndrome (Ross et al 1998). Some of these effects are very rapid, but result from the actions of either 17α-E2 or 17β-E2 and thus, it is not clear whether an ER is important. Alternatively, the CNS ER may be structurally altered to accommodate both isoforms of E2. Recently, it has been shown that excitotoxicity-induced neuronal necrosis could be prevented by 17β-E2 signalling through the ERK member of the MAP kinase family (Singer et al 1999); this is likely to be due to a membrane ER action.

Other membrane effects of E2

E2 appears to interact with other membrane based proteins to effect cellular actions. Oestrogen-induced uterine cell proliferation is partly mediated through

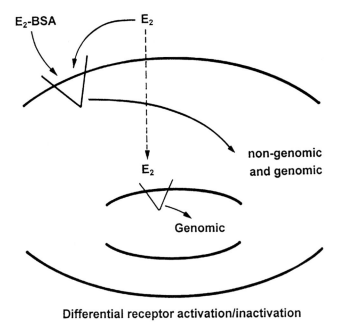

Differential receptor activation/inactivation

FIG. 2. Effects of oestradiol (E2) to activate both membrane and nuclear receptors, leading to the cell biological actions of the steroid. E2-BSA, BSA-conjugated oestradiol.

epidermal growth factor (EGF) action (Gametchu 1987) and may involve the membrane ER utilizing the EGF receptor as a signal transduction scaffold molecule. This is analogous to the dependence of a wide variety of G protein-coupled receptors to utilize the EGF receptor to effect signalling (Nelson et al 1991, Daub et al 1996). In turn, EGF and other growth factors can phosphorylate and therefore activate the oestrogen receptor via MAP kinase phosphorylation of Ser118 of ER (Bunone et al 1996); this residue has been identified to be important for ER activation, perhaps in allowing coactivators to associate with the target DNA. A comparable interaction is seen when the HER-2 tyrosine kinase receptor is overexpressed in human breast cancer cells (Pietras et al 1995). This leads to the steroid-independent phosphorylation and activation of ER by HER-2 ligands, and promotion of cell growth. E2 may interact with additional membrane receptor–ligand complexes. In the prostate and other reproductive tissues, sex hormone binding globulin (SHBG) binds its putative receptor on cell plasma membranes. In conjunction with some form of E2 action at the prostate cell membrane, the SHBG–receptor complex can stimulate adenylate cyclase, and protein kinase A, resulting in the secretion of prostate specific antigen (Nakhla et al 1997).

Conclusions and perspectives

Emerging results have begun to identify discrete actions of E2 that are mediated through the membrane ER. As we learn more about the structure of this protein, it is likely that additional signal transduction and cross-talk pathways will be elucidated, suggesting additional effects of this steroid receptor. In some situations, the membrane ER will act in conjunction with the nuclear receptor to augment and effects cellular actions (Fig. 2). However, it is likely that the rapid, discrete effects of E2 largely results from membrane interactions, and therefore are potentially able to be modified by synthetic agonists or antagonists that do not penetrate the cell. Molecules of this type could serve as the second generation of oestrogen response modifiers.

References

Aronica SM, Kraus WL, Katznellenbogen BS 1994 Estrogen action via the cAMP signaling pathway: stimulation of adenylate cyclase and cAMP-regulated gene transcription. Proc Natl Acad Sci USA 91:8517–8521

Behl C, Widmann M, Trapp T, Holsboer F 1995 17-β estradiol protects neurons from oxidative stress-induced cell death *in vitro*. Biochem Biophys Res Commun 216:473–482

Blackmore PF, Neulen J, Lattanzio F, Beebe SJ 1991 Cell surface-binding sites for progesterone mediate calcium uptake in human sperm. J Biol Chem 266:18655–18659

Bunone G, Briand PA, Miksicek RJ, Picard D 1996 Activation of the unliganded estrogen receptor by EGF involves the MAP kinase pathway and direct phosphorylation. EMBO J 15:2174–2183

Chen Z, Yuhanna IS, Galcheva-Gargova Z, Karas RH, Mendelsohn ME, Shaul PW 1999 Estrogen receptor α mediates the nongenomic activation of endothelial nitric oxide synthase by estrogen. J Clin Invest 103:401–406

Daub H, Weiss FU, Wallasch C, Ullrich A 1996 Role of transactivation of the EGF receptor in signalling by G-protein-coupled receptors. Nature 379:557–560

Farhat MY, Abi-Younes S, Dingaan B, Vargas R, Ramwell PW 1996 Estradiol increases cyclic adenosine monophosphate in rat pulmonary vascular smooth muscle cells by a nongenomic mechanism. J Pharmacol Exp Ther 276:652–657

Fiorelli G, Gori F, Frediani U et al 1996 Membrane binding sites and non-genomic effects of estrogen in cultured human pre-osteoclastic cells. J Steroid Biochem Mol Biol 59:233–240

Gametchu B 1987 Glucocorticoid receptor-like antigen in lymphoma cell membranes: correlation to cell lysis. Science 236:456–461

Gridley KE, Green PS, Simpkins JW 1998 A novel, synergistic interaction between 17β-estradiol and glutathione in the protection of neurons against Aβ 25-35-induced toxicity *in vitro*. Mol Pharmacol 54:874–880

Halachmi S, Marden E, Martin G, MacKay H, Abbondanza C, Brown M 1994 Estrogen receptor-associated proteins: possible mediators of hormone-induced transcription. Science 264:1455–1458

Hayashi T, Yamada K, Esaki T et al 1995 Estrogen increases endothelial nitric oxide by a receptor-mediated system. Biochem Biophys Res Commun 214:847–855

Janknecht R, Ernst WH, Pingoud V, Nordheim A 1993 Activation of ternary complex factor Elk-1 by MAP kinases. EMBO J 12:5097–5104

Kaji H, Sugimoto T, Kanatani M, Nasu M, Chihara K 1996 Estrogen blocks parathyroid hormone (PTH)-stimulated osteoclast-like cell formation by selectively affecting PTH-responsive cyclic adenosine monophosphate pathway. Endocrinology 137:2217–2224

Le Mellay V, Grosse B, Lieberherr M 1997 Phospholipase C beta and membrane action of calcitriol and estradiol. J Biol Chem 272:11902–11907

Lieberherr M, Grosse B, Kachkache M, Balsan S 1993 Cell signaling and estrogens in female rat osteoblasts: a possible involvement of unconventional nonnuclear receptors. J Bone Min Res 8:1365–1376

Migliaccio A, Di Domenico M, Castoria G et al 1996 Tyrosine kinase/p21 ras/MAP-kinase pathway activation by estradiol-receptor complex in MCF-7 cells. EMBO J 15:1292–1300

Morey AK, Pedram A, Razandi M et al 1997 Estrogen and progesterone inhibit human vascular smooth muscle proliferation. Endocrinology 138:3330–3339

Morey AK, Razandi M, Pedram A, Hu R-M, Prins B, Levin ER 1998 Oestrogen and progesterone inhibit the stimulated production of endothelin-1. Biochem J 330:1097–1105

Nakajima T, Kitazawa T, Hamada E, Hazama H, Omata M, Kurachi Y 1995 17β-estradiol inhibits the voltage-dependent L-type Ca^{2+} currents in aortic smooth muscle cells. Eur J Pharmacol 294:625–635

Nakhla AM, Romas NA, Rosner W 1997 Estradiol activates the prostate androgen receptor and prostate-specific antigen secretion through the intermediacy of sex hormone-binding globulin. J Biol Chem 272:6838–6841

Nelson KG, Takahashi T, Bossert NL, Walmer DK, McLachlan JA 1991 Epidermal growth factor replaces estrogen in the stimulation of female genital-tract growth and differentiation. Proc Natl Acad Sci USA 88:21–25

Nemere I, Dormanen MC, Hammond MW, Okamura WH, Norman AW 1994 Identification of a specific binding protein for 1α,25-dihydroxyvitamin D3 in basal-lateral membranes of chick intestinal epithelium and relationship to transcaltachia. J Biol Chem 269:23750–23756

Pagès G, Lenormand P, L'Allemain G, Chambard J-C, Meloche S, Pouysségur J 1993 Mitogen-activated protein kinases p42 [mapk] and p44[mapk] are required for fibroblast proliferation. Proc Natl Acad Sci USA 90:8319–8323

Pappas TC, Gametchu B, Yannariello-Brown J, Collins TJ, Watson CS 1994 Membrane estrogen receptors in GH3/B6 cells are associated with rapid estrogen-induced release of prolactin. Endocrine 2:813–822

Pappas TC, Gametchu B, Watson CS 1995 Membrane estrogen receptors identified by multiple antibody labeling and impeded-ligand binding. FASEB J 9:404–410

Panickar KS, Purushotham K, King MA, Rajakumar G, Simpkins J 1998 Hypoglycemia-induced seizures reduce cyclic AMP response element binding protein levels in the rat hippocampus. Neuroscience 83:1155–1160

Pietras R, Szego CM 1977 Specific binding sites for oestrogen at the outer surfaces of isolated endometrial cells. Nature 265:69–72

Pietras RJ, Szego CM 1980 Partial purification and characterization of oestrogen receptors in subtractions of hepatocyte plasma membranes. Biochem J 191:743–760

Pietras, RJ, Arboleda J, Reese DM et al 1995 HER-2 tyrosine kinase pathway targets estrogen receptor and promotes hormone-independent growth in human breast cancer cells. Oncogene 10:2435–2446

Prevot V, Croix D, Rialas CM et al 1999 Estradiol coupling to endothelial nitric oxide stimulates gonadotropin-releasing hormone release from rat median eminence via a membrane receptor. Endocrinology 140:652–659

Razandi M, Pedram A, Greene GL, Levin ER 1999 Cell membrane and nuclear estrogen receptors (ERs) originate from a single transcript: studies of ERα and ERβ expressed in Chinese hamster ovary cells. Mol Endocrinol 13:307–319

Ross JL, Roeltgen D, Feuillan P, Kushner H, Cutler GB Jr 1998 Effects of estrogen on nonverbal processing speed and motor function in girls with Turner's syndrome. J Clin Endocrinol Metab 83:3198–3204

Rubanyi GM, Freay AD, Kauser K et al 1997 Vascular estrogen receptors and endothelium-derived nitric oxide production in the mouse aorta. Gender difference and effect of estrogen receptor gene disruption. J Clin Investig 99:2429–2437

Schmidt R, Fazekas F, Reinhart B et al 1996 Estrogen replacement therapy in older women: a neuropsychological and brain MRI study. J Am Ger Soc 44:1307–1313

Schwartz Z, Gates PA, Nasatzky E et al 1996 Effect of 17β-estradiol on chondrocyte membrane fluidity and phospholipid metabolism is membrane-specific, sex-specific, cell maturation-dependent. Biochim Biophys Acta 1282:1–10

Shi J, Zhang YQ, Simpkins JW 1997 Effects of 17β-estradiol on glucotransporter 1 expression and endothelial cell survival following focal ischemia in the rats. Exp Brain Res 117:200–206

Simpkins JW, Rajakumar G, Zhang YQ et al 1997a Estrogens may reduce mortality and ischemic damage caused by middle cerebral artery occlusion in the female rat. J Neurosurg 87:724–730

Simpkins JW, Green PS, Gridley KE, Singh M, de Fiebre NC, Rajakumar G 1997b Role of estrogen replacement therapy in memory enhancement and the prevention of neuronal loss associated with Alzheimer's disease. Am J Med 103:19S–25S

Singer CA, Figueroa-Masot XA, Batchelor RH, Dorsa DM 1999 The mitogen-activated protein kinase pathway mediates estrogen neuroprotection after glutamate toxicity in primary cortical neurons. J Neurosci 19:2455–2463

Stampfer MJ, Willett WC, Colditz GA, Rosner B, Speizer FE, Hennekens CH 1991 Postmenopausal estrogen therapy and cardiovascular disease: ten year follow up from the Nurse's Health Study. N Engl J Med 325:757–762

Sylvia VL, Hughes T, Dean DD, Boyan BD, Schwartz Z 1998 17β-estradiol regulation of protein kinase C activity in chondrocytes is sex-dependent and involves nongenomic mechanisms. J Cell Phys 176:435–444

Tesarik J, Mendoza C 1995 Nongenomic effects of 17β-estradiol in maturing human oocytes: relationship to oocyte developmental potential. J Clin Endocrinol Metab 80:1438–1443

Watters JJ, Campbell JS, Cunningham MJ, Krebs EG, Dorsa D 1997 Rapid membrane effects of steroids in neuroblastoma cells: effects of estrogen on mitogen activated protein kinase signalling cascade and c-fos immediate early gene transcription. Endocrinology 138:4030–4033

Wehling M 1995 Nongenomic aldosterone effects: the cell membrane as a specific target of mineralocorticoid action. Steroids 60:153–156

White RE, Darkow DJ, Lang JL 1995 Estrogen relaxes coronary arteries by opening BKCa channels through a cGMP-dependent mechanism. Circ Res 77:936–942

Yallampalli C, Byam-Smith M, Nelson SO, Garfield RE 1994 Steroid hormones modulate the production of nitric oxide and cGMP in the rat uterus. Endocrinology 134:1971–1974

DISCUSSION

McEwen: In collaboration with Dr Teri Milner (Cornell Medical School), Dr Steven Alves and I have done electron microscopic immunogold localization of the ERα in the rat hippocampus using several anti-ERα antibodies. These results are very clean and we have seen them with both antibodies, but we still lack the definitive blocking experiment, which is underway. There is immunoreactivity associated with the dendritic spine apparatus. Spines like these are induced to form by oestrogens in the female hippocampus in the CA1 region. Besides the spines, Milner and Alves have seen the ERα receptor in cell nuclei of inhibitory interneurons.

Baulieu: If we believe that in the same cells there might be both membrane and nuclear receptors, could you define their respective and possibly coordinated activities in terms of cellular physiology?

Levin: One suggestion would be that the rapid effects which are related to signal transduction are occurring through the membrane receptor, whereas the more prolonged effects act through the nuclear receptor. However, this is probably too simplistic.

Baulieu: In terms of function, do you see a collaboration, cooperation and/or reciprocal regulation between the activity of receptors and ligands in the nucleus and at the membrane?

Levin: If you look at Fos transcription, this can be activated rapidly by the membrane receptor signalling through ERK, and then when Fos needs to stay on for longer, the nuclear receptor takes over. In this way, the cell has an exquisite regulation of how long it wants to maintain certain functions.

Baulieu: In molecular terms, do you think that the ligand-binding domain (LBD) of the membrane receptor itself is very similar to that of either ERα or ERβ?

Levin: Yes, I think since the H222 antibody recognizes the LBD with the same affinity, it will be extremely similar.

Baulieu: Does it have the DNA-binding domain (DBD) portion and the rest of the molecule (N-terminal segment)?

Levin: We will only know by isolating it and sequencing it.

Watson: I have done studies using nine different antibodies recognizing seven different epitopes all across the molecule. It seems to be closely related to ERα. All domains seem to be there, but remember that those domains were named because of their description in the nuclear receptor system, so whether or not the DBD in a receptor sitting in a membrane is functional is unclear.

Baulieu: Have you any idea of what determines whether a given molecule binds to the membrane receptor or reaches the nucleus? And for receptor molecules themselves, is there a mechanism to distribute the receptor to the nucleus or membrane location?

Levin: That is a critical question. My view is that the cell, by some novel mechanism, dices out a portion of the proteins that result from the single transcript, and post-translationally modifies the protein to allow it to move to the membrane. We have only just begun to understand what these modifications are, and I have no understanding about how the cell could parcel out a small proportion of receptors to do that.

Murphy: In your immunolabelling, where you used the FITC–BSA conjugate, I noticed that the staining avoided the nucleus. It also appeared to be punctate: could this indicate a clustering of receptors?

Levin: Some of that may be an artefact of preparation, but I think there is clustering. The staining avoiding the nucleus is just an artefact of preparation.

Murphy: If there is receptor clustering, how is the receptor translocating to the membrane? Presumably the ER is cytosolic and is not associated with any conventional membrane trafficking elements inside the cell.

Levin: We think there are lipid modifications. Gametchu has shown for the glucocorticoid receptors, for instance, that there are lipid modifications of that receptor that allows it to move to the membrane. Studies are underway in our lab to see whether there is a facilitated vesicular transport. The other critical issue is how the ER could burrow through the membrane to externalize its LBD. This is a huge hole in mammalian cell biology. In bacteria and yeast there are enzymes that are known to do that.

Simpkins: In view of the preference of oestrogens for lipid membrane, is it possible that this putative membrane receptor is really a transport molecule for oestrogens? It could be exported from the nucleus and goes to the membrane, picks up ligand and then just moves it back. This raises the question: when you saw the membrane receptor associated with the ligand, was that early after exposure to the ligand?

Levin: All our signalling studies are rapid, as are our binding studies. What you propose is reasonable and possible, but given the way this moiety signals, it really looks like a G protein-coupled receptor: the FGF-2/mannose-6-phosphate receptor is G protein linked and it is a single transmembrane spanner. But we won't know for sure until we get the sequence.

Simpkins: If you labelled oestradiol, picked up membrane labelling and did a time-course, would you expect that label to move into the cell?

Levin: It clearly moves into the cell and binds nuclear receptors. This is why we work with pure membrane preparations. A lot of the work I showed you in the CHO cells is with pure membrane preps, so we could exclude a lot of the problems. This doesn't eliminate, for instance ER membranes from Golgi on RER, but there has been no signal induction from these cytosolic sites by anyone's understanding to date.

Behl: Did you do some screening with your FITC method for other cell lines?

Levin: We have used this for astrocytes, endothelial cells and vascular smooth muscle. These are rat, bovine or human primary cultures. A variety of cells show these membrane receptors. I would put it the other way round: show me a cell that has an ER and doesn't have this kind of membrane labelling.

Watson: Depending on how you do the immunocytochemistry, these membrane receptors might not be seen. I would also add that several lipid modifications actually get put on in the endoplasmic reticulum–Golgi compartment, and therefore could end up tethering the protein on the outside of the cell.

Levin: That's tethering in the endoplasmic reticulum–Golgi, not tethering at the plasma membrane.

Watson: Yes, but it could travel on out to the cell surface as many proteins in these compartments do.

I want to make a comment about the oestradiol–BSA labelling of the membrane oestrogen receptors and the possibility of these being clustered by the labelling reagent. So far just about every reagent we have used is multivalent, whether they are antibodies or oestradiol–BSA (which has 15 or 30 oestrogen molecules stuck on a single BSA molecule). So although membrane ERs may be clustered, there is still the potential artefact of the reagents causing the clustered appearance.

Becker: I'm having trouble reconciling the picture that I saw of oestrogen binding to the ER inside of the receptor, with these pictures of oestrogen–BSA sitting on the outside of the cell. What do you think is going on?

Levin: We need the structure. We need isolation of the sequence and then crystallization of the membrane receptor.

Becker: In Dr Levin's transfected cells, I assume that the ER being bound at the cell membrane is the same receptor as the already cloned ERα. I make this assumption because Dr Levin transfected ERα into the cells, and this conferred oestrogen binding to the cells. This suggests that the binding domain described by Dr Kushner and the domain to which oestrogen–BSA is binding in Dr Levin's studies must be similar.

Simpkins: The majority of the energy in the binding of the oestrogen molecule to receptor is that initial A ring in the oestrogen molecule interacting. The BSA preparations are tethered to the 6, 4, 11 or 17 carbon, so that that 3 position is available, and there can therefore be interaction with the receptor (Anstead et al 1997).

Levin: If this thing is within the membrane bilayer I can't see how it would structurally change in response to oestrogen like the nuclear receptor does.

Becker: There are many examples where a receptor binds a ligand, internalizes and changes its 3D conformation.

Levin: This will also endocytose and retrocytose as a GPCR should, but I still don't think that this is going to be the same as the nuclear receptor binding an agonist.

Baulieu: I'm not sure I understood whether or not you have taken your purified membrane preparation and seen an *in vitro* clear biochemical effect of oestrogen, in the same way that a long time ago we saw the inhibition of adenylate cyclase activity by progesterone in *Xenopus laevis* oocytes (Finidori-Lepicard et al 1981).

Levin: When we did the G protein activation studies, those were in membrane preparations. The ability of oestradiol to stimulate the $G_\alpha q$ or $G_\alpha s$ binding of

GTPγS is in the membrane. The ability to activate adenylate cyclase is in the membrane preparations.

Behl: Can you exclude the possibility that the binding of the FITC fluorescence assay is not due to non-specific binding to other receptors? There are many reports of interactions between oestrogens and progesterone, for instance with GABA receptors, and this binding can be displaced by the corresponding receptor ligands such as GABA.

Levin: Probably the biggest conundrum for these studies is that it looks like the progesterone receptor can bind everything and anything. We have looked at a variety of peptides, vasoactive substances and steroids, and the binding is non-competable.

Watson: If the protein is tethered in the membrane by some post-translational lipid modification, it might well be free to undergo conformational changes that we know and expect of the nuclear receptor.

Levin: I think once it moves into the interior of the cell or is tethered to the interior layer of the membrane bilayer, there are all manner of conformational changes that take place.

Watson: It could be tethered on the outside and still make contact.

Levin: It could be.

Gandy: Have the predicted secondary structures of the molecule been revisited in light of the new evidence for cell surface signalling? Has the idea that it exists solely as a globular molecule been re-examined? Have transmembrane alternatives been considered using thermodynamic predictions?

Levin: No. We are doing a variety of classical receptor studies, but we haven't done thermodynamics.

Watson: Every time I talk to a structural biologist about analysing a lipid-attached or membrane-associated protein they usually grind their teeth. We are not anywhere close to being able to manipulate these proteins for such studies, at least in the appropriate lipid environment.

Gandy: I was thinking more of playing with computer modelling rather than immediately doing an experiment; then, perhaps you could design a cell biological experiment based on this.

Gibbs: A lot of your work focused on ERα but I think you showed that you can get membrane receptors with transfected ERβ. You also mentioned that progesterone alone can inhibit the activation of ERK. So all of these, ERα, ERβ and PR can independently activate the ERK pathway.

Levin: Oestradiol activates ERK in traditional target organs and turns it off in our cardiovascular models. In trophic cells it has been shown that oestradiol or oestradiol–BSA can activate ERK and I think progesterone has been shown to activate ERK.

Gibbs: Is there a difference in the efficacy of ERα versus ERβ?

Levin: We don't have any knockout animals, but in the CHO cell models, ERα seems to be a little more potent than ERβ, but then we started to see some differential signalling, which is very exciting. ERα depressed JNK, for example, whereas ERβ activated it. This is what led us to the whole scenario with the MCF7 cells. If this holds up in other cell types it will be exciting.

McEwen: One final comment, as a bookmark: there are effects of oestrogens that we're going to be hearing about tomorrow which defy the specificity of ER, where 17α-oestradiol is at least as potent as 17β-oestradiol. Apparently, unless the receptor specificity changes, which is doubtful, there is some other kind of membrane site that we are going to have to consider.

References

Anstead GM, Carlson KE, Katzellenbogen JA 1997 The estradiol pharmacophore: ligand structure–estrogen receptor binding affinity relationships and a model for the receptor binding site. Steroids 62:268–303

Finidori-Lepicard J, Schorderet-Slatkine S, Hanoune J, Baulieu E 1981 Progesterone inhibits membrane-bound adenylate cyclase in *Xenopus laevis* oocytes. Nature 292:255–257

Novel sites and mechanisms of oestrogen action in the brain

C. Dominique Toran-Allerand

Departments of Anatomy & Cell Biology, and Neurology, and Centers for Neurobiology & Behavior, and Reproductive Sciences, Columbia University College of Physicians and Surgeons, 630 West 168th Street, New York, NY 10032, USA

Abstract. We are investigating novel, non-transcriptionally mediated mechanisms that may contribute to the differentiative effects of oestrogen in developing forebrain neurons. Recent findings in the cerebral cortex document that 17α- and 17β-oestradiol elicit rapid and sustained activation of the Ras–Raf–MAP kinase cascade, a major growth factor signalling pathway. Using oestrogen receptor (ER) α knockout (ERKO) mice, we addressed the identity of the receptor mediating activation of the MAP kinase cascade. 17β-oestradiol increased B-Raf activity and MEK-dependent ERK phosphorylation in explants of wild-type and ERKO cerebral cortex. Although neither the ERα-selective ligand, 16α-iodo-17β-oestradiol (16α-IE2) nor the ERβ-selective ligand, genistein, elicited ERK phosphorylation, as little as 0.1 nM 17β-oestradiol did so. Moreover, 16α-IE2 acted as an inhibitory modulator of ERK activation, and the ER antagonist ICI 182 780 blocked oestradiol action only in wild-type cultures. These data suggest that neither ERα nor ERβ mediate activation of the MAP kinase cascade. A putative, novel, oestradiol-sensitive and ICI 182 780-insensitive receptor, designated ER-X may, rather, be involved. Association of ER-X with flotillin, the neuronal homologue of the caveolar protein, caveolin, places ER-X within plasma membrane caveolae and supports the hypothesis that a membrane-associated ER may mediate rapid oestrogen activation of the MAP kinase cascade.

2000 Neuronal and cognitive effects of oestrogens. Wiley, Chichester (Novartis Foundation Symposium 230) p 56–73

Both the gonadal steroid hormone 17β-oestradiol and the neurotrophin family of growth factors, such as nerve growth factor (NGF), brain-derived neurotrophic factor (BDNF), NT-3 and NT-4/5, have been implicated in the differentiation, survival, plasticity and ageing of mammalian forebrain neurons that subserve cognitive functions. My laboratory first showed that the oestrogen and neurotrophin receptor systems co-localize extensively within these neurons, suggesting that interactions between oestrogen and the neurotrophins may underlie the differentiative effect of oestrogen in this region (see Toran-Allerand 1996 for references and review). I first demonstrated in 1976 that 17β-oestradiol elicits the selective enhancement of axon and dendrite (neurite) growth and differentiation in cultured slices of developing rodent forebrain regions (Figs 1A,B) (Toran-Allerand et al 1980). Oestrogen stimulation of neurite growth in the rodent is developmentally regulated and not normally seen in the adult female brain. However, following loss of trophic support, whether induced by injury or oestrogen deprivation, responsiveness to oestrogen returns, and oestrogen can again be shown to influence the growth and differentiation of neurite-derived structures such as axons, dendrites, dendritic spines and synapses (see Toran-Allerand 1996 for references and review).

In the traditional view, oestrogen receptors (ERs) (ERα and ERβ) are intranuclear receptors which act as ligand-inducible, transcription factors, regulating the expression of target genes on binding to cognate response elements (Landers & Spelsberg 1992 for review). However, this classical view inadequately explains the complete and extensive range of oestrogen's effects in the brain, including its ability to regulate many genes that do not exhibit an apparent consensus response element and its very rapid (milliseconds to minutes) effects. While such a rapid time-course appears inconsistent with transcriptional modulation via an intranuclear ER, it could be explained by the existence of ERs within the plasma membrane that may be coupled to signal transduction pathways typically associated with activation by growth factors. Putative membrane ERs have been described since the mid-1970s (Anuradha et al 1994, Pietras & Szego 1977, Watson et al 1999). However, it remains unclear whether membrane ERs exist as a small subpopulation of both ERα (Watson et al 1999) and ERβ, or in fact represent novel members of the ER family (Das et al 1997, Gu et al 1999). It has even been suggested that the unoccupied (unliganded) goat uterine ER may be a unique, membrane-associated, oestrogen-binding site, which, while cross-reacting immunologically with nuclear ERs, is not only different structurally (Anuradha et al 1994) but, like growth factor receptors, also exhibits tyrosine kinase activity (Karthikeyan & Thampan 1996). On the other hand, Razandi et al (1999) have proposed that nuclear and plasma membrane ERα and ERβ may originate from a single transcript, based on their studies of Chinese hamster ovary (CHO) cells which had been transiently transfected with ERα or ERβ cDNA. However, since CHO cells do not normally express ERs, the extent to which such findings are applicable to cells which *normally* express ERs is unknown.

However, the very concept of functional membrane ERs remains a source of much scepticism. Although oestradiol/BSA, an immobilized oestradiol covalently linked to BSA (E2/BSA) has been used repeatedly to show that there are membrane receptors for oestrogen, Stevis et al (1999) have recently demonstrated that E2/BSA is *not* biologically equivalent to oestradiol. Following dialysis of E2/BSA, which removed large amounts of immunoassayable oestrogen, Stevis et al (1999) showed that, although dialyzed E2/BSA was now unable to bind

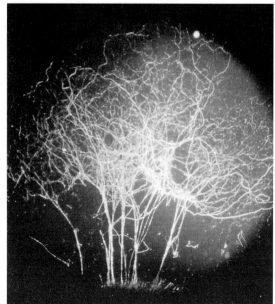

A

B

nuclear ER, it yet remained capable of eliciting rapid and persistent MAP kinase activation in neuroblastoma cells. This response, which typically follows growth factor activation of their membrane receptors, had been attributed originally to oestrogen action in BSA-treated neuroblastoma cells (Watters et al 1997). On the contrary, however, Stevis et al (1999) specifically documented in such neuroblastoma cells the very inability of oestradiol itself to activate MAP kinase.

Oestrogen and neurotrophin signalling

The chain or cascade of molecular events in the brain which follows oestrogen binding to its receptors is poorly understood. In contrast, signals from polypeptide growth factors, such as the neurotrophins, are propagated by an essentially linear flow of intracellular events, characterized by sequential protein phosphorylation (by kinases) and dephosphorylation (by phosphatases). These intracellular protein cascades serve to funnel, amplify and propagate signals generated at the cell surface into complex biological responses. For example, exposure of neural tissue to the neurotrophins elicits activation of their cognate Trk receptors, leading to activation of multiple signal transduction pathways, including, in particular, the mitogen-activated protein (MAP) kinase cascade (Fig. 2), a major pathway involved in cell proliferation, differentiation and cell survival. The initial activation of MAP kinase (also known as extracellular signal-regulated kinase, or ERK) requires the small G protein Ras, but its activation is sustained by the small G protein Rap1, which forms a stable complex with B-Raf. Rap1 activation of B-Raf is followed by sequential phosphorylation and activation of MEK (MAP kinase/ERK kinase) and the MAP kinase isoforms, also known as ERK1 and ERK2 (Marshall 1995). Activated ERK either translocates to the nucleus directly or first activates intermediary signalling proteins such as Rsk which also translocates. Nuclear translocation is the means by which many signals, arising at the plasma membrane, activate transcription factors such as CREB and Elk-1 and immediate early genes such as c-*fos* and c-*jun*. The nature of the cellular response to activation of the MAP kinase cascade depends upon (i) the *ligand itself* (e.g. in PC12 cells, epidermal growth factor [EGF] elicits mitosis; NGF

FIG. 1.   Neurite-promoting effects of oestradiol in hypothalamic explant cultures, 19 d *in vitro*. Photomicrographs of right and left homologous coronal halves of a Holmes' silver-impregnated pair of explants from the preoptic area. Darkfield microscopy, ×125. (A) Control exposed only to oestrogens endogenous to the horse serum component of the nutrient medium. The silver-impregnated neurofibrils (bundles of neuron-specific, neurofilament proteins) course outward from the margin of the explant. (B) Exogenous 17β-oestradiol. There is a significant enhancement of neurite growth from the same region of the homologous explant half, with extensive arborization of neurites in the outgrowth (reprinted by permission from Toran-Allerand et al 1980).

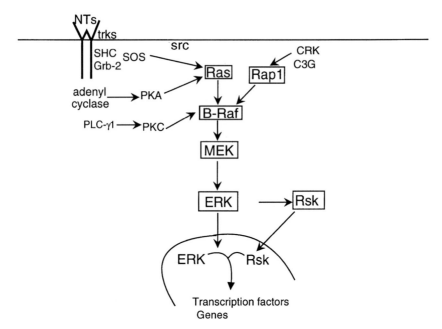

FIG. 2. The MAP kinase cascade. Treatment with neurotrophins elicits dimerization and activation of the cognate Trk receptor through tyrosine autophosphorylation. Tyrosine autophosphorylation regulates interactions of activated Trk with multiple intracellular proteins that either have enzymatic functions, such as phospholipase C (PLC)-γ1 and c-Src, or that are docking proteins without intrinsic enzymatic activity, such as Shc and Grb2. Shc and Grb2 with the GTP/GDP exchange protein SOS connect activated Trk to the initial signalling enzyme Ras. Although the initial activation of MAP kinase/ERK requires the small G protein Ras, its activation is sustained by the small G protein Rap1 which forms a stable complex with B-Raf. Rap1 activation of B-Raf is followed by sequential phosphorylation and activation of MEK and the MAP kinase isoforms, ERK1 and ERK2. Activated ERK either translocates to the nucleus directly or first activates intermediary signalling proteins such as Rsk which also translocates. Nuclear translocation is the means by which many signals, arising at the plasma membrane, activate intranuclear transcription factors and genes.

elicits differentiation); (ii) *cell context* (e.g. mitotic versus postmitotic); and (iii) *duration of the activation* (e.g. rapid onset with transient activation of ERK *without* nuclear translocation is a mitogenic signal; rapid onset with sustained ERK activation, *accompanied* by nuclear translocation, is differentiative) (Marshall 1995).

Convergence of oestrogen and neurotrophin signalling

We have proposed (Singh et al 1999) that oestrogen and neurotrophin receptor co-localization might lead to the sharing of similar, if not overlapping, sequences of intracellular biochemical events, through convergence or

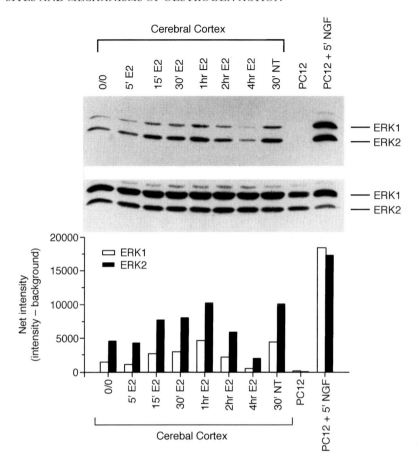

FIG. 3. Oestradiol-induced ERK phosphorylation. Lysates derived from cerebral cortical explants were probed with a phosphospecific ERK1/ERK2 antibody. Shown is a time-course for oestradiol-induced ERK phosphorylation (top panel) and the reprobed blot for ERK1 and ERK2 protein (middle panel). Explants also were treated with a 100 ng/ml neurotrophin cocktail (NGF, BDNF, NT-3, and NT-4/5) for a single 30 min time point that served as the experimental positive control. Note the similarity in the intensity of the response to oestrogen and the neurotrophins in the cortical explants. Untreated PC12 and NGF-treated PC12 cells served as negative and positive methodological controls, respectively. Densitometric representations of the relative intensities of the phosphorylated ERK1 and ERK2 bands are provided also (bottom panel). (Reprinted by permission from Singh et al 1999.)

cross-coupling of their signal transduction pathways. Singh et al (1999) showed for the first time in the brain that 17β-oestradiol elicits activation of the MAP kinase cascade. Oestradiol rapidly (within 5–15 min) elicited the sustained (at least 2–3 h) phosphorylation and activation of ERK1 and ERK2 (Fig. 3). Oestradiol-induced sustained activation of ERK in the cortical explants was accompanied by nuclear accumulation of phosphoERK and a striking level of cytoplasmic phosphoERK expression, as shown by immunohistochemistry, using phospho-specific ERK antibodies (G. Sétáló Jr & C. D. Toran-Allerand, unpublished observations 1999). This response was comparable to that seen with the positive control, the neurotrophin BDNF, and was significantly greater than that of the inactive baseline control. Since rapid onset of sustained ERK activation with nuclear translocation has been associated with neurotrophin-induced neuronal differentiation, a similar pattern seen following oestrogen exposure would appear to be consistent with its observed differentiative actions on neurites (Fig. 1).

Furthermore, 17α-oestradiol, the stereoisomer of 17β-oestradiol with 100-fold lower affinity for ERα, also elicited the rapid and sustained activation of ERK to a level perhaps even more intense than that elicited by 17β-oestradiol. The effect of 17β-oestradiol on ERK phosphorylation was not affected by either inhibitors of transcription or of protein synthesis. The inability of either of these inhibitors to prevent 17β-oestradiol induction of ERK activation, coupled with 17α-oestradiol's ability to elicit ERK phosphorylation, strengthens the likelihood that these oestrogens may be acting upstream of DNA, perhaps via ERs localized to the plasma membrane.

Oestrogen receptor subtypes mediating ERK activation

In order to identify the ER subtype mediating oestrogen activation of ERK (i.e. whether ERα, ERβ or perhaps some other ER), we initiated studies on cerebral cortical explants derived from the ERα knockout (ERKO) mouse in which the ERα gene is disrupted (Lubahn et al 1993). Preliminary studies (Singh et al 2000) document that oestradiol not only elicited ERK activation in ERKO cortical explants, but this increase in ERK phosphorylation was 29-fold above baseline by quantitative densitometry, a level significantly greater than that seen with the wild-type control which was increased only fourfold.

To further evaluate the individual contributions of ERα and ERβ to ERK activation, we analysed the responses to the ERα-selective ligand, 16α-iodo-17β-oestradiol (16α-IE2) (Hochberg & Rosner 1980) and the ERβ-selective ligand, genistein (Witkowska et al 1997). 16α-IE2, which has a selective affinity for ERα at concentrations below 10 nM (Shughrue et al 1999), failed to elicit ERK phosphorylation in cerebral cortical explants at concentrations of 0.1, 1 and 10 nM. In fact, at the 0.1 nM dose of 16α-IE2, for example, ERK was reduced approximately 70% from baseline levels of phosphorylation. This response not only provides evidence that 16α-IE2 was active in the explants but also that ERα may act as an inhibitory modulator of ERK phosphorylation. Moreover, since both 16α-IE2 and 17β-oestradiol have similar affinities for ERα (Hochberg &

Rosner 1980), the inability of 16α-IE2 to stimulate ERK phosphorylation is probably not due to inadequate binding to the ER, particularly, since a concentration as low as 0.1 nM 17β-oestradiol was sufficient to activate ERK. Similarly, genistein, the ERβ-selective ligand, did not influence ERK phosphorylation at concentrations ranging from 0.1 nM to 100 nM, suggesting that ERβ may not be involved in this oestrogen action either. However, at concentrations at which genistein acts as a tyrosine kinase inhibitor (100 μM), 17β-oestradiol-induced ERK phosphorylation was successfully inhibited, verifying the activity of the genistein used.

The ability of ICI 182 780 to block 17β-oestradiol-induced ERK phosphorylation only in the wild-type cultures, where the normal complement of ERs exist (i.e. ERα, ERβ and perhaps as yet unidentified ERs) but not in ERKO was surprising. This observation suggests that, in the absence of ERα (in ERKO), no such modulation could occur despite ICI's ability to bind to both ERα and ERβ with approximately equal affinity. But while ERα may mediate the inhibitory effect of the ICI compound, there must also exist a novel, oestradiol-sensitive and ICI 182 780-insensitive oestrogen receptor which actually elicits activation of ERK. In support of our findings, experiments in the ERKO mouse uterus (Das et al 1997) and hippocampus (Gu et al 1999) have also suggested the existence of novel, ICI-insensitive ERs that are neither classical ERα nor ERβ, as mediators of oestrogen's short-term, rapid and non-transcriptional actions. Because preliminary studies (G. Sétáló Jr, M. Singh, X. Guan & C. D. Toran-Allerand, unpublished observations 1999) suggest that our putative receptor may be different not only from classical ERα and ERβ, but also from those described in the ERKO uterus (Das et al 1997) and hippocampus (Gu et al 1999), we have provisionally designated it 'ER-X'.

One possible mechanism for the failure of ERKO cultures to respond to the ICI compound lies in the existence of normal heterodimerization between ERα and ERβ (Ogawa et al 1998), which may result in the ability of one receptor (ERα) to modulate the effect of its associated receptor (ERβ or even ER-X). In ERKO, heterodimers of ERα would not exist, precluding any inhibitory influence of ICI 182 780. And while ERα may not be necessary for oestrogen-induced activation of the MAP kinase cascade, ERα may be crucial in mediating or modulating regulation of this pathway, perhaps via a complex interplay between ERα and ER-X (Singh et al 2000). Further analysis in mice lacking the ERβ receptor will undoubtedly help test this hypothesis. Whether ER-X is a novel receptor or an ER variant is unclear. ERα and ERβ splice variants and mutations have been documented in both normal and neoplastic extra-neural oestrogen target tissues (Murphy et al 1997) and possibly the ERKO brain as well (Moffatt et al 1998) and may have a role in tissue- or region-specific oestrogen responses.

Potential pathways of oestrogen-induced ERK activation

What pathways might oestrogen use to activate ERK in the brain? Although oestrogen reportedly elicits tyrosine phosphorylation of the EGF receptor, there was no evidence of oestrogen-induced phosphorylation of the cortical neurotrophin receptor Trk (Singh et al 1999). On the other hand, in explants of wild-type and ERKO mouse cerebral cortex, 17β-oestradiol increased B-Raf kinase activity to a level comparable to that elicited by the neurotrophins (Singh et al 2000). However, in both wild-type and ERKO cortical cultures, oestrogen- and neurotrophin-induced phosphorylation of ERK was inhibited to baseline levels in the presence of the MEK inhibitor PD 98059 (Singh et al 1999, 2000). Taken together, these findings suggest that convergence of oestrogen and neurotrophin signalling pathways must occur at least at the level of MEK or perhaps further upstream, at the level of B-Raf, Rap1 or Src, but certainly downstream of Trk.

An oestrogen receptor-containing complex

Singh et al (1999) have recently identified by co-immunoprecipitation and Western blotting an ER-containing, multi-molecular complex in cerebral cortical explants. This complex consists (thus far) of an ER physically associated with Hsp90 and signalling kinases such as Src, B-Raf, MEK and ERK, as well as the amyloid precursor protein (APP) (X. Guan, M. Singh, G. Sétáló Jr & C. D. Toran-Allerand, unpublished observations 1999). The importance of the ER/B-Raf association was further emphasized by demonstrating that a significant level of B-Raf kinase activity remained associated with the ER/B-Raf co-immunoprecipitate and that the co-precipitated band observed was, in fact, kinase-active B-Raf (Singh et al 1999). On the other hand, this complex did not contain c-Raf (Raf-1) or Trk receptors. The lack of association with Trk and c-Raf was correlated with the inability of oestradiol to activate either of these proteins. In contrast, Src, B-Raf, and ERK, which are complexed with the ER, have all been shown to become rapidly activated by oestradiol, providing functional correlates to their association. In contrast, the neurotrophins increased the activity of both B-Raf and c-Raf strongly. This finding, whose significance is unknown, represents the first difference found between the cortical responses to oestrogen and the neurotrophins.

Caveolae of the plasma membrane

These observations have led us to propose that the ER-containing multi-molecular complex may be localized to the neuronal plasma membrane in association with caveolar-like structures (Toran-Allerand et al 1999) and that its ER may be the putative 'ER-X'. Caveolae are small vesicular invaginations of the plasma

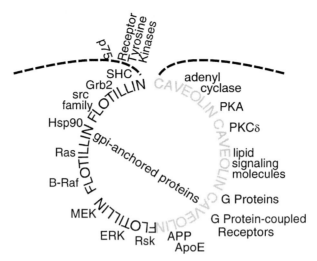

FIG. 4. Caveolae of the plasma membrane. Caveolae of the plasma membrane are highly enriched in caveolin (non-neuronal cells) or flotillin (neurons) both of which are integral membrane protein components. These proteins form multi-valent scaffolding onto which many signalling molecules assemble to generate pre-assembled signalling complexes and which may also functionally regulate their activation state. Concentrated within these 'crowded little caves' (Schlegel et al 1998) are, for example, receptor tyrosine kinases (e.g. the neurotrophin, insulin, EGF and PDGF receptors); the low affinity neurotrophin p75[NTR]; the adaptor proteins Shc and Grb2; signal transduction molecules such as Ras, B-Raf, the Src family of tyrosine kinases, MEK, ERK; adenylate cyclase, protein kinase A and protein kinase C; G proteins and G protein-coupled receptors; lipid signalling molecules; APP; and glycosyl-phosphatidylinositol-anchored proteins, among many others. The physical association of an ER ('ER-X') with flotillin proves its membrane localization.

membrane that have been implicated in signal transduction and lipid/protein trafficking (Schlegel et al 1998) (Fig. 4). Caveolae are highly enriched in caveolin, an integral membrane protein which forms a multivalent scaffold onto which many signalling molecules can assemble to generate pre-assembled signalling complexes. Caveolar localization of certain inactive signalling molecules may provide a compartmental basis for their regulated activation and explain extensive cross-talk between different signalling pathways (Okamoto et al 1998).

Some of the proteins concentrated within these aptly named 'crowded little caves' (Schlegel et al 1998) include, among many others: (i) receptor tyrosine kinases (e.g. the neurotrophin, insulin, EGF and platelet-derived growth factor [PDGF] receptors); (ii) the low affinity neurotrophin receptor p75[NTR]; (iii) Hsp90; (iv) the Src family of tyrosine kinases; (v) the docking/adaptor proteins Shc and Grb2; (vi) signal transduction molecules such as members of the MAP kinase cascade, (Ras, B-Raf (Rap1), MEK and ERK); adenylate cyclase, protein kinase A and protein kinase C; (vii) G proteins and G protein-coupled receptors;

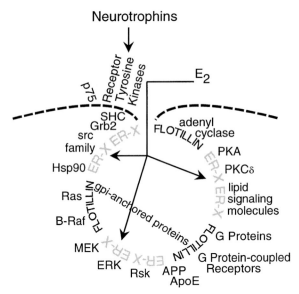

FIG. 5. The membrane ER as a caveolar complex. The initial route taken to elicit rapid oestrogen activation of ERK may involve direct activation of the putative membrane ER-X associated with multimolecular complexes of signalling kinases. Following neuronal exposure to oestradiol (E$_2$), dissociation of the oestrogen receptor (ER) from the complex, consequent to oestradiol-induced tyrosine phosphorylation of Src, the ER, and Hsp90, may elicit conformational changes within the rest of complex and trigger sequential phosphorylation and activation of physically associated kinases such as B-Raf and ERK. Conversely, neurotrophin activation of Trk receptors, followed by activation of the MAP kinase cascade, would, in turn, provide a reciprocal signalling pathway by which the unliganded ER, bound through its physical associations with the rest of the complex could become tyrosine phosphorylated by the neurotrophins. In this manner, multimeric complexes of the ER, members of the MAP kinase cascade and Src, could serve as intracellular junctions which link the oestrogen and neurotrophin signalling pathways. (Re-drawn from Toran-Allerand et al 1999.)

(viii) lipid signalling molecules; (ix) nitric oxide synthase (eNOS); (x) APP (Brouillet et al 1999); and (xi) glycosyl-phosphatidylinositol (GPI)-anchored proteins. In the brain, where mRNA species for known caveolin gene family members have been detected only in astrocytes and microglia, flotillin, which is abundantly expressed in neurons and co-localizes within their caveolae, is considered the neuronal counterpart of caveolin (Bickel et al 1997).

Further support for caveolar localization of ER-X in the brain comes from recent preliminary studies which demonstrate co-immunoprecipitation of an ER with flotillin itself (X. Guan & C. D. Toran-Allerand, unpublished observations 1999). These findings support the hypothesis that a caveolar-associated, hence membrane-associated, ER would be uniquely positioned to mediate rapid, oestrogen-induced activation of the MAP kinase cascade. As diagrammed in

Fig. 5, rapid activation of ERK by oestradiol may involve initial activation of ER-X within caveolar-like structures, leading to rapid phosphorylation of Src, the ER and Hsp90 which causes dissociation and conformational changes within this complex. This in turn may trigger sequential phosphorylation and activation of physically associated kinases, such as the B-Raf/Rap1 complex, leading to ERK activation. Conversely, neurotrophin activation of Trk receptors, followed by activation of the MAP kinase cascade, would, in turn, provide a reciprocal pathway by which the unliganded ER, bound through its associations within the caveolar complex, could not only become tyrosine phosphorylated by the neurotrophins (Toran-Allerand et al 1996) but also serine phosphorylated by insulin-like growth factor (IGF)-I and EGF, as others have shown (see Toran-Allerand et al 1999 for references). Association of an ER with caveolar-like complexes of signalling kinases could mediate the rapid effects of oestrogen and provide both a mechanism and location at which convergence between oestrogen and neurotrophin signalling may occur.

Alzheimer's disease and the MAP kinase cascade

Oestrogen activation of the MAP kinase cascade may be particularly relevant for cognition and its disorders. Oestrogen exposure (Jaffe et al 1994, Xu et al 1998) and activation of the MAP kinase cascade (Mills et al 1997, Desdouits-Magnen et al 1998) are involved in the processing of the APP, a caveolar-associated protein (Brouillet et al 1999) of considerable focus in Alzheimer's disease (AD) pathophysiology (Selkoe 1994). APP is cleaved physiologically by α-secretase to the non-amyloidogenic, (non-toxic) soluble form, but is also cleaved by alternative pathways to form amyloid β-peptide (Aβ), the principle constituent of the amyloid plaques associated with AD (Sisodia 1992, Selkoe, 1994). Oestrogen (Jaffe et al 1994, Xu et al 1998) and MAP kinase activation up-regulate α-secretase activity, favouring formation of soluble APP and may (Mills et al 1997) or may not (Desdouits-Magnen et al 1998) inhibit generation of toxic Aβ. Thus, oestrogen activation of the MAP kinase cascade may be very important not only for neuronal differentiation but may also explain how, through regulation of APP metabolism, oestrogen could exert some of its known neuroprotective effects in neurodegenerative disorders such as AD (Tang et al 1996 for references and review).

Acknowledgements

Many thanks are due Drs Xiaoping Guan, György Sétáló Jr. and Meharvan Singh for their hard work and significant contributions to the work described in this paper; to Mr Matthew Warren and Ms Hae-Jung Chung for their expert technical assistance; and to Drs Lloyd A. Greene and Neil J. MacLusky (Columbia University) for constructive criticisms and helpful discussions throughout. Original work in the author's laboratory was supported in part by grants from

NIH (NIA), NIMH, NSF, the Alzheimer's Association/Burks B. Lapham grant, the Bader Foundation, and an NIMH Research Scientist Award to C. D. T-A.

References

Anuradha P, Khan SM, Karthikeyan N, Thampan RV 1994 The nonactivated estrogen receptor (naER) of the goat uterus is a tyrosine kinase. Arch Biochem Biophys 309:195–204

Bickel PE, Scherer PE, Schnitzer JE, Oh P, Lisanti MP, Lodish HF 1997 Flotillin and epidermal surface antigen define a new family of caveolae-associated integral membrane proteins. J Biol Chem 272:13793–13802

Brouillet E, Trembleau A, Galanaud D et al 1999 The amyloid precursor protein interacts with Go heterotrimeric protein within a cell compartment specialized in signal transduction. J Neurosci 19:1717–1727

Das SK, Taylor JA, Korach KS, Paria BC, Dey SK, Lubahn DB 1997 Estrogenic responses in estrogen receptor-α deficient mice reveal a distinct estrogen signaling pathway. Proc Natl Acad Sci USA 94:12786–12791

Desdouits-Magnen J, Desdouits F, Takeda S et al 1998 Regulation of secretion of Alzheimer amyloid precursor protein by the mitogen-activated protein kinase cascade. J Neurochem 70:524–530

Gu Q, Korach KS, Moss RL 1999 Rapid action of 17β-estradiol on kainate-induced currents in hippocampal neurons lacking intracellular estrogen receptors. Endocrinology 140:660–666

Hochberg RB, Rosner W 1980 Interaction of 16α-[^{125}I]iodo-estradiol with estrogen receptor and other steroid-binding proteins. Proc Natl Acad Sci USA 77:328–332

Jaffe A, Toran-Allerand CD, Greengard P, Gandy S 1994 Estrogen regulates metabolism of Alzheimer amyloid β precursor protein. J Biol Chem 269:13065–13068

Karthikeyan N, Thampan RV 1996 Plasma membrane is the primary site of localization of the nonactivated estrogen receptor in the goat uterus: hormone binding causes receptor internalization. Arch Biochem Biophys 325:47–57

Landers JP, Spelsberg TC 1992 New concepts in steroid hormone action: transcription factors, proto-oncogenes and the cascade model for steroid regulation of gene expression. Crit Rev Eukaryotic Gene Expression 2:19–63

Lubahn DB, Moyer JS, Golding TS, Couse JF, Korach KS, Smithies O 1993 Alteration of reproductive function but not prenatal sexual development after insertional disruption of the mouse estrogen receptor gene. Proc Natl Acad Sci USA 1993 90:11162–11166

Marshall C 1995 Specificity of receptor tyrosine kinase signaling: transient versus sustained extracellular signal-regulated kinase activation. Cell 80:179–185

Mills J, Laurent Charest D, Lam F et al 1997 Regulation of amyloid precursor protein catabolism involves the mitogen-activated protein kinase signal transduction pathway. J Neurosci 17:9415–9422

Moffatt CA, Rissman EF, Shupnik MA, Blaustein JD 1998 Induction of progestin receptors by estradiol in the forebrain of estrogen receptor-α gene-disrupted mice. J Neurosci 18: 9556–9563

Murphy LC, Leygue E, Dotzlaw H, Douglas D, Coutts A, Watson PH 1997 Oestrogen receptor variants and mutations in human breast cancer. Ann Med 29:221–234

Ogawa S, Inoue S, Watanabe T et al 1998 The complete primary structure of human estrogen receptor β (hER β) and its heterodimerization with ER α in vivo and in vitro. Biochem Biophys Res Commun 243:122–126

Okamoto T, Schlegel A, Scherer PE, Lisanti MP 1998 Caveolins, a family of scaffolding proteins for organizing 'pre-assembled signaling complexes' at the plasma membrane. J Biol Chem 273:5419–5422

Pietras R J, Szego CM 1977 Specific binding sites for oestrogen at the outer surfaces of isolated endometrial cells. Nature 265:69–72

Razandi M, Pedram A, Greene GL, Levin ER 1999 Cell membrane and nuclear estrogen receptors (ERs) originate from a single transcript: studies of ERα and ERβ expressed in Chinese hamster ovary cells. Mol Endocrinol 13:307–319

Schlegel A, Volonte D, Engelman J A et al 1998 Crowded little caves: structure and function of caveolae. Cell Signal 10:457–463

Selkoe D J 1994 Normal and abnormal biology of the β-amyloid precursor protein. Annu Rev Neurosci 17:489–517

Shughrue P J, Lane MV, Merchenthaler I 1999 Biologically active estrogen receptor-β: evidence from *in vivo* autoradiographic studies with estrogen receptor α-knockout mice. Endocrinology 140:2613–2620

Singh M, Sétáló Jr G, Guan X, Warren M, Toran-Allerand CD 1999 Estrogen-induced activation of mitogen activated protein kinase in cerebral cortical explants: convergence of estrogen and neurotrophin signaling pathways. J Neurosci 19:1179–1188

Singh M, Sétáló Jr G, Guan X, Frail DE, Toran-Allerand CD 2000 Estrogen-induced activation of the MAP kinase cascade in the cerebral cortex of estrogen receptor-α knockout (ERKO) mice. J Neurosci 20:1694–1700

Sisodia SS 1992 Secretion of the β-amyloid precursor protein. Ann NY Acad Sci 674:53–57

Stevis PE, Deecher DC, Frail DE 1999 Differential effects of estradiol and estradiol-BSA conjugates. Endocrinology 140:5455–5458

Tang MX, Jacobs D, Stern Y et al 1996 Effect of oestrogen during menopause on risk and age at onset of Alzheimer's disease. Lancet 348:429–432

Toran-Allerand CD 1996 The estrogen/neurotrophin connection during neural development: is co-localization of estrogen receptors with the neurotrophins and their receptors biologically relevant? Dev Neurosci 18:36–48

Toran-Allerand CD, Gerlach JL, McEwen BS 1980 Autoradiographic localization of ^{3}H-estradiol related to steroid responsiveness in cultures of the newborn mouse hypothalamus and preoptic area. Brain Res 184:517–522

Toran-Allerand CD, Mauri E, Leung C, Warren M, Singh M 1996 Activation of MAP kinases (ERKs) by estradiol in cerebral cortical explants: cross-coupling of the estrogen and neurotrophin signaling pathways. Soc Neurosci Abstr 22:555

Toran-Allerand CD, Singh M, Sétáló Jr G 1999 Novel mechanisms of estrogen action in the brain: new players in an old story. Front Neuroendocrinol 20:97–121

Watson CS, Norfleet AM, Pappas TC, Gametchu B 1999 Rapid actions of estrogens in GH3/B6 pituitary tumor cells via a plasma membrane version of estrogen receptor-α. Steroids 64:5–13

Watters J J, Campbell JS, Cunningham M J, Krebs EG, Dorsa DM 1997 Rapid membrane effects of steroids in neuroblastoma cells: effects of estrogen on mitogen activated protein kinase signalling cascade and c-fos immediate early gene transcription. Endocrinology 138:4030–4033

Witkowska HE, Carlquist M, Engstrom O et al 1997 Characterization of bacterially expressed rat estrogen receptor β ligand binding domain by mass spectrometry: structural comparison with estrogen receptor α. Steroids 62:621–631

Xu H, Gouras GK, Greenfield JP et al 1998 Estrogen reduces neuronal generation of Alzheimer β-amyloid peptides. Nat Med 4:447–451

DISCUSSION

McEwen: Were the studies you have shown done in whole cells or in membrane preparations?

Toran-Allerand: All the work that I have described was done on explant slices 360 μm thick from the developing rodent cerebral cortex, usually taken on the day after birth. These are coronal sections that are maintained for one week in organotypic culture. I have not yet worked with a membrane preparation or single cell preparations. We have, however, also analysed the complex in studies of the whole postnatal day 7 brain and are studying the complex in membrane preparations.

McEwen: Ellis Levin, your finding of the inhibition of kinase activities through ERα would seem to be consistent with the conclusion Dominique Toran-Allerand is drawing about its possibly inhibitory role. Do you think they are related?

Levin: I think they're very much related. Why is oestrogen a trophic, positive signalling factor in the one, and an inhibitory influence in another cell? I think these kinds of interactions will be important to explain this.

I would like to ask a question. Flotillin is generally thought to be a negative regulatory signalling molecule. If this is true, are there changes in the association of ER with flotillin and this other complex? I guess you don't know whether this is indirect or direct, given the model you are looking at. It could be that ER is associating with flotillin through three other proteins in this protein gemisch.

Secondly, how would you reconcile that aspect with the positive signalling actions of oestrogen in your model? The corollary of this is, is there a difference between the unliganded and the liganded ER and its association with any of these molecules?

Toran-Allerand: We've only recently started looking at flotillin. According to Michael Lisanti who has done much of the work on flotillin, it is not yet clear whether it behaved as caveolin (Bickel et al 1997). There have only been seven papers published on flotillin. No one has yet investigated its role as a positive or negative signalling molecule.

In terms of the association between ER and flotillin and whether this is direct or indirect, I think everything is binding to flotillin in my system. In other systems everything may be binding to caveolin. But in our case the common denominator is probably flotillin. For caveolin at least, the signalling complexes are pre-assembled in quiescent cells through binding to caveolin. Exposure of quiescent cells to a growth factor or other signal, for example, generates the signal.

Levin: But these are inhibitory proteins.

Toran-Allerand: There is no evidence that flotillin is inhibitory in an active manner. It appears to be an integral membrane protein to which signalling proteins are attached in the quiescent state. It is inhibitory in the sense that the kinases are.

Levin: From what is known about them, caveolin and flotillin are inhibitory proteins that inhibit signalling in the basal state. But the question I wanted an answer to was whether the ER was physically associating with flotillin differently in the ligated versus unligated state?

Toran-Allerand: We don't know that yet, but we are actively pursuing this question. With respect to the binding of the ER to flotillin, we don't know much. Moreover, we are dealing with neurons — most of the of the studies that have been done have been done in PC12 cells that have been transfected with caveolin, whereas I'm trying to look at the wild-type condition where we're not altering anything. All I can tell you is that if you take the tissue, isolate ER by co-precipitation and probe for all the supposed members of caveolae, you find a physical association. Whether this is because the ER binds physically to Raf or ERK or whatever, I don't know. My feeling is that if flotillin, like caveolin, is a scaffolding protein, this is a kind of common denominator, and when you immunoprecipitate for the ER and probe for another member of the complex you are finding it because everything is bound to this multivalent protein. We are currently looking at whether or not the presence or absence of the ligand affects this. In the studies I have shown, we have not added ligands, and since these are explants that are grown in steroid-deficient conditions I can't answer the other part of the question.

Watson: I thought the first part of your paper was leading to the conclusion that there was a unique oestrogen receptor protein in the membrane that you called 'ER-X'. However, in the latter half when you described immunoprecipitations, were those ERα or ERβ antibodies you were coprecipitating with?

Toran-Allerand: That's a good question. I think it is a different receptor. It has a ligand-binding domain (LBD) which may be common, because obviously we're getting an effect with oestradiol. It doesn't have to have a DNA-binding domain; it doesn't really move anywhere. All it has to do is to allow conformational changes to take place which then trigger other responses. We're currently sequencing the protein and trying to clone the gene, but I don't know the results yet. We are using antibodies to ERα, but these are antibodies which recognize either only the LBD or are generated by the full length protein.

Watson: With regard to the pharmacological profile of this receptor, if it is ERα sitting in a membrane lipid environment, it might be conformationally different: its ligand binding pocket geometry might be slightly altered. This is a potential explanation for why most people find a slightly different pharmacological profile for these receptors.

Simpkins: Do you see the transient nature of the ERK phosphorylation as part of the biology of the cell, rather than the pharmacology of oestrogens in the system? I presume you put the oestrogen in and leave it in.

Toran-Allerand: We pulse for different time periods.

Simpkins: So when you describe a response for 2 h, have the cells been exposed for 2 h?

Toran-Allerand: We deprive the cells of steroid for 36 h, then we pulse, and we then wait for varying periods after the oestradiol is washed out. At least for our purposes, it is important that the ERK activation seems to last at least 2–3 h, because this is more related to differentiative effects then the rapid short-lasting effects that are more associated with mitogenic activity.

McEwen: Does the activation of ERK last longer than in the ERKO animals, which lack ERα?

Toran-Allerand: It's stronger, but I don't know about longer. We have had problems obtaining enough animals to do these experiments, since we are still establishing the colony.

Kushner: Don't give up on the ERKOs because of your results with 17α-oestradiol. We just tested whether it would allow interactions between receptor and coactivator. It worked just fine. I looked up some recent studies that Jan-Åke Gustafsson has done and he got the same results. Your intriguing result that the ERKOs have this potent activity really puts the finger of suspicion on ERβ, which is there and is not going to be subject to some of the opposing influences of ERα.

Then you did the genistein experiment. Genistein is going to require more than 100 nM concentration to bind efficiently to and saturate ERβ.

Toran-Allerand: You have to be very careful with genistein because it's a receptor tyrosine kinase inhibitor at high concentrations.

Kushner: I know, and it's a very sharp peak. We have done reporter gene studies in which we get this activation, saturate the receptor, and right after that you start to get the tyrosine kinase inhibition. It is tricky to find the concentration.

Bethea: My understanding of 17α-oestradiol is not that it's some kind of inert compound, but that it just has a faster degradation in whole animals and is cleared quickly. However, if you give it to cells in culture, or in situations where it is not rapidly degraded, then it can act effectively at oestrogen receptors.

Toran-Allerand: First, I hope I didn't say it was inert; I just said it had a lower affinity. Neil MacLuskey who has done extensive binding studies tells me that it has 100-fold lower affinity for the receptor than 17β-oestradiol.

Simpkins: If you load animals up with high levels of 17α-oestradiol, you can keep it occupying oestrogen receptors. But it is in the range of 100–1000-fold lower in potency than 17β-oestradiol. This is in a variety of cell types.

Pfaff: I was watching carefully for evidence against the notion that molecular events in the nucleolus and the nucleus, some of which are strongly influenced by oestrogen (Cohen & Pfaff 1981, 1992, Jones et al 1990), are involved in the growth response that you showed in your first slide. I didn't see any evidence against that.

Toran-Allerand: I didn't say that I had evidence against it. I'm interested in various mechanisms by which oestrogen could be stimulating the growth and arborization of axons or dendrites, a differentiative effect of oestrogen I first described in 1976 (Toran-Allerand 1976).

Oestrogen can be survival-promoting not just by influencing caspases or Bcl-2, or any of the various apoptotic mechanisms. One of the important ways by which neurons survive is to be either afferented or connected. If one has a substance such as oestrogen which allows or enhances the formation of contacts between neurons, one will then get survival of those neurons; if one deprives them of those contacts, those cells will die. This is a different mechanism than influencing apoptosis pathways, and it is something that people don't consider too often. But I think that, particularly during development, it's an extremely important mechanism of cell survival.

I have not at all dismissed the possibility of influences at the genomic level influencing growth. Traditionally there is ample evidence that one can block the growth of axons and dendrites by giving inhibitors of the oestrogen receptor. However, I'm interested in other mechanisms, and I've always been particularly interested in interactions of oestrogen with neurotrophins. This is why I happen to focus on cross-coupling of their signalling pathways. But it certainly is not the sole mechanism by which oestrogen elicits the growth and differentiation of neurons, since neurite growth and differentiation consist of many components (e.g. initiation, elaboration, arborization). It is entirely conceivable that rapid oestrogen signalling may mediate the early steps of neurite growth which may not depend solely on transcription, but that later events do require events occurring at the genomic level.

References

Bickel PE, Scherer PE, Schnitzer JE, Oh P, Lisanti MP, Lodish HF 1997 Flotillin and epidermal surface antigen define a new family of caveolae-associated integral membrane proteins. J Biol Chem 272:13793–13802

Cohen RS, Pfaff DW 1981 Ultrastructure of neurons in the ventromedial nucleus of the hypothalamus in ovariectomized rats with or without estrogen treatment. Cell Tissue Res 217:451–470

Cohen RS, Pfaff DW 1992 Ventromedial hypothalamic neurons in the mediation of long-lasting effects of estrogen on lordosis behavior. Prog Neurobiol 38:423–453

Jones K, Harrington C, Chikaraishi D, Pfaff DW 1990 Steroid hormone regulation of ribosomal RNA in rat hypothalamus: early detection using *in situ* hybridization and precursor-product ribosomal DNA probes. J Neurosci 10:1513–1521

Toran-Allerand CD 1976 Sex steroids and the development of the newborn mouse hypothalamus and pre-optic area *in vitro*: implications for sexual differentiation. Brain Res 106:407–412

Oestrogen modulation of noradrenaline neurotransmission

Allan E. Herbison, Sharon X. Simonian, Niren R. Thanky and R. John Bicknell

Laboratory of Neuroendocrinology, The Babraham Institute, Cambridge CB2 4AT, UK

Abstract. Noradrenaline (NA) exerts an important neuromodulatory role within diverse neuronal networks and is also likely to be a target for oestrogen in the brain. Distinct, highly organized sub-populations of brainstem NA neurons express oestrogen receptors (ERs) and some of these display species differences. A number of genes expressed by NA neurons, ranging from transcription factors to co-released neuropeptides, are influenced by oestrogen and may have roles in the predominant enhancement in NA activity in response to oestrogen. The effects of oestrogen on genes involved directly in NA biosynthesis are less clear, although promoter transgenic work suggests oestrogen to have a powerful influence upon tyrosine hydroxylase gene transcription. In addition to direct actions on NA neurons, evidence suggests that oestrogen also regulates adrenergic receptor expression and function within the ER-rich hypothalamus as well as the cerebral cortex. Together, these investigations point to a multifaceted pre- and postsynaptic regulation of NA transmission by oestrogen. While the hypothalamic neuronal networks controlling reproduction remain the principal site of investigation of oestrogen regulated NA transmission, the role of oestrogen and NA and their potential interactions in cortical functioning are becoming of equal interest.

2000 Neuronal and cognitive effects of oestrogens. Wiley, Chichester (Novartis Foundation Symposium 230) p 74–93

Noradrenaline (NA) exerts potent neuromodulatory actions upon the functioning of diverse neuronal networks within the mammalian nervous system. The neurons responsible for these widespread effects of NA have their cell bodies clustered within the pons and medulla of the brainstem where they form several distinct populations (Dahlstrom & Fuxe 1964). Of these, the so-called A1 and A2 groups of the medulla and the A6 population, comprising the locus coeruleus of the pons, are the largest and are believed to provide most of the ascending and descending NA inputs within the nervous system. More specifically, the A1 and A2 neurons are known to project widely within the hypothalamus and associated limbic regions while the NA neurons forming the A6 innervate essentially the whole neuroaxis and are the unique source of NA for the cerebral cortex (Ungerstedt 1971, Swanson & Hartman 1974, Ricardo & Koh 1978). The actions of NA upon single neurons have been found to depend upon region but have often been observed to alter signal-to-noise ratios and 'gate' or 'enable' the responses of networks to afferent stimuli (Foote & Morrison 1986, Waterhouse et al 1988, Manunta & Edeline 1999). These properties have led to NA being regarded as a permissive or neuromodulatory signalling molecule. As the electrical activity of NA neurons in the locus coeruleus is synchronously activated by multiple different sensory stimuli (Aston-Jones & Bloom 1981), NA is thought to be particularly important in determining levels of network arousal in response to external stimuli (Aston-Jones & Bloom 1981, Foote & Morrison 1986, Waterhouse et al 1988).

Thus, the NA system represents one where relatively small populations of discretely localized neurons use highly divergent projection pathways to exert widespread effects throughout the nervous system. As such, the NA neurons would be the ideal substrate through which gonadal steroids might influence brain functioning in a global manner. This paper will review evidence for the sensitivity of the brainstem NA neurons to the gonadal steroid oestrogen, and assess the degree to which NA may be involved in the oestrogen-dependent plasticity of brain function.

Oestrogen receptor (ER) expression by NA neurons

Early receptor autoradiography investigations undertaken in the rat demonstrated that up to 70% of A1 and A2 neurons were able to accumulate radioactive oestradiol (Heritage et al 1977, Sar & Stumpf 1981). Subsequent immunocyto-chemical studies revealed the presence of nuclear-located ERα protein in A1 and A2 neurons (Simonian & Herbison 1997) and identified a clear rostrocaudal topography of ERα expression by NA neurons; A1 and A2 neurons located in the caudal medulla expressed ERα while those in the rostral medulla did not. Intriguingly, this rostrocaudal topography was the inverse of that observed for neuropeptide Y synthesis by NA cells, and triple labelling studies revealed that A1 and A2 neurons expressed either ERα or NPY but not both (Simonian & Herbison 1997).

It is of note that only 30% of caudal A1 cells were found to be immunoreactive for ERα while the earlier autoradiographic studies had reported that 50–70% of A1 cells accumulated oestradiol (Heritage et al 1977, Sar & Stumpf 1981). This discrepancy may be resolved by the finding that scattered cells in the region of the A1 cell group express mRNA for the second ER, ERβ (Shugrhue et al 1997). As such, it is possible that some A1 cells express ERα while others synthesize ERβ. A similar argument is not required for the A2 neurons where the percentage of NA neurons accumulating oestradiol and expressing ERα immunoreactivity are equal (Heritage et al 1977, Simonian & Herbison 1997). There is, nevertheless,

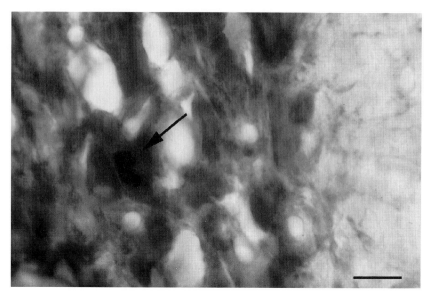

FIG. 1.   High-power photomicrograph of the locus coeruleus double stained for dopamine-β-hydroxylase (grey) and ERα (black). A single dual-labelled cell is indicated by the arrow. Scale bar represents 10 μm.

substantial ERβ mRNA expression in the region of the A2 neurons (Shugrhue et al 1997) and it remains possible that these neurons express both ERs.

Alongside the demonstration of ERs in A1 and A2 NA neurons, the autoradiography experiments also reported that 30–40% of A6 neurons accumulated oestradiol (Sar & Stumpf 1981). In the course of our dual-labelling experiments involving the A1 and A2 neurons, we also detected ERα immunoreactivity in A6 neurons (S. X. Simonian & A. E. Herbison, unpublished results; Fig. 1) although co-expressing cells were relatively infrequent and much fewer than that suggested by autoradiography. Again, as ERβ transcripts have been described in the locus coeruleus (Shugrhue et al 1997), it is possible that both ERs are present in different populations of A6 neurons.

The above work suggests that all three of the major NA cell groups in the brainstem express classic nuclear-located ERs in the rat. Limited work has been undertaken in other species to determine the generality of these findings within the mammalian brain. Dual-labelling immunocytochemical studies in the ovine brain have identified a rostrocaudal topography in ERα expression by A1 and A2 neurons which is identical to the rat, although no evidence was found for ERα immunoreactivity in any A6 neurons of the sheep brain (Simonian et al 1998). Although methodological problems may result in false-negative findings, it is

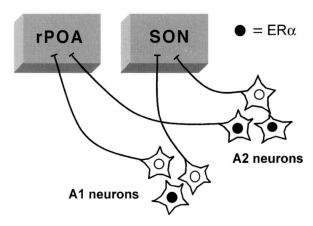

FIG. 2.   Schematic diagram showing which of the A1 and A2 neurons projecting to the rostral preoptic area (rPOA) and supraoptic nucleus (SON) express ERα in the female rat. Although non-NA neurons containing ERα project to the SON, none of the NA inputs synthesize ERα. In contrast, it is only the A2 neurons projecting to the rPOA which express ERα.

interesting that a recent study has similarly failed to detect ERα immunostaining in A6 neurons of the rhesus monkey (Schutzer & Bethea 1997). The presence of ERs in A1 and A2 neurons in primate species has not been investigated. Together, these limited data suggest that no substantial species differences exist with regard to ER expression in the A1 and A2 cell groups but that, at least for ERα, this may not be the case for the A6 neurons.

Projections of ER-expressing NA neurons within the forebrain

As the expression of ERα in brainstem NA neurons exhibited a high degree of specificity both in terms of co-expressed neuropeptides and neuroanatomical topography (Simonian & Herbison 1997), it was of great interest to determine the projection sites of oestrogen-receptive NA neurons. Using triple-labelling immunocytochemical techniques combined with retrograde tracing, we have examined the projection patterns of oestrogen-receptive A1 and A2 neurons to both the rostral preoptic area (rPOA) and the supraoptic nucleus (SON) of the rat hypothalamus (Fig. 2). These studies demonstrated that both A1 and A2 cell groups innervated the rPOA, and that approximately 50% of the A2 afferents expressed ERα compared with none of the A1 neurons (Simonian et al 1999). In contrast, none of the A1 and A2 cells innervating the SON were immunoreactive for ERα even though neurochemically unidentified ERα-expressing neurons immediately adjacent to the A2 cells projected to the SON (Voisin et al 1997). Although these studies have examined the NA innervation of just two hypothalamic brain regions, they do suggest a further degree of functional organization within the oestrogen-receptive NA neurons of the medulla. Despite the quite diffuse projections of A1 and A2 neurons (Ricardo & Koh 1978), it would seem that NA neurons expressing ERα have relatively specific targets within the brain. Clearly, more work will be required to test this hypothesis and learn whether it may also be applicable to the more globally projecting A6 neurons.

Effects of oestrogen on brainstem NA neurons

During development, it would appear that gonadal steroids, and in particular oestrogen, can exert important effects upon the survival of NA neurons as well as their future biosynthetic capacity. The number of A6 neurons in the adult rat depends upon perinatal exposure to testosterone (Guillamon et al 1988), and prenatal oestrogen exposure has been reported to determine the postnatal NA content of specific brain regions such as the cingulate cortex (Stewart & Rajabi 1994).

In the mature animal there is substantial evidence that oestrogen can alter both the biosynthetic and electrical activity of brainstem NA neurons. Electro-physiological studies have revealed that oestrogen can increase the firing rate of A1 neurons (Kaba et al 1983) and also induce expression of transcription factors such as c-Fos and the progesterone receptor in A2 cells (Jennes et al 1992, Haywood et al 1999). *In situ* hybridization studies have also shown that oestrogen can induce the expression of galanin in A6 neurons (Tseng et al 1997). However, the effects of oestrogen upon genes involved directly in NA biosynthesis have been less easy to demonstrate. A study in the male rat found both acute inhibitory and stimulatory effects of oestrogen on mRNA levels of the rate limiting enzyme tyrosine hydroxylase (TH) in the A1 but not elsewhere in the brainstem (Liaw et al 1992). Another found no effect of oestrogen on TH mRNA expression in the locus coeruleus of the female rat (Tseng et al 1997). Furthermore, *in situ* hybridization studies in the female monkey (Schutzer & Bethea 1997) and rat (R. J. Bicknell, unpublished results) have been unable to determine any effect of oestrogen upon NA transporter mRNA levels.

In an attempt to re-examine the potential effects of oestrogen on TH, and in particular TH gene transcription, we have recently used a line of transgenic mice bearing a 9 kb TH promoter–LacZ construct which enables changes in TH gene transcription to be assayed through analysis of β-galactosidase expression in NA neurons (Min et al 1996). These studies have revealed a robust effect of gonadal steroids upon transgene expression in the locus coeruleus A6 neurons of adult female mice (Fig. 3). Long-term ovariectomy was found to dramatically reduce transgene expression while oestrogen replacement for two weeks returned TH promoter-driven transgene expression to levels observed in sham-operated mice (Fig. 3). This result suggests that oestrogen, either directly or indirectly, exerts a powerful role in enhancing TH gene transcription in A6 neurons.

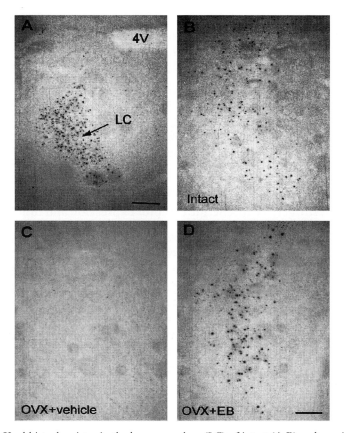

FIG. 3. Xgal histochemistry in the locus coeruleus (LC) of intact (A,B) and ovariectomized vehicle- (C) and oestrogen- (D) treated TH9-LacZ transgenic mice. In these animals 9 kb of tyrosine hydroxylase sequence directs LacZ expression to the noradrenergic phenotype in the LC (A). Long-term ovariectomy results in an almost complete absence of transgene expression (C) while oestrogen replacement for 2 weeks returns transgene expression to normal levels (D). 4V, fourth ventricle. Scale bars represent 100 μm (A) and 50 μm (B–D).

Together, these various investigations raise the possibility that oestrogen acts on NA neurons to alter TH, but not NA transporter, synthesis as well as to modulate the expression of several transcription factors (Fig. 4). How these effects integrate to alter biosynthesis and firing rates of NA neurons is not understood at present. Nevertheless, a consensus exists suggesting that the net effect of oestrogen on NA neurons is to increase the release of neurotransmitter from NA terminals. It is well documented that oestrogen enhances NA release and turnover rates in various hypothalamic regions (see Herbison 1998) and a single report suggests a similar effect in the cerebral cortex (Karkanias et al 1998). Although it is undoubtedly useful to establish the overall quantitative effect of oestrogen on NA release in target sites, it is likely to be equally important in future studies to determine more subtle effects of oestrogen on parameters such as the rate and patterning of NA release at target sites (Brown et al 1994).

Finally, it is worth noting that the actions of oestrogen on NA neurons may also occur in a reciprocal fashion with NA modulating ER expression. Studies by Blaustein and colleagues have shown that the inhibition of either NA synthesis or adrenergic receptor functioning in the brain can reduce ER expression in the hypothalamus (Blaustein et al 1986, Blaustein 1987). Thus NA may serve to transmit oestrogen status within the brain while also maintaining the receptivity of its targets to oestrogen.

Effects of oestrogen on adrenergic receptor expression and function

Pharmacological studies have long demonstrated that oestrogen influences both the binding and responsiveness of different brain regions to adrenergic agonists (see Wagner et al 1979, Weiland & Wise 1987, 1989). More recently, the work of Etgen and colleagues has provided a detailed dissection of the manner in which oestrogen regulates α adrenergic receptor expression in the hypothalamus and preoptic area of the female rat. Oestrogen was found to exert actions on two specific receptor subtypes; down-regulation of the presynaptic $\alpha_{2A/D}$ receptor reduced NA autoregulation at the nerve terminal and led to enhanced NA release, while up-regulation of the postsynaptic α_{1B} receptor mRNA and protein resulted in the indirect potentiation of β adrenergic-stimulated cAMP levels (Karkanias & Etgen 1993, Karkanias et al 1996). Both of these actions should enhance NA signalling at as yet unspecified neuronal populations within the hypothalamus. In support of this, electrophysiological studies have demonstrated that oestrogen enhances the responsiveness of populations of hypothalamic neurons to NA and α_1 adrenergic receptor agonists (Kim et al 1987, Condon et al 1989). However, it is important to note the region- and neuron-specific nature of oestrogen action, as evidence from other studies indicates that oestrogen may equally repress the consequences of NA signalling in relation to specific neuronal phenotypes of the hypothalamus (Herbison et al 1990, Condé et al 1996). In this light it is important to note that alterations in the signal-to-noise ratio within any network may result from enhancement of the specific signal and/or a reduction in background firing. Thus, oestrogen's ability to both enhance and suppress the NA responses of different neuronal elements would be entirely compatible with its ability to change signal-to-noise ratios within a specific network.

It is also relevant and interesting to note that oestrogen influences adrenergic receptor expression within the cerebral cortex. In this region, mRNA expression of both the $\alpha_{2A/D}$ and α_{1B} receptors is down-regulated by oestrogen although only the former is found to result in altered functional receptor binding (Karkanias et al

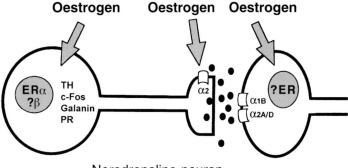

Oestrogen   Oestrogen   Oestrogen

Noradrenaline neuron

FIG. 4.   Schematic diagram showing the potential sites of oestrogen action upon NA signalling within the brain. Specific populations of NA neurons express ERα, and possibly ERβ, and these receptors may mediate known effects of oestrogen upon tyrosine hydroxylase (TH), c-Fos, galanin and progesterone receptor (PR) expression. Equally, ERs in undefined target neurons may mediate known actions of oestrogen on the expression of specific adrenergic receptors.

1996, 1997). These findings, in association with oestrogen-induced alterations in the β adrenergic responsiveness of the cerebral cortex (Wagner et al 1979), have extended the spectrum of oestrogen-modulated NA signalling to areas outside of the hypothalamus. In the case of the cerebral cortex, any direct genomic influence of oestrogen is likely to be occurring through ERβ (Shugrhue et al 1997, McEwan & Alves 1999). These studies also highlight the importance of oestrogen in modulating both pre- and post-synaptic components of the NA network (Fig. 4).

Oestrogen, NA and neuronal plasticity

It is clear that oestrogen exerts an influence upon brainstem NA neurons. Whether this actually occurs through the nuclear-located ERs synthesized by many of these neurons, or through a non-classical mechanism of oestrogen action in the brain (McEwan & Alves 1999), remains to be established. Nevertheless, at present, the simplest manner of defining oestrogen-receptive NA neurons is through their expression of ERs, and on this basis, it would appear that specific species differences are likely to exist. Furthermore, it seems that the brainstem NA neurons with their highly divergent and widespread afferent projections may not be used by oestrogen to influence brain functioning in a global manner. Based on the limited tract tracing data available, the projections of NA neurons expressing ERα have turned out to be highly target-specific (Fig. 2). Thus, the NA neurons may function more as a neural substrate co-ordinating oestrogen actions amongst specific yet diverse neuronal networks.

Many studies have examined which hypothalamic networks might be influenced by oestrogen-dependent NA neurotransmission and it is perhaps not surprising to find that most evidence points to those regulating reproductive function (see Etgen et al 1993, Herbison 1998). For example, in terms of the gonadotropin-releasing hormone (GnRH) neurons, it is thought that the ER-expressing A2 neurons play a role in mediating oestrogen's ability to alter GnRH biosynthesis and secretion (Herbison 1998). In this case, it is suggested that NA exerts a permissive or enabling role within the GnRH network and that fluctuating oestrogen levels through the ovarian cycle drive fluctuations in NA signalling which, at their peak, enable other neurotransmitters to massively activate the GnRH neurons to induce ovulation (Herbison 1998).

Outside of the hypothalamus, relatively little research has been undertaken to examine the networks which may be influenced by oestrogen-dependent NA inputs. The one region that has achieved growing interest is the cerebral cortex where oestrogen has now been shown to influence adrenergic receptor expression as well as NA release in the female rat (Karkanias et al 1997, 1998). There is increasing evidence that oestrogen exerts subtle actions upon a range of cognitive processes in rodents and primates and that it may also be related to negative affect in depression (see McEwan & Alves 1999). Equally, adrenergic receptors, and in particular the α_2 receptor, are thought to be involved in the processes of cognition, selective attention and arousal while reduced levels of NA have been implicated in the pathophysiology of depression (Arnsten & Goldman-Rakic 1985, Steketee et al 1989, Berridge et al 1993, Arnsten et al 1996). Interestingly, there is also evidence for a decline in NA signalling with age and a substantial reduction in the number of A6 neurons in Alzheimer's disease (Iversen et al 1983, Arnsten & Goldman-Rakic 1985). Although it remains necessary to establish oestrogen-dependent alterations in NA signalling in the primate cortex, the potential for oestrogen to enhance NA neurotransmission appears to hold promise as an important means through which cognitive function can be enhanced in the damaged or ageing brain.

Acknowledgements

We are especially grateful to the numerous colleagues past and present of the Laboratory of Neuroendocrinology who have contributed to our work on the brainstem noradrenaline neurons. Work supported by the BBSRC.

References

Arnsten AFT, Goldman-Rakic PS 1985 α_2-adrenergic mechanisms in prefrontal cortex associated with cognitive decline in aged nonhuman primates. Science 230:1273–1276

Arnsten AFT, Steere JC, Hunt RD 1996 The contribution of α_2-noradrenergic mechanisms to pre-frontal cortical cognitive function. Potential significance for attention-deficit hyperactivity disorder. Arch Gen Psychiatry 53:448–455

Aston-Jones G, Bloom FE 1981 Nonrepinephrine-containing locus coeruleus neurons in behaving rats exhibit pronounced responses to non-noxious environmental stimuli. J Neurosci 1:887–900

Berridge CW, Arnsten AF, Foote SL 1993 Noradrenergic modulation of cognitive function: clinical implications of anatomical, electrophysiological and behavioural studies in animal models. Psychol Med 23:557–564

Blaustein JD 1987 The α_1-noradrenergic antagonist prazosin decreases the concentration of estrogen receptors in female rat hypothalamus. Brain Res 404:39–50

Blaustein JD, Brown TJ, Swearengren ES 1986 Dopamine-β-hydroxylase inhibitors modulate the concentration of functional estrogen receptors in female rat hypothalamus and pituitary gland. Neuroendocrinology 43:150–158

Brown DA, Herbison AE, Robinson JE, Marrs RW, Leng G 1994 Modelling the luteinising hormone-releasing hormone pulse generator. Neuroscience 63:869–879

Condé GL, Herbison AE, Fernandez-Galaz C, Bicknell RJ 1996 Estrogen uncouples noradrenergic activation of Fos expression in the female rat preoptic area. Brain Res 735:197–207

Condon TP, Ronnekleiv OK, Kelly MJ 1989 Estrogen modulation of the α-1-adrenergic response of hypothalamic neurons. Neuroendocrinology 50:51–58

Dahlstrom A, Fuxe K 1964 Evidence for the existence of monoamine-containing neurons in the central nervous system. Acta Physiol Scand Suppl 232:1–55

Etgen AM, Ungar S, Petitti N 1993 Estradiol and progesterone modulation of norepinephrine neurotransmission: implications for the regulation of female reproductive behaviour. J Neuroendocrinol 4:255–272

Foote SL, Morrison JH 1986 Extrathalamic modulation of cortical function. Annu Rev Neurosci 10:67–95

Guillamon A, de Blas MR, Segovia S 1988 Effects of sex steroids on the development of the locus coeruleus in the rat. Dev Brain Res 40:306–310

Haywood SA, Simonian SX, van der Beek EM, Bicknell RJ, Herbison AE 1999 Fluctuating estrogen and progesterone receptor expression in brainstem norepinephrine neurons through the rat estrous cycle. Endocrinology 140:3255–3263

Herbison AE 1998 Multimodal influence of estrogen upon gonadotropin-releasing hormone neurons. Endocr Rev 19:302–330

Herbison AE, Heavens RP, Dyer RG 1990 Oestrogen modulation of excitatory A1 noradrenergic input to rat medial preoptic gamma aminobutyric acid neurones demonstrated by microdialysis. Neuroendocrinology 52:161–168

Heritage AS, Grant ID, Stumpf WE 1977 ^{3}H estradiol in catecholamine neurons of rat brain stem: combined localization by autoradiography and formalin-induced fluorescence. J Comp Neurol 176:607–630

Iversen LL, Rossor MN, Reynolds GP et al 1983 Loss of pigmented dopamine-β-hydroxylase positive cells from locus coeruleus in senile dementia of Alzheimer type. Neurosci Lett 39: 95–100

Jennes L, Jennes ME, Purvis C, Nees M 1992 c-fos expression in noradrenergic A2 neurons of the rat during the estrous cycle and after steroid hormone treatments. Brain Res 586:171–175

Kaba H, Saito H, Otsuka K, Seto K, Kawakami M 1983 Effects of estrogen on the excitability of neurons projecting from the noradrenergic A1 region to the preoptic and anterior hypothalamic area. Brain Res 274:156–159

Karkanias GB, Etgen AM 1993 Estradiol attenuates α_2-adrenoceptor-mediated inhibition of hypothalamic norepinephrine release. J Neurosci 13:3448–3455

Karkanias GB, Ansonoff MA, Etgen AM 1996 Estradiol regulation of α_{1b}-adrenoceptor mRNA in female rat hypothalamus-preoptic area. J Neuroendocrinol 8:449–455

Karkanias GB, Li C-S, Etgen AM 1997 Estradiol reduction of α_2-adrenoceptor binding in female rat cortex is correlated with decreases in $\alpha_{2A/D}$-adrenoceptor messenger RNA. Neuroscience 81:593–597

Karkanias GB, Morales JC, Etgen AM 1998 Effects of diabetes and estradiol on norepinephrine release in female rat hypothalamus, preoptic area and cortex. Neuroendocrinology 68:30–36

Kim YI, Dudley CA, Moss RL 1987 A1 noradrenergic input to medial preoptic-medial septal area: an electrophysiological study. Neuroendocrinology 45:77–85

Liaw JJ, He JR, Hartman RD, Barraclough CA 1992 Changes in tyrosine hydroxylase mRNA levels in medullary A1 and A2 neurons and locus coeruleus following castration and oestrogen replacement in rats. Brain Res Mol Brain Res 13:231–238

Manunta Y, Edeline J-M 1999 Effects of noradrenaline on frequency tuning of auditory cortex neurons during wakefulness and slow-wave sleep. Eur J Neurosci 11:2134–2150

McEwan BS, Alves SE 1999 Estrogen actions in the central nervous system. Endocr Rev 20:279–307

Min N, Joh TH, Corp ES, Baker H, Cubells JF, Son JH 1996 A transgenic mouse model to study transsynaptic regulation of tyrosine hydroxylase gene expression. J Neurochem 67:11–18

Ricardo JA, Koh ET 1978 Anatomical evidence of direct projections from the nucleus of the solitary tract to the hypothalamus, amygdala, and other forebrain structures in the rat. Brain Res 153:1–26

Sar M, Stumpf WE 1981 Central noradrenergic neurones concentrate ^{3}H-oestradiol. Nature 289:500–502

Schutzer WE, Bethea CL 1997 Lack of ovarian steroid hormone regulation of norepinephrine transporter mRNA expression in the non-human primate locus coeruleus. Psycho-neuroendorinology 22:325–336

Shugrhue PJ, Lane MV, Merchenthaler I 1997 Comparative distribution of estrogen receptor-α and -β mRNA in the rat central nervous system. J Comp Neurol 388:507–525

Simonian SX, Herbison AE 1997 Differential expression of estrogen receptor and neuropeptide Y by brainstem A1 and A2 noradrenaline neurons. Neuroscience 76:517–529

Simonian SX, Delaleu B, Caraty A, Herbison AE 1998 Estrogen receptor expression in brainstem noradrenergic neurons of the sheep. Neuroendocrinology 67:392–402

Simonian SX, Spratt DP, Herbison AE 1999 Identification and characterization of estrogen receptor α-containing neurons projecting to the vicinity of the gonadotropin-releasing hormone Perikarya in the rostral preoptic area of the rat. J Comp Neurol 411:346–358

Steketee JD, Silverman PB, Swann AC 1989 Forebrain norepinephrine involvement in selective attention and neophobia. Physiol Behav 46:577–583

Stewart J, Rajabi H 1994 Estradiol derived from testosterone in prenatal life affects the development of catecholamine systems in the frontal cortex in the male rat. Brain Res 646:157–160

Swanson LW, Hartman BK 1974 The central adrenergic system. An immunofluorescence study of the location of cell bodies and their efferent connections in the rat utilizing dopamine-β-hydroxylase as a marker. J Comp Neurol 163:467–504

Tseng JY, Kolb PE, Raskind MA, Miller MA 1997 Estrogen regulates galanin but not tyrosine hydroxylase gene expression in the rat locus coeruleus. Brain Res Mol Brain Res 50:100–106

Ungerstedt U 1971 Stereotaxic mapping of the monoamine pathways in the rat brain. Acta Physiol Scand Suppl 367:1–48

Voisin DL, Simonian SX, Herbison AE 1997 Identification of estrogen receptor-containing neurons projecting to the rat supraoptic nucleus. Neuroscience 78:215–228

Wagner HR, Crutcher KA, Davis JN 1979 Chronic estrogen treatment decreases β-adrenergic responses in rat cerebral cortex. Brain Res 171:147–151

Waterhouse BD, Sessler FM, Cheng J, Woodward DJ, Azizi SA, Moises HC 1988 New evidence for a gating action of norepinephrine in central neuronal circuits of mammalian brain. Brain Res Bull 21:425–432

Weiland NG, Wise PM 1987 Estrogen alters the diurnal rhythm of α_1-adrenergic receptor densities in selected brain regions. Endocrinology 121:1751–1758

Weiland NG, Wise PM 1989 Diurnal rhythmicity of beta-1- and beta-2-adrenergic receptors in ovariectomized, ovariectomized estradiol-treated and proestrous rats. Neuroendocrinology 50:655–662

DISCUSSION

Levin: You showed an inverse relationship between NPY and oestrogen receptors in A1 and A2. Is there evidence of regulation in either direction? And, if so, as NPY is so important for certain behaviours, can you postulate some regulatory mechanisms?

Herbison: We initiated that study because, in the oestrogen modulation of the hypothalamus, indices of NPY and adrenergic activity appear to fluctuate together. They both change in response to gonadal replacement or gonadectomy. We therefore thought we should look at NPY in the brainstem A1 and A2 neurons to see whether this was the source of these changes. However, what we have shown is that there is a complete dissociation: noradrenaline cells with oestrogen receptors do not make NPY. In this case, we speculate that the oestrogen-dependent changes in NPY within the hypothalamus are occurring through the hypothalamic arcuate nucleus NPY populations, which do in fact possess oestrogen receptors in the rat and sheep. So it is most likely that the similar regulatory influence of oestrogen on NPY and noradrenaline release occurs through independent neuronal pathways.

McEwen: Has anyone tried injecting orthograde tracer into the rostral or caudal A1 or A2, to find out whether there are differences in projection?

Herbison: This has never been done. There is marked rostro-caudal topography in terms of oestrogen receptor and other biochemical markers in the A1 and A2, and the tracing the we have done is all retrograde. At the least, this shows that the A1 and A2 neurons have very specific projection pathways.

McEwen: There is an old literature on the turnover of catecholamines and serotonin, indicating that there is heterogeneity in the terminal fields in terms of what oestrogens do. This has always been difficult to reconcile with the idea that these types of neuron form a tree with uniform projections. You are saying, however, that it may not be so uniform, and that in fact these terminal areas where there is, say, an effect on oestrogen turnover may be related to the presence of oestrogen receptors in a subset of cell bodies for those particular terminals. Of course, we then have sex differences on top of this. This is really interesting.

Luine: Mary Kritzer did some elegant studies with gonadectomy on both females and males and described changes in tyrosine hydroxylase in the frontal cortex (Kritzer & Kohama 1998). She saw changes immunocytochemically which were confirmed to particular layers. This result suggests that you might not need the cell body but perhaps it is that particular cells within the frontal cortex receive a specific innervation which confers sensitivity to the hormones.

Herbison: I interpret that work as being much more related to the dopaminergic innervation of the frontal cortex. Tyrosine hydroxylase will pick up both dopamine and noradrenaline. It is a great shame that these studies weren't done with a noradrenergic marker also.

Hurd: The rostro-caudal differences for the noradrenergic system are not only evident in their forebrain projection areas, but also in the phenotype of the noradrenergic cells. What about doing single cell PCR and then characterizing the phenotype?

Herbison: We could do that. In fact, we have recently established a single cell RT-PCR protocol using an acute brain slice preparation for evaluating the presence of ERα and ERβ transcripts within single neurons. We have not yet evaluated the noradrenergic neurons but, surprisingly, have shown that most gonadotropin-releasing hormone (GnRH) neurons in the mouse express ERα, while a small number, around 10%, contain ERβ transcripts (Skynner et al 1999). If you look at medial preoptic area neurons, most of these neurons also have ERα and a few have ERβ while, in comparison, we found that approximately 25% of striatal cells had ERα but none expressed ERβ. We think these results are going to be controversial but also very interesting as they suggest that low levels of ER transcripts may exist in neuronal populations not previously thought to express these receptors.

Hurd: In the striatum, you say that about 25% of the cells had ERα and that no ERβ was detected by PCR. We have carried out *in situ* hybridization using human and rat brain specimens, but we have never seen ER mRNA expression in the striatum (Österlund et al 1998, 2000a). Even with the use of PCR we have never detected any signal in the human striatum, thus validating our *in situ* hybridization results (M. K. Österlund, K. Grandien, E. Keller & Y. L. Hurd, unpublished results). With 25% of the cells in the rat striatum being ERα-positive, one would expect to see a signal. Is it just that the human and rat ER expression levels are different? We have detected mRNA expression for other genes in cell populations that make up only 10% of the striatum.

Herbison: This could have something to do with species differences, or possibly the primers that were used for the PCR. We have designed ER primers trying to avoid known splice variants. I agree; if you were just punching out the whole striatum, I would expect you to have seen ERα.

Toran-Allerand: I don't think that you should be discouraged about the distribution pattern of the oestrogen receptors. The use of antibodies to determine which neurons possess oestrogen receptors tends to lead to an underestimation. This is because the antibody will only pick up those neurons that have high levels of the receptor protein. There could be many neurons that are oestrogen receptor positive that one wouldn't see because the protein levels are significantly lower yet are still sufficient to elicit a response.

Herbison: I'm not often discouraged by my own results! Although I agree with you, it is an extremely difficult issue to decide how much oestrogen receptor protein is going to be enough *in vivo* within a neuron to mediate any oestrogen actions.

Toran-Allerand: With respect to the presence of ERs in striatal neurons, it is possible to show using steroid autoradiography with an iodinated oestrogenic ligand significant numbers of oestrogen binding sites within developing striatal neurons (Toran-Allerand et al 1992). The numbers are comparable to those that Allan Herbison was talking about. The confusion on this issue has to do with the fact that the levels may not be excessively high. On the basis of our findings with the iodinated ligands, there are striatal neurons that have oestrogen binding sites.

Gibbs: Using immunocytochemistry and fine-tuning it to make it as sensitive as possible, we were able to detect ER immunoreactivity within striatal neurons in approximately the same percentage of neurons (Gibbs 1996).

Henderson: As you stated, the A6 group projects widely to neocortex and hippocampus, and it is therefore postulated that these neurons have widespread effects. Do you think that these postulated effects are mediated through direct neuronal–neuronal interactions, or might they be mediated indirectly through noradrenergic effects on the cerebral vasculature?

Herbison: Noradrenaline has a large number of very interesting effects at the electrophysiological level. There are sufficient adrenergic receptors on neurons within the cortex to facilitate a direct interaction with noradrenaline. I don't know whether changes in vasculature come into it or not: the substrate is there for direct innervation and activation of the neurons themselves.

McEwen: Is it clear from the histology whether the innervation from any of these areas is both to the vasculature and to the surrounding neurons?

Herbison: I'm not aware of that. All I can say is that if you administer compounds that will detect neurons projecting outside of the blood–brain barrier, some noradrenergic neurons will be labelled.

Kirschbaum: Concerning the interaction of oestrogens and noradrenaline in A1 and A2 cells, is it possible that there differences between effects on an acute, short-term basis versus those on a long-term or chronic basis? I ask this because of findings from human experiments. If we give oestrogens to males via patches for 24–48 h and then stress them, they show an enhanced adrenocorticotropic hormone (ACTH) and cortisol response compared with placebo. But if we go to females and look at response patterns across the menstrual cycle, we don't see any evidence of an association between oestradiol levels and ACTH responsiveness. This should be the case if there is a 1:1 effect of oestrogens on A1 and A2 neurons.

Herbison: I have raised the issue of tyrosine hydroxylase gene expression being regulated by oestrogen. We have done an initial study using a single point oestrogen treatment, but studies that have been done in the arcuate nucleus, for example, have shown a time-dependent down- and up-regulation of tyrosine hydroxylase expression. There is therefore a suggestion that somehow — directly or indirectly — there could be opposite effects of oestrogen on this gene expression. It is certainly not unreasonable to think that a relatively constant level of oestrogen presentation to a cell provides a very different environment to one in which there are changes in oestradiol. The clear sex differences in oestrogen action upon tyrosine hydroxylase gene expression may, however, be part of the phenomenon you describe.

Gandy: Allan Herbison, did you see any behavioural correlates of the changes that you were seeing in tyrosine hydroxylase? I am thinking here of the Alzheimer's disease literature that links depressive clinical phenotypes to levels of catecholamines in the brainstem. The subpopulation of Alzheimer's disease patients who are the most depressed seem to be the ones with the highest depletion of catecholamines in the brainstem.

Herbison: At this point we have no behavioural data for mice treated in this way.

Luine: We have found that ovariectomized rats that have been treated with oestrogen show enhanced working memory. We also found a decrease in noradrenaline levels in the frontal cortex when the rats were sacrificed. The lower levels may indicate increased noradrenergic activity. There is also a correlation between changes in noradrenaline and enhanced cognitive function (Luine et al 1998).

Herbison: The techniques that we use in measuring noradrenaline release are relatively crude.

Baulieu: Do you think that oestrogen acts directly on GnRH neurons, or indirectly by noradrenergic innervation?

Herbison: At present, we have evidence for the presence of ER transcripts, predominantly ERα, inside GnRH neurons. We are very interested in whether or not these ERs are functional. It remains an unresolved question as to whether noradrenergic terminals innervate GnRH neurons directly.

Baulieu: Have you found progesterone receptor (PR) in GnRH neurons?

Herbison: There was a paper by Joan King and Jeff Blaustein suggesting that something like 5% of GnRH neurons in the guinea pig have the PR (King et al 1995). We have not yet conducted single cell RT-PCR experiments on GnRH neurons to determine whether PR transcripts exist.

Becker: In terms of the response to dopamine stimulation, constant or prolonged oestrogen treatment has different effects from acute treatment. Acute oestrogen exposure causes enhancement of stimulated dopamine release and activity. Prolonged exposure to oestrogen causes down-regulation of stimulated dopamine release and also causes dopamine receptor supersensitivity. There are also findings in other systems that support the idea of differential effects depending on the dose, and biphasic effects depending upon the dose: this is seen in osteoblasts, for example. Within a physiological dose range there is an enhancement of the dopamine response in the striatum, but higher doses can be inhibitory. I didn't know what dose of oestrogen you used in your system in terms of the effect on tyrosine hydroxylase, but we have to keep in mind that the response to oestrogen depends upon the dose and duration of treatment.

I also wanted to mention that oestrogen binding sites aren't necessarily oestrogen receptors.

Toran-Allerand: The way steroid autoradiography has been carried out in the past is best suited to show nuclear binding.

Becker: That's the point I'm making.

Toran-Allerand: I never meant to imply that there were not membrane oestrogen receptors; all I meant is that for years we heard that the striatum was an oestrogen-responsive region that had no oestrogen receptors. When you carry out steroid autoradiography with a ligand that is active enough and show nuclear binding, it is fair to assume that there is at least one kind of oestrogen receptor.

Becker: Do you look at oestrogen receptor localization in male rats to see whether there is a sex difference in striatum, or whether there is a sex difference in the effect of oestrogen in noradrenergic neurons?

Herbison: We have not looked for a sex difference in ER expression in noradrenergic neurons. With single-cell RT-PCR we have looked for sex differences in GnRH neurons but not striatal cells.

Gibbs: We've heard today about oestrogen membrane effects that can influence receptor-mediated events; we've also heard a little bit about effects of oestrogen on γ-aminobutyric acid (GABA)ergic neurons. Is there evidence for GABAergic regulation of the noradrenergic neurons, and is there evidence for oestrogen-mediated effects at the adrenergic receptor level?

Herbison: I'll take the second part of your question first. Anne Etgen has really championed that work and has quite clearly demonstrated the oestrogen regulation of adrenergic receptor expression in the hypothalamus. She also showed oestrogen down-regulation of $\alpha2$ receptors in the frontal cortex (Karkanais et al 1997).

With regard to the GABAergic interactions with noradrenaline neurons, we did an experiment some time ago in which we electrically stimulated the A1 noradrenergic cell group and measured GABA and noradrenaline release in the preoptic area using microdialysis (Herbison et al 1990). We found an $\alpha1$ adrenergic activation of preoptic GABA neurons in ovariectomized rats. However, if you treat animals with oestrogen, it completely uncouples the stimulatory effect of A1 neurons on preoptic GABA release.

Gibbs: So that effectively decreases GABAergic tone.

Pfaff: If I remember correctly, your GnRH neurons were sampled from a slice preparation, so their axons would have been cut shortly before the mRNA was sampled. This means that they would be undergoing a chromatolytic reaction in which large numbers of genes are turned on. Is this a theoretical caveat?

Herbison: It is certainly a problem, especially as experimental stroke models have recently been shown to modulate expression of ERs in the cerebral cortex (Dubal et al 1999). But what I didn't say was that the percentage of GnRH neurons expressing ERα transcripts fluctuates with the oestrous cycle. For example, at oestrus about 80% of GnRH neurons have ER transcript, whereas at pro-oestrus it is only about 20%.

Pfaff: The susceptibility could therefore vary with the oestrous cycle.

Is the GnRH protein proven be produced in those cells, or are they are identified by mRNA with RT-PCR?

Herbison: The GnRH neurons are demonstrated by the RT-PCR of individual cells for GnRH transcripts.

Pfaff: I should mention that, on the functional side, Barbara Attardi (unpublished results) finds that in the GnRH-producing GT1 cells, if she transfects ER, oestradiol will turn on transcription by 1000–2000-fold, whereas if she doesn't transfect the ER there's no stimulation by oestrogen at all. If endogenous ER is functional, it is functional to the tune of something less than 0.1%.

Herbison: That's good news in the sense that clearly all the coactivators are present in GnRH neurons. I'm not sure that you would want me to start on the physiological relevance of GT1 cells!

Levin: Is it correct that you have identified the receptors by RT-PCR, but you can't show any protein yet?

Herbison: That's right. Using conventional immunocytochemistry we are unable to detect them.

Levin: Have you tried *in situ* hybridization?

Herbison: We haven't looked at that. It may be interesting to try the iodinated oestradiol ligands that Dominique Toran-Allerand was talking about.

Fillit: Sam Gandy said that about 30% of Alzheimer's patients have catecholaminergic deficits and have depression. If what you're saying is true, then many women who become menopausal and then subsequently oestrogen deficient should be experiencing depression, and yet we know that there's no greater incidence of depression at the time of menopause. I am not questioning the clinical significance of what you're saying, but there must be other variables here that could be related not only to environment but also to the sensitivity of individuals related to how this oestrogen withdrawal state affects adrenergic and other neurochemical systems. Could you speculate on how individual variation might affect the response such that some ultimately do become at risk whereas others don't. If we could identify these kinds of risk factors in individuals, we might be able to develop better therapeutic strategies.

Herbison: That's a very good point. I have perhaps presented it slightly falsely: I said here's a noradrenaline neuron with an oestrogen receptor, give it oestrogen and it will start off firing. This is not the situation. These noradrenaline neurons are very important in sleep–wake cycles and show continuous electrical activity which is modulated by multiple sensory inputs. We think that oestrogen is modulating in someway the baseline level of activity. Consequently, the activity of a noradrenergic neuron depends on many factors. For example, if someone was suffering severe pain, then this factor might be expected to create a level of noradrenaline activation which is way above anything that oestrogen could add. So we are dealing with a neuronal system that is subject to influences from a wide range of sources that will provide a likely wide spectrum of activation within a population. Having said this, however, if we think of noradrenergic neurons as being an integration or nodal point, it is not to say that in a pathological or ageing state that it would not be possible to use oestrogen to override inherent or external deficits in noradrenergic neuron activation.

Bethea: Not all postmenopausal women get depressed, but some do.

Fillit: Not all Alzheimer patients get depressed, but some do — and that's my point. I'm trying to understand why some do and some don't, and what the risk factors are. There is individual variation that we don't understand.

Bethea: You could use genetics to get at this. There are five single nucleotide polymorphisms in the promoter region of the TPH gene, and some of those are linked. There's one polymorphism in intron 7 of the TPH gene that's been linked to impulsive suicide. There are polymorphisms in the serotonin reuptake transporter and in monoamine oxidase, and these are linked with neurotic depression and alcoholism. If you had an individual that was particularly genetically vulnerable for some of these factors, then withdrawal of oestrogen or progesterone could tip them over the edge.

Resnick: That may be the case for individuals but large population-based studies don't show any increase in depressive symptoms during the postmenopausal years.

Bethea: My understanding was the incidence of depression increases in women at menopause. David Rubinow and Peter Scmhidt (NIMH) have a fairly large population of postmenopausal women that exhibit depression (Schmidt et al 1999).

Resnick: They select those people for depression, though.

Gibbs: If there is no correlation with age, why do we talk about 'late-life mood disorder'?

Resnick: I think we are talking about different things. The late-life mood disorder is interesting, because one of the questions that surrounds it is the extent to which it is an early sign of disease in people who are developing dementia. A substantial number of individuals who have a first-onset depression after age 70 go on to develop dementia within a few years. Late-life depression may be an early manifestation of dementia.

Bethea: But for menopausal depression we would be looking at people in their 50s.

Resnick: I'm not denying that there are individuals who have depressive symptoms related to menopause, but if you look at population-based studies, there is no increase in depressive symptoms. Several large studies have looked at this.

Sherwin: Mood changes around the time of menopause are extremely common. Their intensity is such that most would not meet criteria for diagnosis of a major depressive disorder; it is subclinical depression but very common. If these women are then treated with oestrogen, it usually stabilizes their mood. Women who have premenstrual syndrome (PMS), or who had PMS, or who had prior episodes of depression are more vulnerable for the development of a major depressive disorder at the time of menopause. We talk about them as having a vulnerability factor but no one knows what the nature of this factor might be.

Henderson: To reinforce what Dr Resnick has said, from the epidemiological literature there is no strong indication that the time of menopause is associated with an excess of major depression, although depression is certainly much more common in women compared with men throughout the lifespan. On the other hand, postpartum depression is a real clinical entity, and oestrogen does seem to have an effect on mood in this disorder (Gregoire et al 1996).

As an aside regarding Alzheimer's disease, there are epidemiological data that indicate that early episodes of depression may increase the risk of later developing Alzheimer's disease (Speck et al 1995).

Bethea: I think the epidemiologists need to go back and look at a subclinical measure of depression.

Hurd: The problem is not with the epidemiology, it is with the diagnosis and clinical measures used in grouping the subjects. Genetics and polymorphism data are increasingly more available in the clinic, and these markers should be used in addition to the DSM IV criteria. There are clearly individual differences in affective disorders that are quite intriguing. Even post-mortem when we look at gene expression of oestrogen in human brain specimens, we see individual differences that are not related with sex or post-mortem interval (M. Österlund & Y. L. Hurd, unpublished results).

Sherwin: There is some evidence that the serotonergic system is involved in the affective symptoms that women complain of. We have looked at differences in tritiated imipramine binding sites on platelets and found that the density of binding sites, which is a serotonin transport mechanism, was increased in oestrogen treated postmenopausal women but not in placebo-treated women, and it correlated with mood score (Sherwin & Suranyi-Cadotte 1990). Therefore, the administration of oestrogen to postmenopausal women up-regulated the serotonergic system and enhanced mood.

Hurd: We have looked in a rat model of depression, and these animals do show abnormalities of the serotonin system at the mRNA level which can be normalized by oestradiol (Österlund et al 2000b).

References

Dubal DB, Shughrue PJ, Wilson ME, Merchenthaler I, Wise PM 1999 Estradiol modulates Bcl-2 in cerebral ischaemia: a potential role for estrogen receptors. J Neurosci 19:6385–6393

Gibbs RB 1996 Expression of estrogen receptor-like immunoreactivity by different subgroups of basal forebrain cholinergic neurons in gonadectomized male and female rats. Brain Res 720:61–68

Gregoire AJP, Kuman R, Everitt B, Henderson AF, Studd JWW 1996 Transdermal oestrogen for treatment of severe postnatal depression. Lancet 347:930–933

Herbison AE, Heavens RP, Dyer RG 1990 Oestrogen modulation of excitatory A1 noradrenergic input to rat medial preoptic γ-aminobutyric acid neurones demonstrated by microdialysis. Neuroendocrinology 52:161–168

Karkanais GB, Li CS, Etgen AM 1997 Estradiol regulation of α_2 adrenoreceptor binding in female rat cortex is correlated with decreases in $\alpha_{2A/D}$ adrenoreceptor messenger RNA. Neuroscience 81:593–597

King JC, Tai PW, Hanna IK et al 1995 A subgroup of LHRH neurons in guinea pigs with progestin receptors is centrally positioned within the total population of LHRH neurons. Neuroendocrinology 61:265–272

Kritzer MF, Kohama SG 1998 Ovarian hormones influence the morphology, distribution, and density of tyrosine hydroxylase immunoreactive axons on the dorsolateral prefrontal cortex of adult rhesus monkeys. J Comp Neurol 395:1–17

Luine VN, Richards ST, Wu VY, Beck KD 1998 Estradiol enhances learning and memory in a spatial memory task and affects levels of monoaminergic neurotransmitters. Horm Behav 34:149–162

Österlund M, Kuiper GGJM, Gustafsson J-Å, Hurd YL 1998 Differential distribution and regulation of estrogen receptor-α and -β mRNA within the female rat brain. Brain Res Mol Brain Res 54:175–180

Österlund MK, Keller E, Hurd YL 2000a The human forebrain has discrete estrogen receptor α messenger RNA expression: high levels in the amygdaloid complex. Neuroscience 95:333–342

Österlund MK, Overstreet DH, Hurd YL 2000b The Flinder's sensitive rat line, a genetic model of depression, show abnormal serotonin receptor mRNA expression in the brain that is reversed by 17β-estradiol. Brain Res Mol Brain Res 10:158–166

Schmidt P, Berman K, Roca C et al 1999 Sex and human behaviour. Proceedings of Annual Meeting of the Endocrine Society, vol 81. June 12–15, 1999, San Diego, CA, p 25 (abstr S7–3)

Sherwin BB, Suranyi-Cadotte BE 1990 Up-regulatory effect of oestrogen on platelet ^{3}H-imipramine binding sites in surgically menopausal women. Biol Psychiatry 28:339–348

Skynner MJ, Sim JA, Herbison AE 1999 Detection of estrogen receptor α and β messenger ribonucleic acids in adult gonadotropin-releasing hormone neurons. Endocrinology 140:5195–5201

Speck CE, Kukull WA, Brenner DE et al 1995 History of depression as a risk factor for Alzheimer's disease. Epidemiology 6:366–369

Toran-Allerand CD, Miranda RC, Hochberg RB, MacLusky NJ 1992 Cellular variations in estrogen receptor mRNA translation in the developing brain: evidence from combined ^{125}I-estrogen autoradiography and non-isotopic *in situ* hybridization histochemistry. Brain Res 576:25–41

Oestrogen and the cholinergic hypothesis: implications for oestrogen replacement therapy in postmenopausal women

Robert B. Gibbs

University of Pittsburgh School of Pharmacy, 1004 Salk Hall, Pittsburgh, PA 15261, USA

Abstract. Cholinergic deficits in the basal forebrain, hippocampus and cortex are thought to contribute to the risk and severity of cognitive decline associated with ageing and Alzheimer's disease. Work in our laboratory has demonstrated that in rats, basal forebrain cholinergic neurons are affected by physiological fluctuations in circulating oestrogen and progesterone, and that long-term loss of ovarian function produces decreases in cholinergic parameters and nerve growth factor receptor (*trkA*) mRNA beyond the effects of normal ageing. Conversely, short-term treatment with oestrogen or oestrogen plus progesterone produces increases in cholinergic parameters and *trkA*, as well as increases in potassium-stimulated acetylcholine release, that are consistent with an increase in basal forebrain cholinergic function. These findings are consistent with recent studies showing the ability of oestrogen and progesterone replacement to enhance spatial memory and reduce performance deficits associated with hippocampal cholinergic impairment. We hypothesize that similar effects of the ovarian hormones on basal forebrain cholinergic neurons in humans may contribute to the effects of hormone replacement on cognitive processes that have recently been described, and to the ability of oestrogen replacement to reduce the risk and severity of Alzheimer's-related dementia in postmenopausal women.

2000 Neuronal and cognitive effects of oestrogens. Wiley, Chichester (Novartis Foundation Symposium 230) p 94–111

Increasing evidence suggests that oestrogen replacement therapy can influence cognitive processes in women and decrease the risk and severity of Alzheimer's-related dementia (see Henderson 1997, Sherwin 1998, Yaffe et al 1998 for reviews). The mechanisms that underlie these effects are unknown; however, basic research has provided increasing evidence for significant oestrogen effects on neuronal anatomy, biochemistry and function within regions of the brain that are known to play an important role in cognitive processes. Cholinergic neurons in the medial septum (MS), diagonal band of Broca (DBB) and nucleus basalis magnocellularis (NBM) have long been known to play an important role in cognitive processes (see Gibbs 2000a for review). These neurons are the primary source of cholinergic input to the hippocampus and cortex and have a significant impact on hippocampal and cortical functioning. These neurons are also lost or impaired in association with ageing and Alzheimer's disease and are thought to contribute to ageing- and Alzheimer's-related cognitive decline (see Gibbs 2000a for review). Drugs that impair cholinergic activity produce cognitive deficits similar in many ways to those observed in conjunction with ageing and Alzheimer's disease, whereas drugs that enhance cholinergic function have been shown to enhance cognitive processes in normal individuals and to reduce the severity of Alzheimer's-related cognitive decline.

We and others have demonstrated that loss of ovarian function and oestrogen replacement has significant effects on cholinergic neurons projecting to the hippocampus and cortex. In particular, the findings suggest that oestrogen is involved in the normal maintenance and physiological regulation of the cholinergic projections, and that oestrogen replacement can enhance the functional status of these cholinergic projections with a corresponding enhancement of cognitive performance. On the basis of these findings we hypothesize that similar effects in humans could contribute to the effects of oestrogen on cognitive processes and help to reduce the risk and severity of cognitive decline in postmenopausal women. This paper summarizes our recent data which show that oestrogen and progesterone have significant effects on basal forebrain cholinergic neurons and behaviour, and discusses the implications of these results with respect to oestrogen replacement therapy in postmenopausal women.

Effects of oestrogen and progesterone on choline markers

Measures of choline acetyltransferase (ChAT) and high affinity choline uptake (HACU) have been used for many years as indicators of effects on cholinergic cell survival and function. ChAT is present throughout the cell bodies, axons and terminals of the basal forebrain cholinergic neurons, whereas HACU is enriched in the terminals of these neurons where it provides for the rapid re-uptake of choline at cholinergic synapses. While there is continuing debate over the relative contributions of changes in ChAT activity and HACU to changes in cholinergic activity, it is well demonstrated that agents that enhance the functional status of basal forebrain cholinergic neurons (e.g. nerve growth factor, brain-derived neurotrophic factor, fibroblast growth factor) produce increases in ChAT activity and HACU in the basal forebrain, hippocampus and cortex which correlate with increased levels of acetylcholine release (see below). Conversely, losses of basal forebrain cholinergic neurons (e.g. by experimental lesions, ageing and

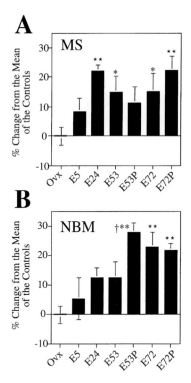

FIG. 1. Relative levels of ChAT mRNA detected in the MS (A) and NBM (B) of ovariectomized animals at different times following a single injection of 17β-oestradiol, and following treatment with oestradiol followed 48 h later with a single injection of progesterone. Animals were sacrificed 5, 24, 53 and 72 h after receiving oestradiol (E), and 5 and 24 h after receiving oestrogen plus progesterone (EP). Bars represent percent change relative to the ovariectomized (Ovx) controls ± SEM. **$P < 0.01$ relative to the controls. *$P < 0.05$ relative to the controls. †$P < 0.05$ comparing E53 with E53P. Adapted from Gibbs (1996a).

Alzheimer's disease) produce decreases in ChAT and HACU in the basal forebrain, hippocampus, and cortex (Gibbs 2000a). These findings suggest that changes in ChAT and HACU are meaningful indicators of physiologically relevant changes in the survival and function of basal forebrain cholinergic neurons.

Recent studies have shown that relative levels of ChAT mRNA and protein, ChAT activity, and HACU are decreased following ovariectomy and/or increased in response to treatment with oestrogen or oestrogen plus progesterone. Luine (1985) originally reported that 10 d of continuous oestradiol treatment (via subcutaneous capsule) produced a significant increase in ChAT activity in the medial aspect of the horizontal limb of the DBB, the frontal cortex and CA1 of the dorsal hippocampus. O'Malley et al (1987) subsequently reported that three weeks following ovariectomy there was a significant decrease in HACU in the frontal cortex relative to gonadally intact controls, and that oestrogen replacement produced a threefold increase in HACU relative to the ovariectomized animals. More recent studies have demonstrated significant increases in the number of ChAT-immunoreactive neurons (Gibbs 1997, Gibbs & Pfaff 1992), as well as increases in ChAT mRNA (Gibbs 1996a, Gibbs et al 1994), in the MS and NBM of ovariectomized animals treated with oestradiol.

One of our goals has been to define the dynamics of these effects as they occur in gonadally intact animals as well as in ovariectomized animals receiving different doses and regimens of hormone replacement. One recent study (Gibbs 1996a) showed that, in ovariectomized animals, a single injection of 17β-oestradiol (10 μg in oil, s.c.) resulted in significant increases in ChAT mRNA in the MS and NBM, with peak effects observed at 24 h (MS) and 72 h (NBM) following oestrogen administration (Fig. 1A and 1B). In contrast, peak oestrogen levels were achieved within hours after oestrogen administration and declined substantially over the next 24–72 h. This demonstrates that acute oestrogen replacement produces significant increases in ChAT mRNA within specific basal forebrain nuclei, and that there is a significant delay between elevated levels of oestrogen and increased levels of ChAT mRNA. In this same study, progesterone administered 48 h after oestrogen appeared to maintain the increase in ChAT mRNA in the MS (Fig. 1A), and to accelerate the increase in ChAT mRNA in the NBM (Fig. 1B). A parallel study conducted with gonadally intact animals revealed that relative levels of ChAT mRNA in the MS and NBM fluctuate over the course of the oestrous cycle and peak at approximately 36 h (MS) and 72 h (NBM) following peak circulating levels of oestradiol. Collectively, these data suggest that oestrogen and progesterone play a role in the physiological regulation of ChAT expression in the MS and NBM, and demonstrate that acute hormone replacement produces increases in ChAT mRNA and protein that are consistent with the physiological fluctuations in ChAT that occur over the course of the oestrous cycle.

More recently we examined the effects of oestrogen on cholinergic parameters as a function of different doses, durations and regimens of oestrogen replacement. In one study (Gibbs 1997) animals received subcutaneous injections containing 0, 2, 10, 25 or 100 μg 17β-oestradiol every other day for 1, 2 or 4 weeks. Sections through the brain were then processed for immunohistochemical detection of ChAT, and the number of ChAT-positive neurons in the MS and NBM were quantified and compared. The results demonstrated a dose-related increase in the number of ChAT-positive profiles in the MS following treatment with 2, 10 and 25 μg oestradiol every other day for 1 week (Fig. 2A). A significant increase in the number of ChAT-positive profiles was also detected in the NBM following treatment with 10 μg oestradiol every other day for 1 week, and following treatment with 2 μg oestradiol every other day for 2 weeks (Fig. 2B). Treatment

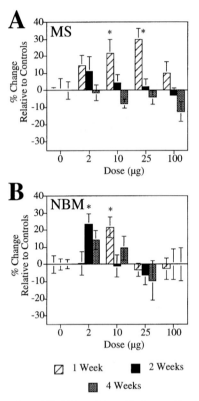

FIG. 2.   Changes in the number of ChAT-positive cells detected in the MS (A) and NBM (B) in response to different doses of repeated oestrogen administration for 1, 2 and 4 weeks. Bars represent percentage change in the number of labelled cells relative to the ovariectomized, vehicle-treated controls. *$P < 0.05$ relative to the corresponding controls. Adapted from Gibbs (1997).

with higher doses of oestradiol, or with lower doses for longer periods had no significant effect on the number of ChAT-positive profiles in the MS and NBM. We interpret these increases in the number of ChAT-immunoreactive cells as an increase in ChAT protein within existing cholinergic neurons that previously contained levels of ChAT below the level of detection. These data are consistent with the increases in ChAT mRNA previously detected, as well as with increases in ChAT-IR previously detected following 1–2 weeks of continuous oestrogen replacement (Gibbs & Pfaff 1992).

To determine whether the effects of oestrogen on cholinergic neurons are affected by the method of treatment, we recently examined the effects of oestrogen and oestrogen plus progesterone on HACU as a function of continuous vs. repeated oestrogen replacement. Ovariectomized animals received hormone

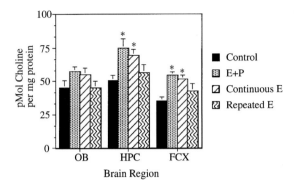

FIG. 3. Effects of different regimens of hormone replacement on HACU in synaptosomes prepared from the olfactory bulbs (OB), hippocampus (HPC) and frontal cortex (FCX). Hormone-treated animals received either repeated administration of oestrogen (E) followed by progesterone (P), continuous oestrogen alone, or repeated administration of oestrogen alone, for 2 weeks (see text for details). Bars represent pmol of HACU per mg protein ± SEM. $N = 4$–6 per group. *$P < 0.05$ relative to ovariectomized, vehicle-treated controls.

High affinity Choline uPtake - HACU

replacement for two weeks. Animals received either continuous oestrogen replacement (a 3 mm silastic capsule containing 17β-oestradiol crystals implanted s.c.), repeated oestrogen replacement (10 μg oestradiol in oil injected every 3 d s.c.), or repeated oestrogen replacement plus progesterone (10 μg oestradiol s.c. followed 48 h later with 500 μg progesterone injected s.c.). Controls received injections of vehicle on the same schedule as the animals receiving repeated oestrogen plus progesterone. HACU in the hippocampus, frontal cortex and olfactory bulbs was then quantified and compared. As shown in Fig. 3, significant increases in HACU were detected in the hippocampus and frontal cortex, but not the olfactory bulbs, following repeated oestrogen replacement combined with progesterone, and following continuous oestrogen replacement. Repeated treatment with oestrogen alone was without effect. Furthermore, the greatest increases in HACU were observed after repeated administration of oestrogen and progesterone. These results demonstrate that in addition to being dependent on the dose and duration of treatment, the effects of oestrogen on cholinergic parameters is also dependent on the method of treatment, with repeated oestrogen administration combined with progesterone being the most effective, and repeated administration of oestrogen alone being the least effective.

Potential mechanisms for hormonal effects on basal forebrain cholinergic neurons: effects of oestrogen and progesterone on neurotrophins and neurotrophin receptors

The mechanisms by which oestrogen influences basal forebrain cholinergic neurons are still largely unknown. Initial autoradiographic and immunohistochemical studies suggest that the majority of cholinergic neurons in the MS and NBM contain high affinity oestrogen binding sites (Toran-Allerand et al 1992) and oestrogen receptor (ER)-like immunoreactivity (Gibbs 1996b), suggesting that oestrogen may directly influence the cholinergic neurons via binding to intracellular receptors followed by direct steroid-mediated effects on gene transcription; however, these initial findings have been difficult to verify in other laboratories using different antibodies and staining techniques. *In situ* hybridization studies have detected ERα mRNA in approximately 41%, 32%, and 4% of the cholinergic neurons in the MS, DBB and NBM, and ERβ mRNA in approximately 2% of these neurons (I. Merchenthaller, personal communication); however, no one has yet examined whether cholinergic neurons in the basal forebrain contain a functional ER that is capable of stimulating expression of an oestrogen-responsive gene. Consequently, the question of whether oestrogen influences basal forebrain cholinergic neurons directly via the activation of a functional ER is currently unknown.

Cholinergic neurons in the MS, DBB, and NBM are influenced by neurotrophins including nerve growth factor (NGF) and brain-derived neurotrophic factor (BDNF). Effects of the neurotrophins on basal forebrain cholinergic neurons include increases in ChAT and ChAT activity (Fusco et al 1989, Knüsel et al 1992, Lorenzi et al 1992, Nonomura & Hatanaka 1992), increases in HACU (Rylett et al 1993), and increases in acetylcholine synthesis and release (Rylett et al 1993). Therefore, changes in the expression of the neurotrophins or their receptors could result in changes in basal forebrain cholinergic function. Evidence for ER-mediated cross-talk with neurotrophin-regulated signalling pathways has also recently been reported (Singh et al 1999) and could influence oestrogen-mediated effects on the cholinergic neurons.

McMillan et al (1996) recently demonstrated a significant decrease in *trkA* (NGF receptor) mRNA in the NBM and horizontal limb of the diagonal band of Broca (HDB) of young adult rats at 10 d following ovariectomy. Oestrogen replacement (10 μg/d s.c. for 3 d) restored *trkA* mRNA levels to control levels. In addition, we recently demonstrated a significant increase (30.4%) in relative levels of *trkA* mRNA in the MS of ovariectomized following a single injection of 17β-oestradiol (10 μg in oil s.c.) (Gibbs 1998a). As with the effects on ChAT mRNA described above, peak levels of *trkA* mRNA in the MS were detected 24 h following oestrogen administration. Levels of *trkA* mRNA subsequently declined, but were maintained in part by the added administration of progesterone.

We and others have also recently demonstrated a significant up-regulation of BDNF mRNA in the hippocampus following acute treatment with oestrogen or oestrogen plus progesterone (Gibbs 1998a, 1999a, Singh et al 1995). In one study *in situ* hybridization techniques were used to detect and quantify relative levels of BDNF mRNA within different regions of the hippocampus following treatment with oestrogen and progesterone (Gibbs 1998a). Animals that received $10 \mu g$ oestrogen s.c. followed 48 h later by $500 \mu g$ progesterone s.c. showed a 73.4%, 28.1% and 76.9% increase in relative levels of BDNF mRNA in the dentate gyrus, CA1 and CA3/4 regions of the hippocampus relative to ovariectomized vehicle-treated controls. Treatment with oestrogen alone was much less effective. In addition, no significant effects on the levels of NGF mRNA in the hippocampus were detected.

In a subsequent study, quantitative reverse transcriptase PCR and ELISA techniques were used to examine changes in BDNF mRNA and protein within different regions of the brain following different regimens of hormone replacement (Gibbs 1999a). The results showed that treatment with oestrogen plus progesterone produced a 72.5% increase in BDNF mRNA in the hippocampus, and a 79.6% increase in BDNF mRNA in the pyriform cortex. Treatment with oestrogen alone also produced significant increases in BDNF mRNA in the hippocampus and pyriform cortex, but was less effective than treatment with oestrogen plus progesterone. Treatment with oestrogen plus progesterone also resulted in a 138.9% increase in BDNF protein in the pyriform cortex; however, a decrease (-34.7%) in BDNF protein was detected in the hippocampus. Analysis of the septal/diagonal band region revealed very low levels of BDNF mRNA and a 59.1% increase in BDNF protein following oestrogen treatment. One possibility, therefore, is that hormone replacement increases BDNF mRNA in the hippocampus, but that levels of BDNF in the hippocampus decrease as a result of increased transport of the BDNF protein from the hippocampus to the septum and diagonal band. This could, in turn, contribute to the effects of oestrogen on ChAT and HACU described above.

Effects of oestrogen on acetylcholine release and cognitive performance

An important question is whether the changes in ChAT and HACU produced by oestrogen administration reflect significant changes in acetylcholine synthesis and release, with corresponding effects on cognitive function. We recently examined the effects of oestrogen replacement on basal and evoked acetylcholine release in the hippocampus and overlying cortex using sensitive *in vivo* microdialysis techniques (Gibbs et al 1997). Ovariectomized animals received either continuous oestrogen replacement (3 mm capsule implanted s.c.) or sham surgery. After 10 d of treatment, the levels of basal and potassium-evoked acetylcholine release in the hippocampus and overlying cortex were quantified and compared. Oestrogen had no significant effect on basal acetylcholine release; however, a 40.3% increase in the percent change in release induced by potassium over a 90 min period was detected in the oestrogen-treated animals vs. the non-oestrogen-treated controls

TABLE 1   Effects of oestrogen replacement on basal and potassium-stimulated acetylcholine release in the hippocampus and overlying cortex

	Ovariectomized	*Oestrogen-treated*
Average basal release (pmol/20 min)	6.2 ± 0.6	5.0 ± 0.7
Average percent increase in release during 90 minute period of increased potassium	97.5 ± 12.6	$137.8 \pm 12.6*$

$*P < 0.05$ relative to ovariectomized controls.
Adapted from Gibbs et al (1997).

(Table 1). This suggests that continuous oestrogen replacement can help to maintain elevated levels of acetylcholine release in response to a sustained depolarizing stimulus.

Subsequent behavioural studies suggest that the effects of oestrogen and progesterone on basal forebrain cholinergic neurons may also be relevant to effects on cognitive function. In one recent study, we used a delayed non-matching to position (DMP) T-maze task to assess the effect of oestrogen replacement on learning, as well as the ability of oestrogen replacement to reduce performance deficits produced by scopolamine, a muscarinic cholinergic receptor antagonist (Gibbs 1999b). Young adult, ovariectomized rats received either continuous oestrogen replacement (3 mm capsule implanted s.c.) or sham surgery. Two months later, the animals were trained on the DMP task. After reaching criterion, animals were tested with increasing intertrial delays, as well as following systemic and intrahippocampal injections of either scopolamine or vehicle. The results indicate that the oestrogen-treated animals reached criterion significantly sooner (11.0 d vs. 15.0 d) and acquired the DMP task at a faster rate (Fig. 4A) than the non-oestrogen-treated controls. In addition, intrahippocampal scopolamine (20 μg/side administered 5 min prior to testing) produced a significant performance deficit in the non-oestrogen-treated controls, but failed to produce a performance deficit in the oestrogen-treated animals (Fig. 4B). This suggests that in addition to enhancing the rate of acquisition, continuous oestrogen replacement resulted in a significant reduction in performance deficit produced by hippocampal cholinergic inhibition. Other studies showing similar effects of oestrogen on the acquisition and performance of a reinforced T-maze task and on the working memory component of a radial arm maze task, as well as the ability of oestrogen to attenuate performance deficits produced by systemic and intrahippocampal scopolamine, have also recently been reported (Bimonte & Denenberg 1999, Daniel et al 1998, Fader et al 1998, 1999, Luine 1997). These findings demonstrate that oestrogen replacement can significantly affect performance on spatial learning tasks, and are consistent with the idea that oestrogen-mediated

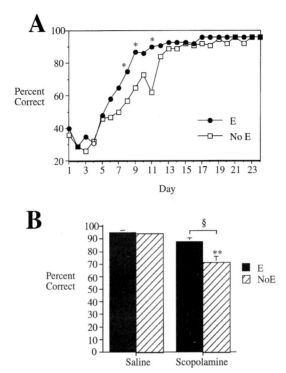

FIG. 4. (A) Learning curves summarizing acquisition of the DMP spatial memory task by ovariectomized oestrogen (E)-treated and non-E-treated animals. Values represent the mean percentage correct on each day of testing. Note that after the first 5 d of training E-treated animals improved at a greater rate than the non-E-treated controls. *$P < 0.05$ relative to non-E-treated controls. (B) Effects of intrahippocampal scopolamine administration on DMP performance by E-treated and non E-treated animals. Bars represent the mean percentage of correct choices. Note that intrahippocampal scopolamine administered 5 min prior to testing significantly reduced performance accuracy in the non-E-treated animals, but not in the E-treated animals, and that E-treated animals performed significantly better than the non-E-treated controls following intrahippocampal scopolamine administration. **$P < 0.001$ by paired t-test relative to corresponding saline-treated animals. §$P = 0.01$ by independent t-test comparing E-treated vs. non-E-treated animals given scopolamine. Adapted from Gibbs (1999b).

enhancements of basal forebrain cholinergic function may contribute to effects on cognitive performance.

Changes in cholinergic parameters associated with long-term loss of ovarian function

The data presented suggest that oestrogen- and progesterone-mediated effects on basal forebrain cholinergic neurons can help to decrease cognitive impairments

TABLE 2 Effects of ovariectomy on relative levels of ChAT and *trkA* mRNA in the medial septum and nucleus basalis magnocellularis in older animals relative to age-matched, gonadally intact controls

	ChAT mRNA	*TrkA mRNA*
Medial septum		
16 month old animals		
Controls	0.0 ± 13.3	0.0 ± 12.8
Ovx 3 Mo	−1.0 ± 7.4	−11.3 ± 12.0
19 month old animals		
Controls	0.0 ± 11.5	0.0 ± 14.3
Ovx 6 Mo	−38.9 ± 12.9*	−34.0 ± 9.4*
Ovx 6 Mo+oestrogen	−15.1 ± 7.9	−37.6 ± 14.3*
Nucleus basalis		
16 month old animals		
Controls	0.0 ± 4.1	0.0 ± 8.8
Ovx 3 Mo	−11.9 ± 7.2	−4.1 ± 14.5
19 month old animals		
Controls	0.0 ± 11.1	0.0 ± 6.3
Ovx 6 Mo	−33.5 ± 3.5*	−32.3 ± 7.9*
Ovx 6 Mo+oestrogen	−37.5 ± 6.1	−24.2 ± 15.3

All Ovx animals were ovariectomized at 13 months of age and were sacrificed either 3 or 6 months later. Values represent percent change from corresponding gonadally intact age-matched controls ± SEM. *$P < 0.05$ relative to controls.
Adapted from Gibbs (1998b).

associated with a hippocampal cholinergic deficit. This raises the possibility that similar effects in humans could help to reduce cognitive deficits associated with basal forebrain cholinergic impairment. Such effects may, in fact, contribute to the recent reports suggesting the ability of oestrogen replacement to reduce the risk and severity of Alzheimer's-related dementia in postmenopausal women.

The findings also raise the interesting question of whether the decreases in oestrogen and progesterone following menopause have negative effects on the cholinergic neurons which, over time, contribute to age-related decreases in basal forebrain cholinergic function. Few studies have examined the effects that the long-term loss of ovarian function has on brain ageing and basal forebrain cholinergic function.

Recently, we examined the effects that long-term loss of ovarian function has on ChAT and *trkA* expression in the MS and NBM as a function of ageing (Gibbs

1998b). Sprague–Dawley rats (13 months old) were either ovariectomized or received sham surgery. Ovariectomized animals were sacrificed at 16 and 19 months of age and compared with age-matched gonadally-intact controls. The results demonstrated significant decreases in the levels of ChAT and *trkA* mRNA in both the MS and NBM of animals sacrificed 6 months, but not 3 months, following ovariectomy relative to age-matched gonadally intact controls (Table 2). This suggests that long-term loss of ovarian function results in decreases in both ChAT and *trkA* expression in the MS and NBM beyond the effects of normal ageing.

Whether these effects can be prevented or reversed by oestrogen and progesterone administration is currently unknown. Short-term (3 d) treatment with 17β-oestradiol initiated 6 months following ovariectomy partially restored the levels of ChAT mRNA in the MS and *trkA* mRNA in the NBM (Table 2), but not to the extent predicted by the effects of oestrogen replacement in young animals. In addition, Singer et al (1998) recently reported that short-term oestrogen replacement (100 μg oestradiol injected s.c. every other day for 1 week) produced a modest but significant increase in ChAT mRNA in the HDB, as well as increases in both ChAT and *trkA* mRNA in the NBM, of aged (24 months old) ovariectomized rats. We have also recently determined that long-term (> 10 months) treatment with oestrogen plus progesterone results in significant enhancements in DMP performance by aged animals relative to ovariectomized, untreated controls (Gibbs 2000b). Whether these effects correlate with increased measures of cholinergic function will need to be determined.

Conclusions

There is increasing evidence that oestrogen and progesterone play a role in the physiological regulation of basal forebrain cholinergic neurons, and that oestrogen and progesterone replacement can enhance the functionality of basal forebrain cholinergic projections to the hippocampus and cortex with corresponding effects on cognitive function. Whether similar effects occur in humans is still unknown. Nevertheless, we speculate that similar effects in humans may contribute to the effects of hormone replacement on cognitive function, as well as on the risk and severity of Alzheimer's-related dementia in women, that have recently been reported. Given the recent results, it seems likely that consideration of the dose, duration and methods of hormone replacement will become important issues in defining the potential of hormone replacement therapy to reduce the risk and severity of cognitive decline in older women.

References

Bimonte HA, Denenberg VH 1999 Estradiol facilitates performance as working memory load increases. Psychoneuroendocrinology 24:161–173

Daniel JM, Fader AJ, Spencer AL, Dohanich GP 1998 Estrogen enhances performance of female rats during acquisition of a radial arm maze. Horm Behav 32:217–225

Fader AJ, Hendricson AW, Dohanich GP 1998 Estrogen improves performance of reinforced T-maze alternation and prevents the amnestic effects of scopolamine administered systemically or intrahippocampally. Neurobiol Learn Mem 69:225–240

Fader AJ, Johnson PEM, Dohanich GP 1999 Estrogen improves working but not reference memory and prevents amnestic effects of scopolamine on a radial-arm maze. Pharmacol Biochem Behav 62:711–717

Fusco M, Oderfeld-Nowak B, Vantini G et al 1989 Nerve growth factor affects uninjured, adult rat septohippocampal cholinergic neurons. Neuroscience 33:45–52

Gibbs RB 1996a Fluctuations in relative levels of choline acetyltransferase mRNA in different regions of the rat basal forebrain across the estrous cycle: effects of estrogen and progesterone. J Neurosci 16:1049–1055

Gibbs RB 1996b Expression of estrogen receptor-like immunoreactivity by different subgroups of basal forebrain cholinergic neurons in gonadectomized male and female rats. Brain Res 720:61–68

Gibbs RB 1997 Effects of estrogen on basal forebrain cholinergic neurons vary as a function of dose and duration of treatment. Brain Res 757:10–16

Gibbs RB 1998a Levels of *trkA* and BDNF mRNA, but not NGF mRNA, fluctuate across the estrous cycle and increase in response to acute hormone replacement. Brain Res 787:259–268

Gibbs RB 1998b Impairment of basal forebrain cholinergic neurons associated with aging and long-term loss of ovarian function. Exp Neurol 151:289–302

Gibbs RB 1999a Treatment with estrogen and progesterone affects relative levels of brain-derived neurotrophic factor mRNA and protein in different regions of the adult rat brain. Brain Res 844:20–27

Gibbs RB 1999b Estrogen replacement enhances acquisition of a spatial memory task and reduces deficits associated with hippocampal muscarinic receptor inhibition. Horm Behav 36:222–233

Gibbs RB 2000a Effects of estrogen on basal forebrain cholinergic neurons and cognition: implications for brain aging and dementia in women. In: Morrison M (ed) Hormones, aging, and mental disorders. Cambridge University Press, Cambridge, in press

Gibbs RB 2000b Long-term treatment with estrogen and progesterone enhances acquisition of a spatial memory task by ovariectomized aged rats. Neurobiol Ageing 21:107–116

Gibbs RB, Pfaff DW 1992 Effects of estrogen and fimbria/fornix transaction on p75[NGFR] and ChAT expression in the medial septum and diagonal band of Broca. Exp Neurol 116:23–39

Gibbs RB, Wu D, Hersh LB, Pfaff DW 1994 Effects of estrogen replacement on relative levels of choline acetyltransferase, TrkA and nerve growth factor messenger RNAs in the basal forebrain and hippocampal formation of adult rats. Exp Neurol 129:70–80

Gibbs RB, Hashash A, Johnson DA 1997 Effects of estrogen on potassium-stimulated acetylcholine release in the hippocampus and overlying cortex of adult rats. Brain Res 749:143–146

Henderson VW 1997 The epidemiology of estrogen replacement therapy and Alzheimer's disease. Neurology 48 (suppl 7):S27–S35

Knüsel B, Beck KD, Winslow JW et al 1992 Brain-derived neurotrophic factor administration protects basal forebrain cholinergic but not nigral dopaminergic neurons from degenerative changes after axotomy in the adult rat brain. J Neurosci 12:4391–4402

Lorenzi MV, Knusel B, Hefti F, Strauss WL 1992 Nerve growth factor regulation of choline acetyltransferase gene expression in rat embryo basal forebrain. Neurosci Lett 140:185–188

Luine VN 1985 Estradiol increases choline acetyltransferase activity in specific basal forebrain nuclei and projection areas of female rats. Exp Neurol 89:484–490

Luine VN 1997 Steroid hormone modulation of hippocampal dependent spatial memory. Stress 2:21–36

McMillan PJ, Singer CA, Dorsa DM 1996 The effects of ovariectomy and estrogen replacement on trkA and choline acetyltransferase mRNA expression in the basal forebrain of adult female Sprague-Dawley rat. J Neurosci 16:1860–1865

Nonomura T, Hatanaka H 1992 Neurotrophic effect of brain-derived neurotrophic factor on basal forebrain cholinergic neurons in culture from postnatal rats. Neurosci Res 14:226–233

O'Malley CA, Hautamaki RD, Kelley M, Meyer EM 1987 Effects of ovariectomy and estradiol benzoate on high affinity choline uptake, ACh synthesis, and release from rat cerebral cortical synaptosomes. Brain Res 403:389–392

Rylett RJ, Goddard S, Schmidt BM, Williams LR 1993 Acetylcholine synthesis and release following continuous intracerebral administration of NGF in adult and aged Fischer-344 rats. J Neurosci 13:3956–3963

Sherwin BB 1998 Estrogen and cognitive functioning in women. Proc Soc Exp Biol Med 217:17–22

Singer CA, McMillan PJ, Dobie DJ, Dorsa DM 1998 Effects of estrogen replacement on choline acetyltransferase and trkA mRNA expression in the basal forebrain of aged rats. Brain Res 789:343–346

Singh M, Meyer EM, Simpkins JW 1995 The effects of ovariectomy and estradiol replacement on brain-derived neurotrophic factor messenger ribonucleic acid expression in cortical and hippocampal brain regions of female Sprague-Dawley rats. Endocrinology 136:2320–2324

Singh M, Sétáló G Jr, Guan X, Warren M, Toran-Allerand CD 1999 Estrogen-induced activation of mitogen-activated protein kinase in cerebral cortical explants: convergence of estrogen and neurotrophin signaling pathways. J Neurosci 19:1179–1188

Toran-Allerand CD, Miranda RC, Bentham WDL et al 1992 Estrogen receptors colocalize with low-affinity nerve growth factor receptors in cholinergic neurons of the basal forebrain. Proc Natl Acad Sci USA 89:4668–4672

Yaffe K, Sawaya G, Lieberburg I, Grady D 1998 Estrogen therapy in postmenopausal women: effects on cognitive function and dementia. JAMA 279:688–695

DISCUSSION

McEwen: In your co-release studies, do you know anything about the latency after giving oestrogen? I ask this because a study by Mark Packard has shown that acute application of oestradiol in hippocampus facilitates memory functions in the hippocampus related to the cholinergic system (Packard 1998). I was wondering if you can get these effects rapidly, perhaps through some kind of membrane action.

Gibbs: The microdialysis studies were done with 10 day chronic administration of oestradiol. This is the only microdialysis study that we have so far completed. We can see the effects on the ChAT mRNA within 24 h. I know that in Packard's study he did some hippocampal oestrogen administration and saw rapid effects (Packard et al 1996), but I don't recall whether his systemic administrations also produced rapid effects. We haven't done the kind of study you are suggesting.

Toran-Allerand: Ovulation is not only accompanied by the secretion of oestrogen, but also large amounts of testosterone. In comparing the oestrous cycle with ovariectomy and oestrogen replacement with or without progesterone, one is omitting the significant contribution of testosterone which is present at the time of pro-oestrus. Does testosterone influence any of the parameters that you were talking about?

Gibbs: In our initial studies we looked at testosterone as well as oestrogen replacement in males, and we didn't see an effect. I have not given testosterone to the females. I recall that Vicky Luine looked at testosterone in males and didn't really see an effect on ChAT activity.

Luine: Testosterone given to ovariectomized females did not alter ChAT activity in the medial preoptic area (Luine et al 1975).

Sherwin: I think the testosterone issue is really neglected. The ovary makes one-third of all testosterone in a woman's body, and after menopause she is therefore testosterone deficient. Furthermore, if you treat women with oestrogen alone it increases SHBG and thereby lowers the amount of bioavailable testosterone. So her diminished postmenopausal levels are even less biologically active.

Toran-Allerand: Farida Sohrabji and I demonstrated the presence of a functional oestrogen response elment (ERE) in the BDNF gene (Sohrabji et al 1995). EREs are also present in the genes for other members of the neurotrophin family and their receptors. This is a good mechanism by which oestrogen may elicit the responses that you were showing.

Simpkins: At the risk of revealing that I reviewed your last manuscript, I will tell you about an experience with subcutaneous oestrogen injection. We have long believed what we taught our students: if we put steroids in oil, we get a diffuse release. It turns out you get a remarkable increase within 5 min of the oestradiol injection. It goes up to nanogram concentrations, and it falls off with a half-life of about 2 h, so that by the time you measure oestradiol at 2 d or 3 d, it is back to physiological levels. However, it certainly is not physiological soon after you give the injection. This raises the issue about whether or not the temporal pattern of the changes you are seeing are really related to this very marked increase in oestradiol and then it falling off over time.

Gibbs: That's a good point. After 3 d our levels are down to zero, and we're seeing elevated levels of ChAT and TrkA message even three days following administration of oestrogen if we then follow it with progesterone as well. We are therefore seeing elevations at times when oestrogen levels are low. This is also true in the cycling animal. Whether that is related to having superphysiological levels for a short period, I don't know. In the capsule studies where we know we're getting a more gradual release, we still see increases in both ChAT and TrkA. These increases generally are not maintained for months. One of the questions is why we are getting behavioural effects following long-term treatment, when the effects on the specific cholinergic parameters we are measuring tend not to be maintained. It may be that the measures we use do not reflect a sustained increase in the functionality of the neuron. In other words, we may get a short term effect on ChAT activity and HACU which then subsides due to regulatory mechanisms, but the neurons themselves may continue in a primed state, enabling the animals to perform better than they would otherwise.

Luine: No one has examined progesterone in relation to behaviour, so it is great that you have done it. Just reinforcing what you reported, oestrogen was important for acquisition and then progesterone appears to enhance performance. Perhaps oestrogen may be priming the neurons to respond to progesterone similar to its role in female reproductive behaviour.

Gibbs: Progesterone is always given two days after the oestrogen.

Woolley: I am curious about the comparison between the effects of hormone administration on a number of your measures versus effects during the oestrous cycle. Some of the things that we are talking about may be the sources of those differences. One would expect that 52 h of oestrogen plus short-term progesterone is a sort of recapitulation of pro-oestrus, yet this treatment increases various measures, whereas in pro-oestrus the same measures were decreased relative to dioestrus in general. Do you think that androgen or spiking levels of oestrogen following injection is the source of these differences? Those of us that use these hormone treatments hope that we are modelling something that happens in reality, because it is beginning to seem that we are not.

Gibbs: The safest thing to say would be that the hormone replacement regimen is not a perfect reproduction of what happens during the cycle. I think the comments about testosterone are important; this may well be responsible for some of the differences. The delay between hormone administration and the changes we see is worth bearing in mind. In terms of trying to understand the mechanisms and therapeutic approaches, if we want to consider using hormone replacement as a method of treating age-related cognitive decline, we have to think about how it's going to be given. We have to consider the delay between hormone administration and an effect, and how long that effect will be maintained even after hormonal levels decrease.

Watson: You can also get a burst of hormone from capsules, depending on how they are made.

Gibbs: The implanted capsules are filled with powder. We take great care to wash our capsules to make sure there's not powder on the outside of the capsule.

Watson: The oestrogen followed by progesterone effect may involve the induction of progesterone receptor. You might be able to test if you can inhibit activity of the progesterone receptor and block this.

Gibbs: That's a good point. We see effects of oestrogen alone and the progesterone seems to enhance these effects. I am not aware of any studies that have examined the presence of progesterone receptors in the cholinergic neurons. The possibility of neurosteroid effects at γ-aminobutyric acid (GABA)ergic receptors should also be considered.

Gustafsson: You say that you found ER protein: it would be interesting to know whether it is ERα or ERβ, because they have different effects.

Gibbs: István Merchenthaler and colleagues (personal communication) have done studies specifically with ERα versus ERβ within cholinergic neurons in rats and monkeys. They do see expression of ERα in some cholinergic neurons, but they don't see significant expression of ERβ, so I suspect that ERβ is not playing a significant role. However, Tom Clarkson's group has done some work with phytooestrogens, and has demonstrated effects on cholinergic neurons (Pan et al 1999). Phytooestrogens are thought to work through ERβ, so we don't have a good explanation for that, but I suspect that there are indirect effects.

Gustafsson: Now that you have a system that responds to oestrogen in the brain, have you checked the role of catecholoestrogens?

There is also the issue of conjugated oestrogens. At least in humans (in rodents it is less clear), the major portion of circulating oestrogen is in the form of sulfates, which are usually regarded as inactive metabolites. This is probably too simplistic an explanation: they may well have a function. Etienne Baulieu showed many years ago that DHEA sulfate was secreted from the adrenal cortex. Should conjugated oestrogens be re-considered as potentially important for oestrogenic effects?

Gibbs: I would not be surprised if catecholoestrogens and conjugated oestrogens had effects that mimic the potency of oestrogens at the ER. I don't know whether these effects would be direct or indirect with respect to the cholinergic neurons. We have not tested conjugated oestrogens or catechol oestrogens directly. I know that other people are doing work with premarin and are certainly seeing effects.

Gustafsson: Androgen metabolites could also have oestrogenic effects. Some androgen metabolites could trigger the ERs. One of the important metabolites is 5α-androstane-3β,17β-diol, which is a good ligand for ERβ and not such a good ligand (but still a ligand) for ERα. It seems that the brain has a particular system to get rid of this metabolite, which has also been studied by Etienne Baulieu, namely a P450 hydroxylase which specifically hydroxylates this androgen metabolite in the 6 and 7 positions, as if the brain has a protective mechanism to eliminate 5α-androstane-3β,17β-diol with its potential oestrogenic activity.

Gibbs: We have not looked at the androgens. There are certainly studies showing that DHEA and DHEA sulfate can affect acetylcholine release in the hippocampus and can have effects on behaviour. My understanding of that literature is that it is primarily acting through neurosteroid effects on the GABA$_A$ receptor. My personal belief is that effects on GABA are going to turn out to be extremely important, and that many of the effects I am seeing may be related to hormonal effects on GABAergic neurons that are then influencing these other systems.

References

Luine VN, Khylchevskaya RI, McEwen BS 1975 Effects of gonadal steroids on activities of monoamine oxidase and choline acetylase in rat brain. Brain Res 86:293–306

Packard MG 1998 Posttraining estrogen and memory modulation. Horm Behav 34:126–139

Packard MG, Kohlmaier JR, Alexander GM 1996 Post-training intrahippocampal estradiol injections enhance spatial memory in male rats: interaction with cholinergic systems. Behav Neurosci 110:626–632

Pan Y, Anthony M, Clarkson TB 1999 Effect of estradiol and soy phytoestrogens on choline acetyltransferase and nerve growth factor mRNAs in the frontal cortex and hippocampus of female rats. Proc Soc Exp Biol Med 221:118–125

Shorabji F, Miranda RC, Toran-Allerand CD 1995 Identification of a potential estrogen response element in the gene encoding brain-derived neurotrophic factor. Proc Natl Acad Sci USA 92:1110–1114

Ovarian steroid action in the serotonin neural system of macaques

Cynthia L. Bethea, Chrisana Gundlah and Stephanie J. Mirkes

Divisions of Reproductive Sciences and Neuroscience, Oregon Regional Primate Research Center, Beaverton, OR 97006, and Department of Physiology/Pharmacology, Oregon Health Sciences University, Portland, OR 97201, USA

Abstract. The serotonin neural system plays an important role in cognitive, emotional and endocrine processes. If the ovarian hormones, oestrogen and progesterone, alter serotonin neural transmission, then functional changes in all of these systems would follow. Therefore, information on the effects of oestrogen and progesterone at a molecular level in the serotonin neural system was sought using non-human primates. Serotonin neurons express nuclear oestrogen receptor β (ERβ) and progesterone receptors (PRs) which are gene transcription factors. Within serotonin neurons, the regulation of three genes related to serotonin neurotransmission was examined. The mRNA for tryptophan hydroxylase (TPH), the committal enzyme in serotonin synthesis, increased significantly with oestrogen treatment and remained elevated when progesterone was added to the oestrogen regimen. Serotonin reuptake transporter (SERT) mRNA decreased significantly with oestrogen treatment and addition of progesterone had no further effect. 5-HT$_{1A}$ autoreceptor mRNA decreased significantly with oestrogen treatment and addition of progesterone caused a further decrease. Little or no regulation of postsynaptic 5-HT$_{1A}$, 5-HT$_{2A}$ or 5-HT$_{2C}$ receptor mRNAs was observed in hypothalamic target neurons. TPH protein is increased by oestrogen treatment and remains elevated with addition of progesterone in a manner similar to TPH mRNA. Medroxyprogesterone (MPA) blocked the stimulatory effect of oestrogen on TPH protein and tamoxifen reduced TPH protein levels below that observed in spayed monkeys. Together these data indicate that ovarian hormones and their synthetic analogues could modify cognitive and autonomic neural functions by acting on the serotonin neural pathway.

2000 Neuronal and cognitive effects of oestrogens. Wiley, Chichester (Novartis Foundation Symposium 230) p 112–133

There are significant differences between the sexes in the incidence of various psychopathologies (Seeman 1997). Epidemiological studies show that worldwide, the lifetime prevalence of depression in women is twice the rate of men (Weissman & Olfson 1995). Several lines of evidence suggest an association between depression and reproductive function (Pearlstein 1995). A subpopulation of women also exhibit a unique form of psychosis which occurs postpartum, coincident with the precipitous decline in oestrogen and progesterone (Steiner

112

1990). With schizophrenia, a disease associated with clusters of symptoms including flat affect, withdrawal, psychosis, disorganized thought and hallucinations, women fall ill three to four years later than men and show a second peak of onset around menopause, but this varies with genetic load (Loranger 1984, Häfner & an der Heiden 1997). When women develop schizophrenia after age 45, they may suffer less severe negative symptoms than men or women with earlier onset (Lindamer et al 1999) and male schizophrenics manifest greater severity in structural brain abnormalities (Cowell et al 1996, Nopoulos et al 1997). Recently, a case was reported in which a postmenopausal woman with schizophrenia showed an improvement with oestrogen replacement therapy (Lindamer et al 1997).

The serotonin neural system projects to nearly every area of the forebrain and evidence exists that serotonin plays an important role in numerous autonomic and cognitive neural processes (Jacobs & Azmitia 1992). There is substantial serotonergic input to the hypothalamus, temporal lobe including the hippocampus and amygdala, and the prefrontal cortex (Azmitia & Gannon 1986, Azmitia & Segal 1978). These are the areas of the forebrain which regulate mood, affective behaviour and integrative cognition.

Therefore, we have investigated whether ovarian steroids can alter serotonin neural function using non-human primates. Any change in serotonin neurotransmission would impact several categories of cognition including organized thought, associative functions and memory, as well as overall mood or affect.

The ovarian steroids, oestrogen and progesterone, act largely through nuclear receptors, oestrogen receptor (ER) and progesterone receptor (PR), which are gene transcription factors (Tsai & O'Malley 1994). Moreover, the presence of ER or PR defines the cell as a target for the cognate hormone (Katzenellenbogen et al 1996).

Serotonin transmission can be envisioned as the sum of several processes including (but not limited to) serotonin synthesis, reuptake, degradation, neural activity which drives release, and receptor activation. In turn, serotonin synthesis is governed by tryptophan hydroxylase (TPH), the rate-limiting enzyme in the conversion of tryptophan to 5-hydroxytryptophan; serotonin reuptake is accomplished by the serotonin reuptake transporter (SERT); serotonin neural activity is inhibited by the 5-HT_{1A} autoreceptor; serotonin is degraded by monoamine oxidases (MAO), and there are at least seven major types of serotonin receptors on target cells which utilize a variety of intracellular transduction mechanisms. Hence, candidate genes that code for these respective proteins are the TPH gene, the SERT gene, the 5-HT_{1A} gene within serotonin neurons which codes for the autoreceptor, MAO-A and MAO-B genes, as well as the genes that code for the different serotonin receptors in target cells.

This paper will focus on three basic questions:

(1) Do serotonin neurons contain ovarian steroid receptors?
(2) Do ovarian steroids alter the expression of pivotal genes in serotonin neurons or in target neurons?
(3) For TPH, do protein levels reflect gene expression, and do synthetic steroids act like natural steroids?

Experimental procedures

To answer these questions we use the rhesus monkey (*Macaca mulatta*). This primate species has substantial neocortex, a reproductive cycle identical to that of women and complex social behaviours as do humans. We have employed a treatment paradigm composed of three groups of female monkeys: untreated controls, a group treated with oestrogen only and a group treated with oestrogen + progesterone. All of the animals are ovariectomized (OVX) and hysterectomized ('spayed') about 3–6 months prior to assignment to our program. The control group receives an empty Silastic capsule on day 0, implanted subcutaneously in the pericapsular region. The oestrogen and oestrogen + progesterone groups receive an oestrogen-filled capsule on day 0. Then, the oestrogen + progesterone group receives an additional progesterone-filled capsule on day 14. All of the animals are euthanized after 28 days. The level of oestrogen achieved in the serum is between 100–200 pg/ml. The level of progesterone achieved in the serum is between 4–8 ng/ml.

The techniques employed and the details of the protocols have been previously published. Single and double immunocytochemistry for serotonin and PR with computer-assisted image analysis were developed in the author's laboratory (Bethea 1993, 1994). Single *in situ* hybridization and densitometric analysis for TPH, SERT, 5-HT$_{1A}$ autoreceptor, 5-HT$_{1A}$, 5-HT$_{2A}$ and 5-HT$_{2C}$ receptors were conducted by this laboratory according to well-accepted procedures (Pecins-Thompson et al 1996, 1998, Pecins-Thompson & Bethea 1999, Gundlah et al 1999). Double *in situ* hybridization for SERT and ERβ were performed in the author's laboratory according the recommendations of Peterson (Petersen & McCrone 1993). For the *in situ* hybridization studies, all tissues were heavily fixed with 4% paraformaldehyde. Radioimmunoassays for oestrogen and progesterone were conducted in the ORPRC Endocrine Services Laboratory according to the methods of Resko (Resko et al 1975).

Results and discussion

We initially demonstrated by double-label immunocytochemistry that serotonin neurons in the dorsal raphe of macaques contain PR (Bethea 1993). The antibody to nuclear PR was bridged to diaminobenzidine substrate yielding a brown reaction product which was detected only in the nucleus. The antibody to serotonin was bridged to alkaline phosphatase, then developed with a blue chromogen. This label was found only in the cytoplasm. Additional neurons in the dorsal raphe expressed nuclear PR, but were of unidentified phenotype. The antibody to PR was JZB39, an anti-human monoclonal, from Dr Geoffrey Greene, University of Chicago, Chicago, IL.

To determine whether the expression of PR was regulated by ovarian steroids in serotonin neurons, we applied single immunocytochemistry for serotonin and PR to adjacent sections and subjected these sections to computer-assisted image analysis. The computer converted the labelled cells to a binary image and then counted the number of cells in the field. We counted PR-positive cells at four levels of the dorsal and medial raphe nucleus of steroid-treated macaques (Bethea 1994). Table 1 shows that very few cells expressed PR in spayed monkeys. However, with oestrogen treatment there was a significant increase in the number of PR-positive cells and addition of progesterone to the oestrogen regimen did not alter the number of cells expressing PR.

The induction of PR by oestrogen has been ubiquitously attributed to an action of oestrogen through a nuclear ER. Following these experiments, we sought the classical ER, now called ERα, in the dorsal raphe of macaques using both immunocytochemistry and *in situ* hybridization. However, we were unable to detect ERα with either of these procedures.

None the less, a new ER has been recently discovered, called ERβ (Kuiper et al 1996, Enmark et al 1997). Upon publication of the sequence of human ERβ, we applied human primers and RT-PCR to RNA extracted from the rhesus monkey prostate gland and constructed monkey-specific cDNAs to the 5′ and 3′+UT regions of ERβ. Empirically, a combination of both probes yields the best signal to noise ratio in monkey brain (Gundlah et al 2000).

TABLE 1   The effect of hormone replacement for 1 month on progestin receptor immunostaining in the raphe of rhesus macaques

	Treatment			
	Spayed	E	E+P	Anova
PR-positive cells ± SEM (monkeys/group)	4.25 ± 0.99 (3)	72.58 ± 8.16[a] (4)	67.33 ± 5.75[a] (4)	P < 0.001

The number of positive cells was counted in a 540×760 μm area at four levels of the dorsal and medial raphe with computer-assisted image analysis. The data represent the average for all levels for all animals.
P, progesterone; E, oestrogen; PR, progestin receptors.
[a]Different from spayed control group with Duncan's *post hoc* pairwise comparison, P < 0.05.

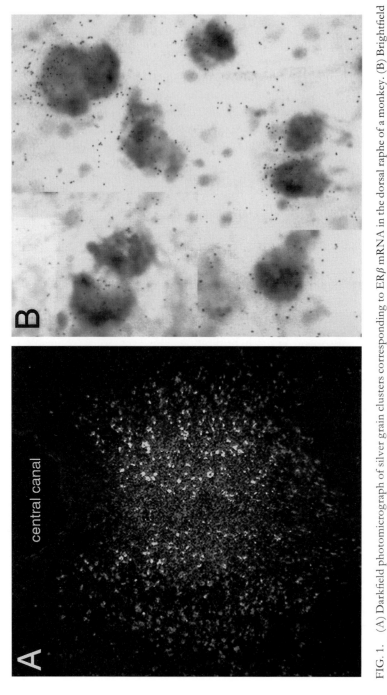

FIG. 1.   (A) Darkfield photomicrograph of silver grain clusters corresponding to ERβ mRNA in the dorsal raphe of a monkey. (B) Brightfield photomicrograph illustrating cells in the dorsal raphe which are double-labelled for SERT mRNA and ERβ mRNA. SERT mRNA is an excellent and specific marker for serotonin cells. The SERT mRNA was hybridized with antisense riboprobes incorporating digoxygenin and developed with a purple-coloured substrate. The ERβ mRNA was hybridized with antisense probe incorporating [^{35}S]UTP and developed with NTB-2 emulsion yielding silver grain clusters over the SERT-positive cells.

ERβ was sought in areas of the midbrain which are rich in serotonin neurons. Figure 1A contains a darkfield photomicrograph of the dorsal raphe region of a rhesus macaque that has been hybridized with ^{33}P-labelled riboprobes for ERβ. Clusters of silver grains corresponding to ERβ mRNA are observed over numerous cells in the dorsal raphe. Similar hybridization patterns were observed in the medial raphe. However, the raphe nuclei contain several different types of neurons, in addition to serotonin neurons.

Therefore, to determine whether serotonin neurons express ERβ, double *in situ* hybridization was applied to the dorsal raphe. Serotonin neurons were identified with a riboprobe against the SERT incorporating digoxygenin-UTP and ERβ was identified with the dual riboprobes incorporating [^{35}S]UTP. Figure 1B illustrates a field of cells in the dorsal raphe of a macaque which has been double labelled. The serotonin cells are labelled with purple alkaline phosphatase reaction product. There are clusters of silver grains corresponding to ERβ over several of these cells. When the purple reaction product is dense, the black grains are difficult to discern, but on the lighter-labelled serotonin cells, there are clusters of ERβ grains which are three to five times higher than background.

From this data we hypothesize that serotonin neurons in higher primates contain ERβ, which when activated by oestrogen, induces expression of PR. Furthermore, oestrogen and progesterone, acting through their respective nuclear receptors, can alter gene expression governing serotonin neurotransmission.

It is then reasonable to ask whether any of the genes which control serotonin neurotransmission are targets for oestrogen or progesterone via their cognate receptors; and if so, how do oestrogen and progesterone regulate those genes? Using riboprobes against the various mRNA species and *in situ* hybridization, we have examined the expression of the TPH, SERT and 5-HT$_{1A}$ autoreceptor genes within serotonin neurons and have also examined expression of the 5-HT$_{1A}$, 5-HT$_{2A}$ and 5-HT$_{2C}$ receptor genes in hypothalamic target neurons.

Riboprobes incorporating [^{35}S]UTP were generated by reverse transcription of cDNA clones of each gene. The monkey-specific TPH cDNA clone contains 249 basepairs (bp) representing the 5′ substrate binding domain of TPH. It was constructed after RT-PCR using RNA extracted from the monkey dorsal raphe and primers based upon the rat TPH sequence (Pecins-Thompson et al 1996). The SERT cDNA was subcloned from the full length human cDNA clone provided by Dr Randy Blakely (Vanderbilt University, Nashville, TN). It contains 253 bp representing the 5′ cytoplasmic tail, a region with low homology to other transporters (Pecins-Thompson et al 1998). The monkey-specific 5-HT$_{1A}$ cDNA clone contains 432 bp corresponding to the region between transmembrane domains V and VI. It was constructed using PCR on rhesus monkey genomic

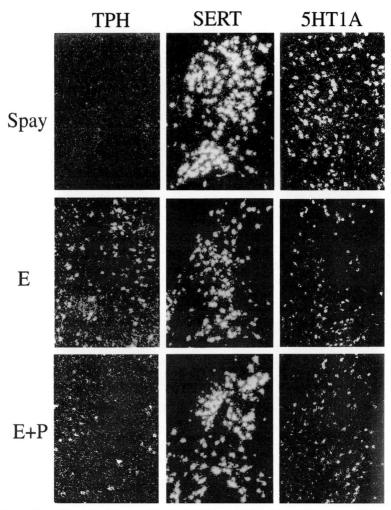

FIG. 2. Darkfield photographs of the right side of the dorsal raphe in representative spayed, oestrogen (E)-treated and oestrogen + progesterone (E+P)-treated monkeys. Sections containing the dorsal raphe were hydridized to antisense riboprobes for TPH, SERT and 5-HT$_{1A}$ autoreceptor mRNAs, dipped in NTB-2 emulsion and then developed. TPH mRNA is virtually undetectable in spayed macaques. There is an obvious increase in silver grains corresponding to TPH mRNA with oestrogen and significant signal remains when progesterone is added to the oestrogen regimen. In contrast, there is robust expression of SERT and 5-HT$_{1A}$ mRNAs in spayed macaques. Then, there is a marked decrease in silver grains corresponding to SERT mRNA and 5-HT$_{1A}$ mRNA with oestrogen and oestrogen + progesterone treatment. To quantitate these changes, film autoradiographs of multiple sections were subjected to densitometric analysis and the results are shown in Table 2. In addition, counts of silver grains on multiple sections such as these confirmed the densitometric analysis.

TABLE 2 The effect of hormone replacement for 1 month on TPH mRNA, SERT mRNA and 5-HT$_{1A}$ autoreceptor mRNA in the dorsal raphe of rhesus macaques as indicated by positive pixel measurements after operator assisted thresholding with NIH Image

	Treatment			
	Spayed	E	E+P	Anova
TPH mRNA-positive pixels ± SEM (monkeys/group)	279 ± 54 (5)	2282 ± 1115[a] (5)	1593 ± 570[a] (4)	P < 0.01
SERT mRNA-positive pixels ± SEM (monkeys/group)	12484 ± 920 (5)	9200 ± 592[b] (5)	9150 ± 816[b] (5)	P < 0.005
5-HT$_{1A}$ mRNA-positive pixels ± SEM (monkeys/group)	4524 ± 677 (5)	2918 ± 635[b] (5)	2579 ± 526[b] (5)	P > 0.02

The data were obtained from adjacent sections of the same animals.
E, oestrogen; P, progesterone.
[a]Different from spayed control group with Dunn's non-parametric *post hoc* pairwise comparison, P < 0.05.
[b]Different from spayed control group with Student Newman Keul's *post hoc* pairwise comparison, P < 0.05.

DNA and primers based upon the human sequence (Pecins-Thompson & Bethea 1999).

Figure 2 contains representative darkfield photographs of the right side of the dorsal raphe from spayed, oestrogen-treated and oestrogen + progesterone-treated macaques hybridized for TPH, SERT and 5-HT$_{1A}$ mRNAs. In spayed monkeys, the TPH mRNA signal is barely detectable, but upon oestrogen treatment, there is a marked induction of TPH mRNA. TPH is still easily detectable when progesterone is added to the oestrogen regimen. In contrast, there is robust expression of SERT mRNA in spayed macaques. Then, the SERT mRNA signal in the dorsal raphe decreases markedly in the oestrogen- and oestrogen + progesterone-treated groups. Also, there is robust expression of 5-HT$_{1A}$ mRNA in the spayed monkey raphe. Upon treatment with oestrogen or oestrogen + progesterone, there is an obvious decrease in the silver grain signal for 5-HT$_{1A}$ mRNA.

In order to quantitate this signal, the NIH Image Program was used to analyse film autoradiographs generated from multiple sections. This software generates the optical density from the grey scale image and it also generates the number of pixels covered by the signal after thresholding. In each monkey, five levels of the dorsal raphe nucleus from a rostral to caudal direction were subjected to image analysis for each gene.

Table 2 contains the average area of positive pixels for the TPH, SERT and the 5-HT$_{1A}$ autoreceptor mRNAs. There was a nearly 10-fold increase in TPH mRNA signal when the monkeys were treated for 28 days with oestrogen. Addition of progesterone to the oestrogen regimen for the last 14 of the 28-day treatment period had no statistical effect. TPH mRNA remained significantly higher in the oestrogen + progesterone-treated group than in the spayed control group, but TPH mRNA was not different between animals treated with oestrogen alone versus animals treated with oestrogen + progesterone. There was a statistically significant, 30%, decrease in SERT mRNA signal with oestrogen treatment and addition of progesterone to the oestrogen regimen did not have any further effect. Image analysis of the autoradiographs for 5-HT$_{1A}$ mRNA signal indicated that there was a significant decrease in 5-HT$_{1A}$ pixel area with oestrogen and oestrogen + progesterone treatment. In the analysis of optical density for 5-HT$_{1A}$ mRNA, an additional and significant decrease in signal was observed when progesterone was added to the oestrogen regimen (data not listed). In addition, silver grains were counted on the emulsion-dipped sections (as illustrated in Fig. 2) which correlated with the densitometric results.

If these changes in gene expression are manifested by changes in protein expression and function, then together, it can be predicted that ovarian steroids would increase serotonin synthesis, decrease serotonin reuptake and decrease autoreceptor inhibition. In concert, these actions would lead to an overall increase in serotonin neural transmission.

To determine whether oestrogen or progesterone regulate serotonin receptor gene expression, we examined the expression of mRNA for 5-HT$_{1A}$, 5-HT$_{2A}$ and 5-HT$_{2C}$ receptors in the hypothalamus. The hypothalamus contains dense populations of cells with ovarian steroid receptors. If oestrogen or progesterone affected the expression of the serotonin receptor genes directly, via their cognate nuclear receptors, then the areas of the hypothalamus containing a high concentration of ER or PR would be expected to manifest that regulation.

Therefore, in addition to the previously described 5-HT$_{1A}$ probe, monkey-specific cDNAs were generated for the 5-HT$_{2A}$ and 5-HT$_{2C}$ receptors using PCR on rhesus monkey genomic DNA and primers based upon the published human sequence. The 5-HT$_{2A}$ construct contains 411 bp corresponding to amino acids 1–135 which span the 5' extracellular tail, the first transmembrane domain, the first intracellular loop and the second transmembrane domain of the receptor. The 5-HT$_{2C}$ construct contains 294 bp corresponding to amino acids 13–110 which are located towards the end of the 5' extracellular tail through the second transmembrane domain (Gundlah et al 1999).

The distribution and regulation of 5-HT$_{1A}$, 5-HT$_{2A}$ and 5-HT$_{2C}$ receptor mRNAs were examined in the hypothalamic terminal field with *in situ* hybridization and densitometry of autoradiographs. Figure 3 contains film autoradiographs which illustrate the distribution of the mRNA for the three serotonin receptors in the hypothalamus. The 5-HT_{1A} receptor mRNA was expressed in the preoptic area (POA), the ventromedial nucleus (VMN), a small contiguous region in the dorsomedial nucleus and posterior hypothalamus (DMN + PH) and the supramammillary region (SMam). The 5-HT_{2A} receptor mRNA was anatomically confined to the paraventricular nucleus (PVN), the supraoptic nucleus (SON) and the mammillary nuclei (Mam). In contrast, the 5-HT_{2C} receptor mRNA was found throughout the hypothalamus.

There was very little overlap in the receptor fields and the 5-HT_{2C} receptor mRNA was, by far, the most prevalent. This agrees well with autoradiographic binding data in rats (Molineaux et al 1989). The two areas where overlap was detected are the POA, which expresses 5-HT_{1A} and 5-HT_{2C}; and the VMN which expresses 5-HT_{1A} and 5-HT_{2C}.

An exhaustive densitometric analysis on each of these receptors, in all of these areas, found little or no regulation by ovarian steroids. This does not rule out the potential for ovarian steroids to alter coupling of these receptors to their intracellular signal cascade, but there does not seem to be a marked effect at the level of gene expression, at least in the hypothalamus. It should be noted that a global analysis such as this includes receptors that are on cells which may or may not contain ovarian steroid receptors. Therefore, a final analysis awaits hybridization and grain counting over various phenotypically identified cells.

Hence, to date, the predominant effects of ovarian steroids appear to be exerted on gene expression within serotonin neurons of macaques. Next, it is important to determine whether the documented changes in gene expression are manifested by changes in protein and/or function. For TPH, a measurement of protein mass could indicate whether translation proceeds from transcription with or without gross modification. TPH is also post-translationally modified by phosphorylation at serine 58 for enzymatic activity (Kuhn et al 1997).

A Western blot analysis was applied to extracts of the dorsal raphe to examine the effects of natural and synthetic steroids on TPH protein mass. Figure 4 illustrates the linearity of the analysis. The NIH Image Program was used to scan the lanes, generate a peak representing the optical density of the band and then provide the area under the peak in arbitrary units. Increasing volumes of homogenates of a pineal gland and a monkey raphe exhibit a linear increase in TPH signal.

Table 3 contains the average optical densities for TPH protein mass of spayed animals that were untreated or treated with natural oestrogen and progesterone in the same manner as the animals used for TPH mRNA analysis. The spay control animals have low, but detectable, TPH protein mass which increases slightly with time after ovariectomy. Upon oestrogen treatment there is a marked induction of

5HT1A   5HT2A   5HT2C

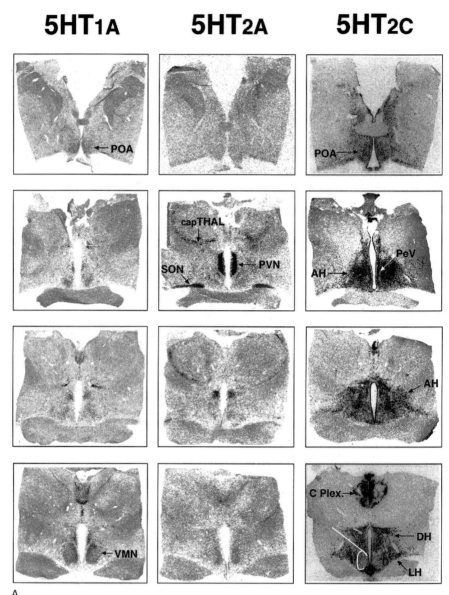

A

TPH protein and a similar increase in TPH protein is detected when progesterone is added to the oestrogen regimen. An additional group of monkeys was primed with oestrogen and then maintained for 4 weeks on progesterone alone. This group showed a similar increase in TPH protein mass over short-term spayed

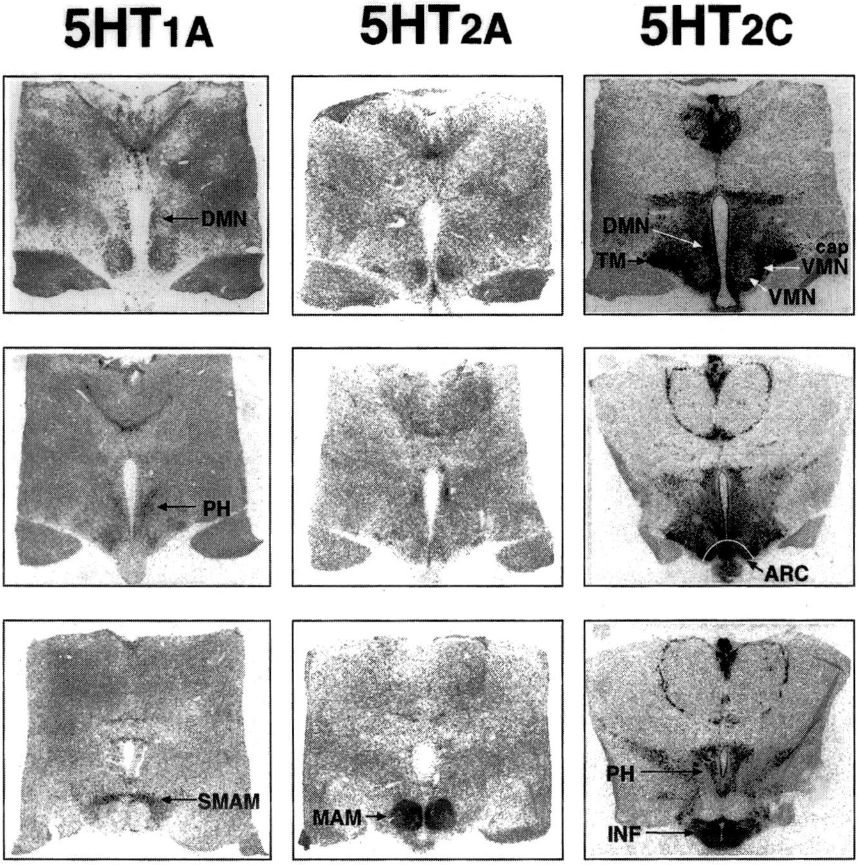

5HT₁ₐ   5HT₂ₐ   5HT₂c

B

FIG. 3. Digitized autoradiographs from representative sections of equivalent levels of the hypothalamus from a spayed monkey illustrating 5-HT_{1A}, 5-HT_{2A} and 5-HT_{2C} receptor mRNA expression as reflected by black pixels. The 5-HT_{1A} receptor signal was localized to the preoptic area (POA), ventromedial nuclei (VMN), a more diffuse contiguous region from the dorsomedial nuclei through the posterior hypothalamus (PH), and a diffuse expression in the supramammillary region (SMAM). The 5-HT_{2A} receptor signal was localized to the paraventricular nuclei (PVN), supraoptic nuclei (SON), the capsule of the thalamus (capTHAL) and more rostrally in the mammillary nuclei (MAM). A variable 5-HT_{2A} receptor signal was observed in the ventromedial nuclei of some animals. In contrast, 5-HT_{2C} receptor signal was much more prevalent, showing dense expression in the preoptic area, anterior hypothalamus (AH), the periventricular region (PeV) of the paraventricular nuclei, the arcuate nucleus (ARC), the area surrounding the ventromedial nuclei, the dorsomedial nuclei (DMN), infundibular area (INF), tuberomammillary nuclei (TM) and the choroid plexus (C Plex.). Moderate 5-HT_{2C} receptor expression was observed within the ventromedial nuclei, lateral hypothalamus (LH) and the dorsal (DH) to posterior hypothalamus (PH). The demarcations used for analysis are illustrated with white lines in some sections for clarity. Black or white arrows indicate the anatomical areas.

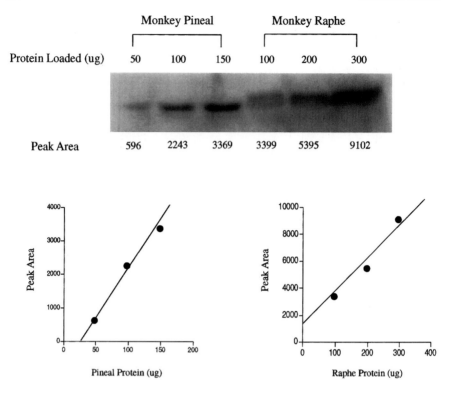

FIG. 4. Demonstration of the linearity of the Western blotting procedure. Increasing concentrations of extracts from the pineal gland and the dorsal raphe produced a linear increase in band density as determined with the NIH Image software subroutine for gel analysis.

controls. There is no statistical difference between oestrogen-treated, oestrogen + progesterone-treated or oestrogen-primed and then progesterone-treated groups. TPH was consistently elevated in each treatment group over the short-term spayed control group. This closely reflects the pattern that we observed with TPH mRNA, suggesting that translation follows transcription with little modification. The treatment of spayed animals with a low dose of progesterone only for 14 days caused an approximately twofold increase in TPH protein mass.

Table 4 contains the average optical densities of TPH protein mass in animals treated with synthetic hormones or tamoxifen, a selective ER modulator (SERM). These data were obtained from cynomolgus macaques treated for 30 months with hormones that are commonly used in hormone replacement therapy in the USA. In addition, the doses of hormones were adjusted to closely approximate clinical

TABLE 3 The effect of natural hormone replacement for up to 1 month on TPH protein mass in the dorsal raphe of rhesus macaques

| | Treatment | | | | | | |
	Spay-short	Spay-long	E	P	E + P	E then P	# of blots
Experiment 1							
OD ± SEM	715 ± 176	1401 ± 167*					1
(monkeys/group)	(3)	(5)					
Experiment 2							
OD ± SEM	498		3367	1450 ± 284			1
(monkey group)	(1)		(1)	(5)			
Experiment 3							
OD ± SEM	736		2794 ± 154		2230 ± 192	2644 ± 621	3
(monkeys/group)	(1)		(3)		(4)	(4)	
Overall ANOVA, $P < 0.003$							
Mean ± SEM	733 ± 145[a,b,c]	1401 ± 167[d,e]	2831 ± 172[a,d,f]	1450 ± 284[f,g]	2230 ± 192[b]	2644 ± 621[c,e,g]	5
(monkeys/treatment)	(3)	(5)	(3)	(5)	(4)	(4)	

The dorsal raphe region was homogenized and TPH protein was determined with western blotting and densitometric analysis of TPH bands on chemiluminescent films. These data are derived from five blots that were run in three experiments over a period of several months. Each blot contained an aliquot of an homogenate of monkey pineal glands as a positive control lane and molecular weight markers. If an animal was run on more than one blot, then the average optical density (OD) of the individual measurements was obtained first. Finally, one OD per animal was used to calculate the overall mean. N is the number of monkeys in each group.
P, progesterone; E, oestrogen.
The same superscript letter indicates there is a difference between the groups with Student Newman Keul's *post hoc* pairwise comparison, $P < 0.05$.
*Different from short-term spayed control group with Student's *t*-test.

TABLE 4 The effect of synthetic hormone replacement for 30 months on TPH protein mass in the dorsal raphe of cynomolgus macaques

| | Treatment | | | | | | |
	Ovx	EE	MPA	EE + MPA	Tamox	# of blots	Anova
Experiment 1							
OD ± SEM	966 ± 243	2299 ± 701*	2370 ± 470*	1295 ± 329#		2	P < 0.01
(monkeys/group)	(5)	(5)	(5)	(5)			
Experiment 2							
OD ± SEM	1089 ± 273	1654 ± 118*			650 ± 172*#	1	P < 0.002
(monkeys/group)	(5)	(5)			(5)		

The dorsal raphe region was homogenized and TPH protein was determined with western blotting and densitometric analysis of TPH bands on chemilumenescent films. These data are derived from three blots that were run in two experiments over a period of three weeks. Each blot contained an aliquot of an homogenate of monkey pineal glands as a positive control lane and molecular weight markers. One optical density (OD) value per animal was used to calculate the overall mean. N is the number of monkeys in each group.

Ovx, ovariectomized; EE, conjugated equine oestrogens; MPA, medroxyprogesterone; Tamox, tamoxifen.

*Different from Ovx group with Student Newman Keul's *post hoc* pairwise comparison, P < 0.05.

#Different from EE group with Student Newman Keul's *post hoc* pairwise comparison, P < 0.05.

doses and administered to the monkeys in their diet. The cynomolgus midbrain tissues were provided by Dr Carol Shively, Department of Comparative Medicine, Bowman Gray School of Medicine.

TPH protein mass was compared between OVX, untreated controls and animals that were treated with Premarin, a preparation of conjugated equine oestrogens, medroxyprogesterone acetate (MPA), a commonly prescribed progestin and a combination of Premarin plus MPA. The conjugated equine oestrogens increased TPH protein mass in a manner similar to natural oestrogen. MPA alone also increased TPH protein mass. But when MPA was added to conjugated equine oestrogen treatment, there was a significant decrease in TPH protein mass from equine oestrogen treatment alone. That is, TPH protein was significantly less in the equine oestrogen + MPA-treated group than in animals treated with equine oestrogen alone or MPA alone, but not different from the OVX controls.

Tamoxifen is a commonly used treatment for women with breast cancer (Horwitz 1995). It is currently in a large-scale clinical trial in the USA for prophylactic use in women at high risk for breast cancer (Yeomans-Kinney et al 1995). There has been some information emerging that women taking tamoxifen are experiencing more depressive symptoms than women not taking tamoxifen (Cathcart et al 1993).

On a separate blot, TPH protein mass was compared in OVX controls, equine oestrogen-treated and tamoxifen-treated monkeys. The equine oestrogen-treated animals reliably demonstrated a significant increase in TPH protein mass over OVX controls. However, there was a significant decrease in TPH protein mass upon treatment with tamoxifen for 30 months. The level of TPH in the tamoxifen group was significantly less than in the equine oestrogen group, and also significantly less than in the control group. There was a low level of oestrogen present in the serum of ovariectomized monkeys (c. 20 pg/ml) which is thought to come from the adrenal gland and fat tissue. Tamoxifen may antagonize the beneficial effect of the small amount of oestrogen that remains in OVX animals.

In conclusion, serotonin neurons of higher primates contain the nuclear receptors, ERβ and PR. The expression of the TPH, SERT and 5-HT$_{1A}$ autoreceptor genes within serotonin neurons is regulated by oestrogen and progesterone in such a manner as to increase serotonin neurotransmission. It currently appears that there is good agreement between TPH gene expression and TPH protein expression. However, there may be a discrepancy in the action of the synthetic progestin, MPA, compared to natural progesterone. Together these data indicate that ovarian hormones and their synthetic analogues could modify cognitive and autonomic neural functions by acting on the serotonin neural pathway (Fig. 5).

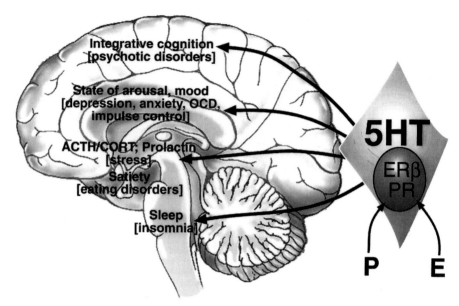

FIG. 5. Diagrammatic summary of how the ovarian steroids may alter cognition, affect and other autonomic neural functions by an action in the serotonin neural system. Serotonin neurons express ERβ, which when activated by its cognate ligand, oestrogen (E), will induce expression of the progestin receptor (PR) gene and the TPH gene. Oestrogen also decreases expression of the SERT and 5-HT$_{1A}$ genes. Progesterone (P) subsequently can act through PR to alter the expression of genes related to serotonin neurotransmission. The molecular mechanism(s) by which oestrogen both increases and decreases gene expression within the serotonin neuron are of significant interest. Moreover, the identity of genes uniquely induced or repressed by PR in the CNS has remained elusive.

Acknowledgements

Supported by NIH grants HD17269 to CLB, T32-HD17133 to CG, P30 Population Center Grant HD18185, and RR00163 for the operation of ORPRC.

References

Azmitia EC, Gannon PJ 1986 The primate serotonergic system: a review of human and animal studies and a report on *Macaca fascicularis*. Adv Neurol 43:407–468

Azmitia EC, Segal M 1978 An autoradiographic analysis of the differential ascending projections of the dorsal raphe and median raphe nuclei in the rat. J Comp Neurol 179:641–667

Bethea CL 1993 Colocalization of progestin receptors with serotonin in raphe neurons of macaque. Neuroendocrinology 57:1–6

Bethea CL 1994 Regulation of progestin receptors in raphe neurons of steroid-treated monkeys. Neuroendocrinology 60:50–61

Cathcart CK, Jones SE, Pumroy CS, Peters GN, Knox SM, Cheek JH 1993 Clinical recognition and management of depression in node negative breast cancer patients treated with tamoxifen. Breast Cancer Res Treat 27:277–281

Cowell PE, Kostianovsky DJ, Gur RC, Turetsky BI, Gur RE 1996 Sex differences in neuroanatomical and clinical correlations in schizophrenia. Am J Psychiatry 153:799–805

Enmark E, Pelto-Huikko M, Grandien K et al 1997 Human estrogen receptor β gene structure, chromosomal localization and expression pattern. J Clin Endocrinol Metab 82:4258–4265

Gundlah C, Pecins-Thompson M, Schutzer WE, Bethea CL 1999 Ovarian steroid effects on serotonin 1A, 2A and 2C receptor mRNA in macaque hypothalamus. Mol Brain Res 63:325–339

Gundlah C, Kohama SG, Mirkes SJ, Garyfallou V T, Urbanski HF, Bethea CL 2000 Distribution of estrogen receptor beta (ERβ) mRNA in hypothalamus, midbrain and temporal lobe of spayed macaque: continued expression with hormone replacement. Brain Res Mol Brain Res 76:191–204

Häfner H, an der Heiden W 1997 Epidemiology of schizophrenia. Can J Psychiatry 42:139–151

Horwitz KB 1995 When tamoxifen turns bad. Endocrinology 136:821–823

Jacobs BL, Azmitia EC 1992 Structure and function of the brain serotonin system. Physiol Rev 72:165–229

Katzenellenbogen JA, O'Malley BW, Katzenellenbogen BS 1996 Tripartite steroid hormone receptor pharmacology: interaction with multiple effector sites as a basis for the cell- and promoter-specific action of these hormones. Mol Endocrinol 10:119–131

Kuhn DM, Arthur R Jr, States JC 1997 Phosphorylation and activation of brain tryptophan hydroxylase: identification of serine-58 as a substrate site for protein kinase A. J Neurochem 68:2220–2223

Kuiper GG, Enmark E, Pelto-Huikko M, Nilsson S, Gustafsson JA 1996 Cloning of a novel estrogen receptor expressed in rat prostate and ovary. Proc Natl Acad Sci USA 93:5925–5930

Lindamer LA, Lohr JB, Harris MJ, Jeste DV 1997 Gender, estrogen, and schizophrenia. Psychopharmacol Bull 33:221–228

Lindamer LA, Lohr JB, Harris MJ, McAdams LA, Jeste DV 1999 Gender-related clinical differences in older patients with schizophrenia. J Clin Psychiatry 60:61–69

Loranger AW 1984 Sex difference in age at onset of schizophrenia. Arch Gen Psychiatry 41:157–161

Molineaux SM, Jessell TM, Axel R, Julius D 1989 5-HT$_{1C}$ receptor is a prominent serotonin receptor subtype in the central nervous system. Proc Natl Acad Sci USA 86:6793–6797

Nopoulos P, Flaum M, Andreasen NC 1997 Sex differences in brain morphology in schizophrenia. Am J Psychiatry 154:1648–1654

Pearlstein TB 1995 Hormones and depression: what are the facts about premenstrual syndrome, menopause and hormone replacement therapy? Am J Obstet Gynecol 173:646–653

Pecins-Thompson M, Bethea CL 1999 Ovarian steroid regulation of 5-HT$_{1A}$ autoreceptor messenger ribonucleic acid expression in the dorsal raphe of rhesus macaques. Neuroscience 89:267–277

Pecins-Thompson M, Brown NA, Kohama SC, Bethea CL 1996 Ovarian steroid regulation of tryptophan hydroxylase mRNA expression in rhesus macaques. J Neurosci 16:7021–7029

Pecins-Thompson M, Brown NA, Bethea CL 1998 Regulation of serotonin re-uptake transporter mRNA expression by ovarian steroids in rhesus macaques. Brain Res Mol Brain Res 53:120–129

Petersen SL, McCrone S 1993 Characterization of the receptor complement of individual neurons using dual-label *in situ* hybrization histochemistry. In: Eberwine JH, Valentino KL, Barchas JD (eds) *In situ* hybridization in neurobiology. Oxford University Press, New York (Adv Methodology) p 78–95

Resko JA, Ploem JG, Stadelman HL 1975 Estrogens in fetal and maternal plasma of the rhesus monkey. Endocrinology 97:425–430

Seeman MV 1997 Psychopathology in women and men: focus on female hormones. Am J Psychiatry 154:1641–1647

Steiner M 1990 Postpartum psychiatric disorders. Can J Psychiatry 35:89–95

Tsai MJ, O'Malley RW 1994 Molecular mechanisms of action of steroid/thyroid receptor superfamily members. Annu Rev Biochem 63:451–486

Weissman MM, Olfson M 1995 Depression in women: implications for health care research. Science 269:799–808

Yeomans-Kinney A, Vernon SW, Frankowski RF, Weber DM, Bitsura JM, Vogel VG 1995 Factors related to enrollment in the breast cancer prevention trial at a comprehensive cancer center during the first year of recruitment. Cancer 76:46–56

DISCUSSION

McEwen: There are important species differences. Dr Steve Alves in our lab has studied the rat, and he has found that there are indeed oestrogen-inducible progestin receptors, but they're not in serotonergic neurons. They seem to be in some kind of interneuron, but we don't yet know the relationship to the serotonin neuron. In mouse, Dr Alves sees inducible progestin receptors in serotonin neurons as well as in non-serotonin neurons. Thus, the mouse is a little bit of the rat and a little bit of the rhesus monkey — it is not simply a small rat! I was wondering also, in your case, about the non-serotonin neurons: do you see them in the rhesus monkey raphe?

Bethea: If I count the serotonin-positive cells versus the PR-positive cells, there is about one and a half times more PR-positive cells. There are more cells with PR than the serotonin neurons, but I have never tried to identify the phenotype of those cells.

McEwen: Steve Alves thinks that some of these are excitatory and not inhibitory, which is even more intriguing in terms of what they may be doing, because there is pretty good body of evidence to show that ovarian hormones do affect serotonin turnover the rat brain. They do it indirectly, rather than the way they do in the rhesus monkey.

Gustafsson: Particularly with reference to the clinical implications, I wonder about these. Is there any indication that obsessive–compulsive disorder (OCD) increases at the time of menopause, for instance? Has anyone tried to implement some kind of oestrogen regimen to treat OCD or anorexia nervosa?

Bethea: I haven't heard of anybody trying to treat OCD with oestrogen, although I found one case of a woman who had a late-onset schizophrenia who was helped by oestrogen replacement therapy.

Gustafsson: We have to remember that compounds other than the classical oestrogens might be acting on ERβ. This may be the key: there may be a physiological ligand out there that is different from oestradiol.

Bethea: Yes, that would be very exciting. It also would not have some of the risks associated with oestradiol. I think oestrogen is now being used as an adjunct to selective serotonin reuptake inhibitor (SSRI) treatment in postmenopausal women, and it does seem to help. There is some evidence that SSRIs are less efficacious in the absence of ovarian steroid hormones.

Toran-Allerand: You and others at this meeting have emphasized the fact that the hormone replacement therapy carefully matches normal circulating levels of oestrogen. One of the problems is that the brain sees oestrogen not only from circulating oestradiol but also from circulating androgen sources because of their aromatization in the brain. Thus, what the brain is exposed to with respect to oestrogen is not only what's circulating around in terms of picograms per millilitre of plasma oestradiol. I wonder how one could take this into account because it might be that some of the responses that one is getting are due to the effects of oestradiol derived from aromatizable androgen sources to which the brain is exposed.

Bethea: We definitely need to examine aromatase activity in the serotonin neurons. Our monkeys still have their adrenal glands, which are a good source of androgens.

Baulieu: What's the distribution of aromatase in the brain?

Toran-Allerand: It depends on the species, but there is quite a bit of aromatase activity in the brain. It is not all over, but it certainly occurs in the basal forebrain, hippocampus, the hypothalamus, during development in the cerebral cortex, and in the midbrain and amygdala. There is enough that it could contribute significantly to the oestrogen that the brain can is exposed to.

McEwen: Then it's all the more important to study the programmed sex differences. There is a lot of evidence to suggest that even if the male is making oestrogen, it is not necessarily doing the same thing.

Behl: Are there any behavioural changes observed?

Bethea: I don't personally do behavioural observations. These animals are in single cages and so their behaviour is somewhat restricted. I have a colleague who does behavioural testing and can discriminate between hippocampal and cognitive functions, so we intend to look at the effects of ovarian hormones on these parameters in both young and old animals.

Behl: So it might be of value to put these animals under stress to imitate HPA dysregulation and then check whether there is a difference between the control and oestrogen-supplemented animals.

Bethea: Yes.

Simpkins: Pattie Green in my lab, working in collaboration with Christian Behl, has used a cell type that as far as we know does not have any oestrogen receptors. She has been able to show that oestrogens are very potent inhibitors of NF-κB activation, which is consistent with what you've seen. These effects are seen with 17β-oestradiol as well as several non-feminizing oestrogens. The oestrogen effect in dorsal raphe may be completely if not substantially receptor-independent.

Kushner: We have also been looking at down-regulation of a test promoter that has NF-κB sites, in collaboration with Pale Leitman and Jan-Åke Gustafsson. Oestrogen and its receptors down-regulate this promoter, and ERβ is very much more potent that ERα.

Bethea: That is wonderful news.

Kushner: I recently attended a meeting on breast and prostate cancer, and it was revealed in this meeting that MPA strongly interacts with the androgen receptor, and that many of these biological effects on women who are treated for breast cancer are probably due to interactions with the androgen receptor. Have you ever looked at androgen receptors? This might explain the paradoxical action of MPA on progestin.

Bethea: I haven't, but I would really like to.

McEwen: I think MPA also has glucocorticoid activity, so this needs to be considered.

Bethea: The reason I think Thomas Clarkson used MPA in his studies is that it is one of the most commonly used progestins in hormone replacement therapy in the USA, and he's trying to get the message out that it blocks whatever beneficial effects oestrogen might have in the cardiovascular system. There are glucocorticoid receptors in serotonin neurons, but to my knowledge no one has looked for androgen receptors, although I imagine that they're there.

Hurd: In your studies, you saw no change in the 5-HT$_{1A}$ receptors in the hypothalamus. We have carried out rat studies on the effects of oestrogen, and we see alterations in the 5-HT$_{1A}$ mRNA in, for example, the amygdala and cortex (Österlund & Hurd 1998, Österlund et al 2000). So, perhaps outside the hypothalamus serotoninergic receptors are altered. Since you do have these spayed monkey brain samples, is it possible to study the extrahypothalamic 5-HT receptors as well? Also, how were the ERα and ERβ mRNAs altered in the spayed monkeys?

Bethea: We have not looked at 5-HT$_{1A}$ in the hippocampus or the amygdala. We've looked at ERβ throughout the hypothalamus, hippocampus, amygdala and midbrain in spayed and hormone-replaced monkeys, and we see almost no regulation: it's there, it stays there. I have looked at ERα in the hypothalamus, and it is expressed very nicely in spayed animals. It continues to be expressed with oestrogen and progesterone. In my hands, I don't see a lot of up- or down-regulation of either ERα or ERβ in the monkey with steroid hormone replacement. I know that it is different in the rat.

Gustafsson: I would like to return to this sex difference issue. What are the levels in male animals of these marker genes that you're studying? If they are in the same range males as they are in non-spayed females, this would indicate that something other than oestrogens would activate this ERβ-dependent system in males. In humans, looking at the sex differences in the pathological conditions you mentioned, anorexia nervosa is much more prevalent in women. With regard to OCD, I don't recall such a sex difference. What are the sex differences for these marker genes?

Bethea: I haven't looked in male monkeys.

Levin: On one of your slides where you showed clearly the protein levels are induced by oestrogen, there is clearly a second band. This suggests phosphorylation to me. Oestrogen might not just increase levels; it may also cause phophorylation, which might affect activity.

Bethea: That enzyme does have to be phosphorylated in order to be active. It would be interesting to see what effects the steroids have on some of the protein kinases and phosphatases.

References

Österlund M, Hurd YL 1998 Acute 17β-estradiol treatment down-regulates serotonin 5-HT$_{1A}$ receptor mRNA expression in the limbic system of female rats. Brain Res Mol Brain Res 55:169–172

Österlund MK, Halldin C, Hurd YL 2000 Effects of chronic 17β-estradiol treatment on the serotonin 5-HT$_{1A}$ receptor mRNA and binding levels in the rat brain. Synapse 35:39–44

Oestrogen effects on dopaminergic function in striatum

Jill B. Becker

Psychology Department, Reproductive Sciences Program and Neuroscience Program, The University of Michigan, Ann Arbor, MI 48109, USA

Abstract. Experiments involving the use of behavioural, neurochemical and electrophysiological methods to explore the mechanisms mediating the effects of oestrogen on dopaminergic activity in the striatum on the female rat are described. Results have shown that oestrogen influences the activity of striatal dopamine terminals and dopamine-mediated behaviours in the female rat. These are rapid effects of oestrogen in the striatum, and are thought to be mediated by a novel membrane-associated receptor. How these novel effects of oestrogen may affect naturally occurring behaviours in the rat will be discussed.

2000 Neuronal and cognitive effects of oestrogens. Wiley, Chichester (Novartis Foundation Symposium 230) p 134–151

The study of the effects of oestrogen on dopamine (DA) function in the striatum by my laboratory began quite by accident over 20 years ago. In experiments conducted in the laboratory of V. D. Ramirez, examining the effects of gonadectomy on DA and noradrenaline release from the mediobasal hypothalamus, we included striatal tissue as control. At the time we thought that the striatum was an excellent control, since it was well known that it does not contain classical oestrogen receptors (Pfaff & Keiner 1973). In these experiments, however, we found that there was a profound effect of ovariectomy (OVX) and subsequent oestrogen replacement on striatal DA release from the striatum (Becker & Ramirez 1981). My laboratory has been investigating how oestrogen acts in the striatum since that time. The experiments to be described use behavioural, neurochemical and electrophysiological methods to explore the mechanisms mediating the effects of oestrogen on dopaminergic activity in the striatum of the female rat.

This neural system lends itself well to both behavioural and neurochemical studies. The ascending DA projection from the midbrain to the dorsal and ventral striatum is a massive projection, within which over 90% of the DA in the brain resides (Lindvall & Björklund 1974). This means that administration of drugs that activate or inhibit DA systems produces behaviours that are due primarily to activity in the ascending DA system. The mesolimbic and nigrostriatal DA systems are contained within the ascending DA projection, and our attention will be focused primarily on the latter. Behaviours that we have employed have been shown to be mediated by the nigrostriatal DA system (Arbuthnott & Crow 1971, Costall & Naylor 1977, Fink & Smith 1980, Ungerstedt 1971, 1974). Two behaviours have been used most frequently in our studies: rotational behaviour and stereotyped behaviours. Each of these behaviours has been demonstrated to depend on activation of the ascending DA system and to be related to the extent of DA activity in the striatum. Thus, these behaviours are used as an index, or behavioural assay, of DA activity in the striatum.

With each method employed, we begin by asking whether DA activity varies with the normal fluctuations in ovarian hormones that occurs during the female rat's reproductive cycle. In other words, do ovarian hormones affect the day-to-day functions of the striatal DA system in the female rat? If this proves to be the case, then we examine which hormones are important by removing the endogenous source of hormones (i.e. OVX) and then we selectively replace hormones to determine which hormones are involved. With this in mind, remember that the female rat has a 4 d oestrous cycle. During this cycle, ovarian hormones are low during the days of metaoestrus (also referred to as dioestrus 1 or D1) and dioestrus 2 (D2). Oestrogen rises rapidly to peak around noon on pro-oestrus (with a 14:10 light:dark cycle, lights on at 6 am). Progesterone peaks 4–6 h later. Behavioural receptivity occurs coincident with ovulation, 4–6 h after the progesterone peak, following the onset of the dark phase of the cycle. In the experiments conducted in my laboratory, animals were tested during the dark phase of the cycle. This means that for animals tested in pro-oestrus, testing occurred prior to the major surges of oestrogen or progesterone. Animals tested during oestrus have just been exposed to high circulating oestrogen and progesterone and are behaviourally receptive.

In one of our first behavioural experiments, we examined rotational behaviour in male and female rats given amphetamine (AMPH) to induce DA release. In this experiment we determined brain and striatal concentrations of AMPH, and gave males a higher dose (1.56 mg/kg AMPH) than females (who received 1.25 mg/kg AMPH) in order to overcome sex differences in liver metabolism of AMPH. When striatal concentrations of AMPH were equalized during the behavioural testing period, male rats turned significantly less than did female rats at all stages of the oestrous cycle except for metaoestrus (Fig. 1; Becker et al 1982). Subsequent experiments demonstrated that gonadectomy resulted in a significant decrease in rotational behaviour in female, but not male rats. We found that OVX produced a decrease in rotational behaviour using either AMPH-induced

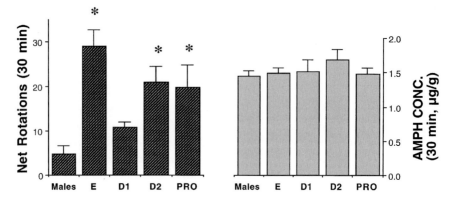

FIG. 1.   Rotational behavior (*left*) and whole brain concentrations of AMPH (*right*) in male and female rats at different stages of the oestrous cycle. Males received 1.56 mg/kg AMPH (i.p.) while females (E = oestrus, D1 = metaoestrus. D2 = dioestrus 2, P = pro-oestrus) received 1.25 mg/kg (i.p.). On the left, the mean (+SEM) number of net rotations made by neurologically intact rats during the 30 min after AMPH injection are plotted. On the right, the mean (+SEM) whole brain concentrations of AMPH 30 min after AMPH. *Significantly different from the male group, E and D2 significantly different from D1 ($P < 0.05$). Redrawn from Becker et al (1982).

rotational behaviour, or rotational behaviour induced by unilateral electrical stimulation of the ascending DA projection to the striatum (Camp et al 1986, Robinson et al 1981). Thus, the effect of OVX to attenuate the behavioural response induced during DA stimulation is not unique to the effect of the drug AMPH, but is apparently a more general effect on the ascending DA system.

Having found that there is oestrous cycle-dependent variation in behaviours induced by DA stimulation, and that OVX attenuates this behavioural response, we then went on to determine whether hormonal replacement could reinstate the response. In these experiments, we found that oestrogen alone is sufficient to enhance rotational behaviour in the OVX female rat (Becker & Beer 1986). Importantly, while rotational behaviour or stereotyped behaviour is enhanced by repeated oestrogen treatment (hormone treatment to mimic the oestrous cycle), a single dose of 5 μg oestradiol benzoate (EB), 30 min prior to AMPH is sufficient to induce a significant increase in AMPH-induced behaviours (Fig. 2; Becker 1990a). Thus, an acute oestrogen treatment rapidly induces an increase in behavioural measures of nigrostriatal DA activity.

Neurochemical techniques have also been used to assess the effects of ovarian hormones on striatal DA activity. Using *in vivo* microdialysis to sample the concentrations of DA in extracellular fluid from the striatum, while simultaneously observing the behaviour of the animals, we see that there is oestrous cycle-dependent variation in the AMPH-induced increase in DA in

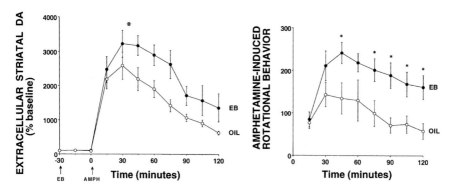

FIG. 2. The rapid effect of acute oestrogen treatment on the AMPH-induced increase in extracellular striatal DA (*left*) and rotational behaviour (*right*) determined during *in vivo* microdialysis. (*Left*) OVX female rats with unilateral nigrostriatal DA depletion received 5 μg oestradiol benzoate (EB) (closed circles) or oil (open circles) at the first arrow labelled EB and 30 min later all animals received 3.0 mg/kg AMPH (i.p.). Dialysis samples were collected at 15 min intervals, DA in dialysate is expressed as a percent of baseline (prior to EB or oil). *There was a significant main effect of EB treatment on the AMPH-induced increase in extracellular DA (F[1,15] = 4.985, P = 0.04). Bars indicate the SEM. (*Right*) AMPH stimulated greater rotational behaviour in OVX animals receiving EB than in animals receiving oil vehicle (P < 0.02). *During these intervals animals that received EB exhibited significantly more rotational behaviour than did oil-treated animals (P < 0.05; Scheffe *post hoc* test). Redrawn from Becker (1990a). Bars indicate the SEM.

striatum. Female rats during oestrus have a significantly greater AMPH-induced increase in extracellular DA in striatum than do female rats during metaoestrus (Becker & Cha 1989). In these same animals, females in oestrus exhibited significantly more stereotyped behaviour than did females in metaoestrus (Becker & Cha 1989), confirming previous behavioural results discussed above.

With *in vivo* microdialysis, we also find that a single dose of 5 μg EB given to OVX rats enhances the AMPH-induced increase in extracellular striatal DA, coincident with enhanced rotational behaviour (Fig. 2; Becker 1990a). We see this effect of acute oestrogen treatment, to enhance AMPH-induced DA in striatum and AMPH-induced behaviours, in OVX female rats but not castrated male rats (Castner et al 1993). Thus, there is a sex difference in the effect of oestrogen in the striatum. Finally, in OVX rats we have compared the effects of acute vs. repeated oestrogen treatments. We find that 30 min after 5 μg EB (i.e. acute EB treatment), AMPH-induced DA release and stereotyped behaviours are significantly greater in OVX rats previously primed for 3 d with EB, than in animals receiving acute EB for the first time, or in animals not receiving acute EB but previously primed with EB. OVX rats treated only with vehicle exhibited significantly less DA release and fewer stereotyped behaviours than did any of the EB treated groups (Fig. 3; Becker & Rudick 1999).

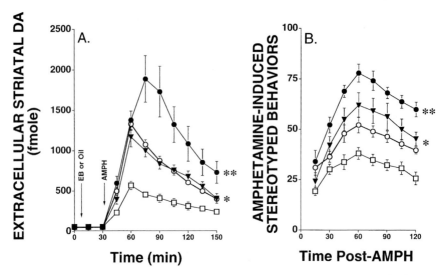

FIG. 3.   The effect of oestradiol benzoate (EB) with and without EB priming on the AMPH-induced (2.5 mg/kg, i.p.) increase in extracellular striatal DA in dialysate (A) and stereotyped behaviours (B). Closed circles: OVX rats received 5 μg EB for 3 d. On the 4th day these animals received 5 μg EB, 30 min before AMPH. Open circles: OVX rats received oil injections for 3 d. On the 4th day these animals received a single injection of 5 μg EB, 30 min before AMPH. Closed triangles: OVX rats received 5 μg EB for 3 d. On the 4th day these animals received oil, 30 min before AMPH. Open squares: OVX rats received oil for 3 d. On the 4th day these animals received oil, 30 min before AMPH. All values are the mean ± SEM. (A) DA concentrations in dialysate expressed in fmole/15 min (mean ± SEM). **Significantly greater than all other groups ($P < 0.05$). *The behaviour of these two groups was significantly greater than the oil control group. (B) AMPH-induced stereotyped movements of the head and forelimbs were counted for 30 s every 7.5 min at times to correspond with the period of the dialysis samples. **Significantly greater than all other groups ($P < 0.05$). *The behaviour of these two groups was significantly greater than the oil control group. Redrawn from Becker & Rudick (1999).

The results of the experiments I have just described demonstrate that oestrogen can affect both behavioural and neurochemical indices of striatal DA function. The question remains, however, whether the effect of oestrogen on DA-mediated behaviours causes the effects on striatal DA release described, or whether the effects of oestrogen on striatal DA release induce the reported effects on behaviour. Experiments conducted examining DA release from striatal tissue, *in vitro*, are important, therefore, to determine causal relations between these events. In experiments using an *in vitro* superfusion system, striatal tissue slices are placed into a chamber through which a Ringer's solution flows continuously. AMPH, oestrogen, and other compounds can then be delivered directly to the tissue and the effluent from the superfusion system can be sampled to measure the release of DA from the striatal tissue. Using this superfusion system, we have confirmed that:

(1) AMPH-induced striatal dopamine release varies with oestrous cycle (Becker 1990b); (2) OVX attenuates AMPH-induced striatal dopamine release while castration of male rats has no effect on this measure (Becker 1990b, Becker & Ramirez 1981); and (3) Oestrogen treatment enhances AMPH-stimulated striatal dopamine release in striatal tissue from female but not male rats (Becker 1990b, Becker & Ramirez 1981). From the results of these experiments we conclude that the effects of oestrogen on striatal DA release occurs in the absence of behaviour. Thus, it is likely that the effects of oestrogen on striatal DA release mediate the changes in behaviour that occur.

The superfusion system is a particularly powerful tool, as it is possible to use this method to determine where oestrogen is acting in order to produce its effects on the striatum. The previous experiments described have used systemic oestrogen treatments to manipulate striatal DA activity, so the results discussed could be due the effects of oestrogen almost anywhere in the brain or body! With the superfusion system, oestrogen can be delivered directly to the striatum, so it is possible to determine whether the effects described above are due to direct effects on the striatum. Thus, we have used the superfusion system to show that oestrogen in physiological concentrations (100 pg/ml) acts directly on the striatum of OVX female rats to enhance AMPH- or K^+-stimulated DA release (Becker 1990b). Furthermore, we find that the pulsatile administration of oestrogen directly stimulates DA release in striatal tissue from female rats, but not male rats (Becker 1990b). We believe, therefore, that the acute effects of oestrogen that we are investigating are due to the direct effects of oestrogen on the striatum.

Interestingly, the pharmacology of the response to oestrogen analogues in the striatum is quite different from the pharmacology at the classical oestrogen receptors. For example, the catecholoestrogens, 2-hydroxyoestradiol and 4-hydroxyoestradiol, also enhance AMPH-induced striatal DA release while 17α-oestradiol, 2-methoxyoestradiol, oestrone, oestriol and diethylstilboesterol are ineffective in this regard (Xiao & Becker 1998). At the classical oestrogen receptor diethylstilboesterol is a potent agonist and oestrone and oestriol are weak agonists, yet we see no effect of these compound even at concentrations that are 10 times higher than the effective dose of oestrogen. Experiments conducted by Li Xiao in my laboratory demonstrated that the effects of oestrogen in the striatum are not blocked by tamoxifen, but are blocked by ICI 182 780 (Xiao & Becker 1997a); both of these agents are active at the classical oestrogen receptor. Thus, the pharmacology of the effects of oestrogen in the striatum indicate that it is steroid-specific with characteristics distinct from oestrogen receptor α (ERα): (1) a rigid steroidal conformation is necessary for efficacy; and (2) hydroxylation of the A ring does not inhibit while modification of the D ring prevents efficacy of a compound in the striatum. Finally, oestrogen conjugated to bovine serum albumin (BSA), to prevent entry into a cell, enhances AMPH-induced striatal DA release

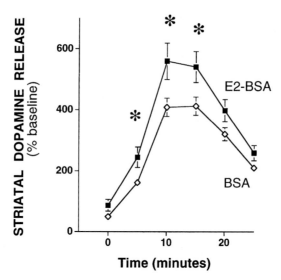

FIG. 4.   Time course of the AMPH-induced increase in DA release from *in vitro* superfused striatal tissue slices obtained from OVX rats. Following a 60 min equilibration period, effluent samples were collected at 5 min intervals, values indicate the mean ± SEM. Closed squares were exposed to 100 pg/ml oestradiol with the oestradiol conjugated to BSA (510 pM E2-BSA), control chambers (open diamonds) were exposed to BSA alone (without oestradiol). There was a main effect of E2-BSA on AMPH (10 μM) induced DA release ($P < 0.05$). *In subsequent pairwise comparisons the AMPH-induced DA release in tissue from the E2-BSA treated group was significantly greater than the control group during the first three intervals after AMPH stimulation. Redrawn from Xiao & Becker (1998).

(Fig. 4; Xiao & Becker 1998). This demonstrates that the mechanism through which oestrogen is acting in the striatum is mediated by an interaction between oestrogen and the exterior of the cell.

Electrophysiological studies using whole cell clamp electrophysiology in acutely dissociated striatal neurons indicate that these rapid effects of oestrogen are mediated by a G protein-coupled membrane receptor. In experiments conducted in collaboration with Paul Mermelstein in Jim Surmeier's laboratory, we found that oestrogen inhibits L-type Ca^{2+} channels in striatal neurons (Mermelstein et al 1996). The effects occur within seconds, reverse as soon as oestrogen delivery ceases, and are seen at physiological concentrations of oestrogen (i.e. picomolar). Furthermore, cells from females show a greater response than males. As we saw with the superfusion system, oestrogen conjugated to BSA is also effective. Interestingly, oestrogen applied internally to cells through the electrode is not effective at reducing Ca^{2+}-currents, nor does it block the effect of 1 pM oestrogen applied externally. Collectively, these results suggest that the effect of oestrogen occurs externally at the membrane surface. In the presence of GTPγS (which prevents inactivation of G protein-mediated events) the effect of 17β-oestradiol does not reverse when hormone delivery ceases. Thus, the effect of oestrogen on striatal neurons is apparently dependent upon a G protein-coupled receptor. Finally, the effect of 17β-oestradiol is stereospecific as 17α-oestradiol does not mimic the modulation, and is steroid-specific as 100 pM oestrone and 3-methoxyoestriol were ineffective while oestriol or 4-hydroxyoestradiol mimic the effect of 17β-oestradiol. It is concluded, therefore, that oestrogen has rapid stereospecific effects on striatal neurons that alter signalling pathways independent of the classical ER (Mermelstein et al 1996).

We then asked what the implications are of these effects of oestrogen on the striatum for the natural behaviour (behaviour not induced by artificial stimulation with drugs, etc.) of rats? In one study, we found that when female rats are trained to traverse a narrow beam, animals make fewer mistakes during oestrus than during other stages of the oestrous cycle (Becker et al 1987). Furthermore, in OVX rats, crystalline 30% oestradiol implanted bilaterally into the striatum enhanced performance on this tack indicating that oestrogen's effects on the striatum were important for enhancing sensorimotor function (Becker et al 1987).

In thinking about the role of the striatum in behaviour, and the more 'ultimate' causes of hormonal modulation of neural activity in the striatum, we have begun to consider an evolutionary view. Evolutionary biologists argue that biological functions that vary with the reproductive cycle are likely to play a role in reproduction. Therefore, my laboratory began to look for a role for the ascending DA systems in sexual behaviour of the female rat. Sexual behaviour in the female rat has typically been studied in the laboratory in a small open arena where the male rat is able to copulate with the female rat at will, resulting in low levels of female proceptive behaviours and a rapid rate of intromissions, leading to ejaculation by the male (e.g. Beach 1976). In semi-natural conditions, however, the female rat will actively control the pace of copulatory behaviour by exhibiting proceptive behaviours and actively withdrawing from the male when there is a place that can be used to hide (McClintock 1984).

It turns out that the preferred rate of intromissions for males and females are different. For the male rat, a rapidly paced series of intromissions (about 1 min between intromissions) is optimal to induce ejaculation in the fewest number of intromissions (Adler 1978). The female rat, on the other hand, requires behavioural activation of a progestational response in order to facilitate pregnancy, which requires that intromissions be spaced further apart than would occur at the male's pace (Adler 1978). A female rat will 'pace' the rate of intromissions if there is a barrier behind which she can escape from the male rat (Erskine 1989, McClintock 1984). In fact within one or two intromissions we find that a naïve sexually receptive female rat will take advantage of a barrier in

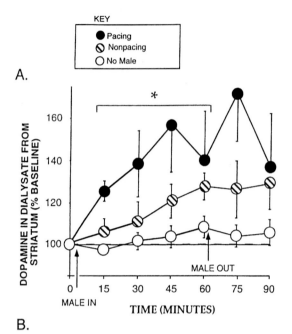

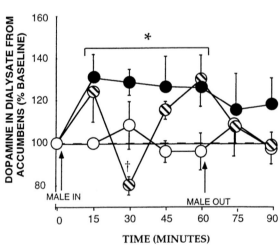

the testing apparatus and begin pacing (personal observations). Two measures are taken during pacing behaviour: (1) percent exits, which is the rate at which the female withdraws from the presence of the male after a copulatory contact; and (2) return latency, which is the time before she returns to the male after a contact. A female that is pacing shows higher rates of percent exits and a longer return latency after an ejaculation than after an intromission or mount.

The role of the ascending DA systems in female rodent sexual behaviour, and pacing behaviour in particular, has been a recent topic of investigation in this laboratory. We find that there is enhanced DA in dialysate from striatum and nucleus accumbens during sexual behaviour in female rats that are pacing sexual behaviour, compared with females that are engaging in sex but not pacing (Fig. 5; Mermelstein & Becker 1995). In the striatum and nucleus accumbens, the increase in DA concentrations in dialysate of oestrogen and progesterone-primed OVX rats pacing copulation is significantly greater than that of non-pacing animals or behaviourally receptive animals tested without a male rat (Mermelstein & Becker 1995). In rats, DA in dialysate from striatum and nucleus accumbens has been found to increase in animals pacing sexual behaviour and in animals where the male rat is introduced by the experimenter at the female's previously determined preferred pace, relative to animals engaging in sexual behaviour but not pacing or those prevented from receiving coital stimulation by the presence of a vaginal mask (Becker et al 1998). These results together support the notion that DA in the striatum and nucleus accumbens is important for coding specific aspects of the coital stimuli received, rather than being related to specific motor behaviours.

Support for the functional dissociation of the roles of the striatum and nucleus accumbens in pacing behaviour comes from studies in which female rats were induced into behavioural oestrus via bilateral ventromedial hypothalamus hormone treatments and then received oestrogen bilaterally into the striatum or nucleus accumbens (Xiao & Becker 1997b). Intrastriatal oestrogen was found to

---

FIG. 5. Extracellular DA concentrations in dialysate from striatum (A) or accumbens (B), expressed as a per cent of baseline (mean±SEM). OVX female rats were primed with oestrogen and progesterone so that they were behaviourally receptive and then underwent microdialysis. A number of conditions were tested, a few of which are shown (for additional data see original source, Mermelstein & Becker 1995). (A) When comparing animals' pacing sexual behaviour (pacing; closed circles) with animals engaging in sexual behaviour who could not pace (non-pacing; hatched circles) and animals primed with hormones but not engaging in sex (no male; open circles) the increase in DA in dialysate from the striatum was significantly greater in the pacing group than the non-pacing or no male group (*P < 0.03). (B) Pacing animals exhibited significantly greater increase in extracellular DA in the nucleus accumbens than did all other groups (*P < 0.05). †During the second interval after the introduction of the male rat there was a significant decrease in extracellular DA in accumbens from non-pacing females relative to baseline and all other groups (P < 0.05).

facilitate percent exits, while intra-nucleus accumbens implants increases the return latency. Conversely, the anti-oestrogen ICI 182 780 applied to the striatum decreased percent exits, while in the nucleus accumbens it decreased return latency (Xiao & Becker 1997b). Thus, the results together suggest that the striatum and nucleus accumbens differentially modulate specific components of pacing behaviour, and that the effects of oestrogen on DA in these brain areas enhances these functions.

Conclusions

The evidence reviewed indicates that oestrogen has rapid, direct effects on the striatum that act to enhance DA release, sensorimotor function and specific components of sexual behaviour in the female rat. It is suggested that hormonal modulation of the striatum may have evolved to facilitate reproductive success in female rats by enhancing pacing behaviour. Electrophysiological studies suggest this effect is indirectly mediated by novel membrane receptors for oestradiol.

Acknowledgements

Supported by the National Science Foundation.

References

Adler NT 1978 On the mechanisms of sexual behavior and their evolutionary constraints. In: Hutchison JB (eds) Biological determinants of sexual behavior. Wiley, New York, p 657–694
Arbuthnott GW, Crow TJ 1971 Relation of contraversive turning to unilateral release of dopamine from the nigrostriatal pathway in rats. Exp Neurol 30:484–491
Beach FA 1976 Sexual attractivity, proceptivity, and receptivity in female mammals. Horm Behav 7:105–138
Becker JB 1990a Estrogen rapidly potentiates amphetamine-induced striatal dopamine release and rotational behavior during microdialysis. Neurosci Lett 118:169–171
Becker JB 1990b Direct effect of 17β-estradiol on striatum: sex differences in dopamine release. Synapse 5:157–164
Becker JB, Beer ME 1986 The influence of estrogen on nigrostriatal dopamine activity: behavioral and neurochemical evidence for both pre- and postsynaptic components. Behav Brain Res 19:27–33
Becker JB, Cha J 1989 Estrous cycle-dependent variation in amphetamine-induced behaviors and striatal dopamine release assessed with microdialysis. Behav Brain Res 35:117–125
Becker JB, Ramirez VD 1981 Sex differences in the amphetamine stimulated release of catecholamines from rat striatal tissue *in vitro*. Brain Res 204:361–372
Becker JB, Rudick CN 1999 Rapid effects of estrogen or progesterone on the amphetamine-induced increase in striatal dopamine are enhanced by estrogen priming: a microdialysis study. Pharmacol Biochem Behav 64:179–187
Becker JB, Robinson TE, Lorenz KA 1982 Sex differences and estrous cycle variations in amphetamine-elicited rotational behavior. Eur J Pharmacol 80:65–72

Becker JB, Snyder PJ, Miller MM, Westgate SA, Jenuwine MJ 1987 The influence of estrous cycle and intrastriatal estradiol on sensorimotor performance in the female rat. Pharmacol Biochem Behav 27:53–59

Becker JB, Rudick CN, Jenkins WJ, Hummer DL 1998 Dopamine in dialysate during pacing of sexual behavior: stimulus or response? Soc Neurosci Abstr 24:356

Camp DM, Becker JB, Robinson TE 1986 Sex differences in the effects of gonadectomy on amphetamine-induced rotational behavior in rats. Behav Neural Biol 46:491–495

Castner SA, Xiao L, Becker JB 1993 Sex differences in striatal dopamine: *in vivo* microdialysis and behavioral studies. Brain Res 610:127–134

Costall B, Naylor RJ 1977 Mesolimbic and extrapyramidal sites for the mediation of stereotyped behavior patterns and hyperactivity by amphetamine and apomorphine in the rat. In: Ellinwood EH, Kilbey MM (eds) Cocaine and other stimulants. Plenum Press, New York, p 47–76

Erskine MS 1989 Solicitation behavior in the estrous female rat: a review. Horm Behav 23: 473–502

Fink JS, Smith GP 1980 Relationships between selective denervation of dopamine terminal fields in the anterior forebrain and behavioral responses to amphetamine and apomorphine. Brain Res 201:107–127

Lindvall O, Björklund A 1974 The organization of the ascending catecholamine neuron systems in the rat brain as revealed by the glyoxylic acid fluorescence method. Acta Physiol Scand Suppl 412:1–48

McClintock MK 1984 Group mating in the domestic rat as context for sexual selection: consequences for the analysis of sexual behavior and neuroendocrine responses. Adv Stud Behav 14:1–50

Mermelstein PG, Becker JB 1995 Increased extracellular dopamine in the nucleus accumbens and striatum of the female rat during paced copulatory behavior. Behav Neurosci 109:354–365

Mermelstein PG, Becker JB, Surmeier DJ 1996 Estradiol reduces calcium currents in rat neostriatal neurons via a membrane receptor. J Neurosci 16:595–604

Pfaff D, Keiner M 1973 Atlas of estradiol-concentrating cells in the central nervous system of the rat. J Comp Neurol 151:121–158

Robinson TE, Camp DM, Becker JB 1981 Gonadectomy attenuates turning behavior produced by electrical stimulation of the nigrostriatal dopamine system in female but not male rats. Neurosci Lett 23:203–208

Ungerstedt U 1971 Striatal dopamine release after amphetamine or nerve degeneration revealed by rotational behavior. Acta Physiol Scand Suppl 367:49–68

Ungerstedt U 1974 Functional dynamics of central monamine pathways. In: Schmitt FO, Worden FG (eds) The neurosciences: third study program. MIT Press, Cambridge, MA, p 979–988

Xiao L, Becker JB 1997a Steroid-specific effects of estrogen agonists and antagonists on amphetamine-induced striatal dopamine released from superfused striatal tissue. Soc Neurosci Abstr 23:403

Xiao L, Becker JB 1997b Hormonal activation of the striatum and the nucleus accumbens modulates paced mating behavior in the female rat. Horm Behav 32:114–124

Xiao L, Becker JB 1998 Effects of estrogen agonists on amphetamine-stimulated striatal dopamine release. Synapse 29:379–391

DISCUSSION

McEwen: Is the Ca^{2+} channel mechanism the same as in smooth muscle? There are oestrogen effects on smooth muscle.

Becker: Yes, the smooth muscle Ca^{2+} channel is also L-type.

Gustafsson: I would urge caution in interpretation of the experiments using BSA-conjugated oestrogen. It is widely accepted that it is not possible to exclude the possibility that there is always some free oestrogen in these preparations. The level of evidence from these experiments is somewhat limited, and one cannot take them as an indication of the presence of membrane receptor.

Becker: That's why we did the second experiment with the oestradiol inside the electrode, applied to the inside of the cell. In this way we showed that there is still an effect with low concentrations of oestrogen outside the cell, even when the inside of the cell is saturated with oestradiol. Together, these two experiments argue that our results with oestrogen conjugated to BSA are not an artefact of a small amount of oestrogen getting inside the cell.

Gustafsson: With a small amount, do you not see an effect, then?

Becker: What I am saying is that there is an effect with oestrogen–BSA; there is also an effect with 100 pM oestrogen inside the cell and then 1 pM applied outside the cell. Thus 1 pM is still effective even when the inside of the cell is saturated with oestradiol.

Levin: That is a great experiment.

Gustafsson: In view of the current knowledge of oestrogen action, it is not necessary to invoke membrane oestrogen receptors to explain your results. There was a time when concepts for steroid hormone action required mRNA production, protein biosynthesis and so on. But as we have heard at this meeting, there are also many cases where oestrogens, glucocorticoids and other steroid hormones act rapidly, according to accepted mechanisms, by interacting with other signal transduction pathways. We have heard today about NF-κB, where the oestrogen receptor interacts with the p65 component. The examples of protein–protein interactions in steroid hormone action are increasing exponentially. One example, which I find particularly tantalizing, is with the 14-3-3 protein. This is a chaperone protein which has two binding sites: it has been suggested that this protein can contact one signal transduction pathway on one side and another pathway on the other side of the molecule. 14-3-3 binds excellently to the glucocorticoid receptor, and ERα and ERβ. Obviously, oestrogen can have rapid effects. Conceptually at least, some of these rapid effects can be explained by activation of the 'conventional' intracellular receptors eliciting rapid effects through various signal transduction pathways. This doesn't require mRNA production. This would be consistent with the fact that you often described quite similar ligand specificities in terms of effects as for ERα and ERβ. For instance, the ICI antagonist has an agonistic effect. Membrane receptors aren't needed here.

Becker: I think you have to invoke a membrane receptor; whether or not it is the classical receptor in the membrane is something we can discuss. The response is being mediated by the effects of oestrogen on the outside of the cell: this is really quite different from the classical response. It is a novel mechanism, even if it is not necessarily a novel receptor.

McEwen: In relation to that, you make a case that the L-type Ca^{2+} channel effect is a necessary and sufficient action of oestrogen to explain the dopamine results. Have you tried to examine the pharmacology of the oestrogen effect to find out if it conforms to a classical receptor in some way?

Becker: I haven't done that experiment, but it needs doing. We have looked at other agonists and antagonists, and there are many similarities with the classical receptor, but there are also some differences. Diethylstilbestrol is not an agonist or antagonist in this system, and tamoxifen is not an antagonist, suggesting there may be something about the steroidal conformation that affects functional activity in the striatum. Modifications at the A and the D ring affect binding, so it seems that the entire three dimensional structure is necessary to maintain function in the striatum.

Levin: Just to give some perspective to this discussion, in the vasculature it is also appreciated that oestrogen can rapidly activate K^+-sensitive Ca^{2+} channels, causing vasodilitation as part of the mechanism. These are rapid effects: Ca^{2+} spikes can be seen in 1 s. There are similar data from constrained progesterone experiments.

There is a pool of cytoplasmic receptors. We know that a number of these move to the nucleus when they are ligated, and we think a small percentage will move to the membrane. There are obviously some other cytoplasmic ERs and we know nothing about their function. Your point that they may hook up with signalling molecules such as 14-3-3 is to be taken in perspective: they may be the unknown pool of receptors that is activating either signal transduction or some other oestrogen actions.

Smith: In comparing the effects on the Ca^{2+} channel with your behaviour and dopamine release experiments, have you looked at the ICI compound to see whether it can block the Ca^{2+} channel effects, or have you looked at oestrogen priming?

Becker: We haven't yet looked at ICI, but tamoxifen doesn't block the Ca^{2+} channels. We haven't looked at oestrogen priming, either.

Smith: What do you think oestrogen priming is doing in the dopamine release experiments?

Becker: I believe that we are activating the intracellular signalling responses. There are well-known effects of G protein-coupled membrane receptors on transcription. My bet is that there are long-term effects of activating specific intracellular messengers that either induce membrane oestrogen receptors or amplify the response intracellularly when oestrogen acts at the membrane.

Smith: You said that you saw sex differences. Could you expand on this?

Becker: There is an effect of oestrogen on L-type Ca^{2+} channels in striatal neurons from male rats, but the response is substantially smaller. In striatal neurons from male rats you produce a small but significant decrease in L-type Ca^{2+} current, but this requires 100-fold higher concentrations of oestradiol than in the female.

Toran-Allerand: I would like to make a comment concerning the membrane receptors as membrane receptors. There is no question that there are cytoplasmic components of the oestrogen receptor, and that there is a shuttling back and forth of the receptor between the cytoplasm and the nucleus. But 14-3-3 and signalling molecules have been shown to be in the membrane in caveolae. They are not all over the cytoplasm, but are structurally localized and poised in such a position that when an activating molecule is exposed to the surface of a cell they are there to respond immediately and set off a chain of biochemical events. It is entirely possible that there can be a population of membrane receptors that are just there for the function of responding to certain aspects of oestrogen action which are dependent on utilization of signalling pathways whose kinases are located within the membrane proper. I would like us to keep an open mind: the issue is not whether it is the same or different receptors, but rather that it is possible that there may be membrane oestrogen receptors that don't do much except stay in the membrane and respond in certain ways to exposure to oestrogen by virtue of their interactions with signalling kinases and molecules like 14-3-3.

Gustafsson: I would agree with Dominique Toran-Allerand: perhaps a lot of this discussion about membrane ERs is just semantics. When we talk about a 'membrane receptor', we invoke that this molecule should have a hydrophobic segment which allows it to be embedded in the membrane in a similar manner to transmembrane receptors. People are looking for these things but have not found them yet. However, if we talk about a 'membrane-associated' receptor, this allows us to think about hydrophilic proteins that may not be hooked up to the membrane themselves by having hydrophobic segments on them, but may be dimerizing or connecting with the plethora of proteins that have been shown to connect with receptors, some of which may be hydrophobic enough to stick to the membrane. Then it is as undramatic to say that there are ERs in the membrane as it is to say that they occur in the nucleus and the cytoplasm. So instead of talking about membrane receptors, which gives the impression that we want to create a new receptor species, perhaps we should talk about membrane-associated receptors instead.

Becker: I absolutely agree.

Watson: One does not have to invoke a transmembrane segment to put a steroid receptor on the membrane. There are plenty of post-translational modifications that could do this. This might explain how ER is attached to the membrane without a transmembrane domain.

Gustafsson: That is a possibility, and dimerization is another possibility. The problem with post-translational modification is that not there is not much evidence for this with ER.

Watson: Recent data suggest that the glucocorticoid receptor has a lipid attachment (Powell et al 1999).

Is there any evidence that oestrogen is the secretagogue for dopamine, in much the same way that in my system oestrogen is a secretagogue for prolactin release. There are probably other compounds that work via a Ca^{2+} elevation to cause dopamine release. Is that possible?

Becker: If you pulse in oestradiol you can induce an increase in dopamine release in striatal tissue. But a constant level of oestrogen doesn't produce the same effect. Constant concentrations seem to give a change in tone or set point, while a pulsatile stimulus, which is what is seen during the oestrous cycle can induce dopamine release.

Watson: The first paper that was published about the secretagogue effect of oestradiol on pituitary cells showed that there was a period of resistance after oestrogen stimulation where the cells are no longer receptive (Dufy et al 1979).

Herbison: How many different types of effects do you think oestrogen is having on the striatum? As I understand it, your *in vitro* work and microdialysis experiments suggests that the effect of oestrogen on dopamine is a presynaptic action. This must be different to the L-type Ca^{2+} channel, so there are at least two effects of oestrogen.

Becker: The model that I believe to be true is that oestrogen is not necessarily acting on the dopamine terminals, but on the γ-aminobutyric acid (GABA) interneurons within the striatum. These are the cells that are being recording from in the experiments examining L-type Ca^{2+} channels. Oestrogen is acting on the GABA neurons, which have recurrent collaterals onto the dopamine terminals. We think that GABA acting on the $GABA_B$ receptors on the dopamine terminals is causing the increase in dopamine release by a release of inhibition. We can mimic the effect of oestrogen with baclofen *in vitro*, suggesting that the release of GABA on the $GABA_B$ receptors is decreased by oestrogen, thus enhancing dopamine release. There may also be an independent effect of oestrogen on the dopamine terminals, because we can still see an effect in the quinolinic acid-lesioned striatum, suggesting that the GABA neurons aren't necessary. We are trying to put all these pieces together, but at present I think there are two processes occurring.

Henderson: I want to leave the world of rats and enter the realm of speculation. Your evidence suggests that oestrogen should ameliorate dopamine-deficient conditions such as Parkinson's disease. The human literature on this point is sparse and messy, but it implies that there may be some effect of oestrogen in Parkinson's disease or perhaps that parkinsonian symptoms in younger women might vary during the menstrual cycle. On the basis of your research, could you speculate what effects oestrogen might have?

Becker: Your interpretation of the human literature is right: it is pretty messy, and I don't know of any nice studies that have been done on the menstrual cycle. There is some suggestion that women have later onset of Parkinson's, and this may be related to an effect of oestrogen to enhance dopamine function pre-menopause, but again there are no data. As far as I know, no one has looked at whether post-menopausal oestrogen replacement ameliorates the symptoms.

Resnick: If I recall, the main thing people have tried to demonstrate is that it does not interfere with standard treatment of Parkinson's disease.

Becker: There are some early-stage Parkinson's disease case studies but we are only talking about one or two individuals where oestrogen has been shown to decrease some of the rigidity but increase some of the side effects of *L*-dopa.

Hurd: There are differences in oestrogen effects depending on the stage of the Parkinson's disease. Although some studies previously showed oestrogen to worsen Parkinsonism, it has been recently published that oestrogen is beneficial in early Parkinson's disease (Saunders-Pullman et al 1999). We have detected ER, primarily ERβ, mRNA in the human subthalamic nucleus (Österlund et al 2000, M. K. Österlund, J.-Å. Gustafsson, E. Keller & Y. L. Hurd, unpublished results), a core structure in basal ganglia function. Oestrogen effects in Parkinson's patients have been previously thought to be mediated in the striatum via indirect mechanisms, but the subthalamic nucleus could be involved in direct oestrogen effects on motor behaviour.

Allan Herbison has already raised the issue of whether your dopamine release was direct or indirect. Was the dopamine released via reversal of transport carrier in the non-amphetamine stage or was it from exocytotic pools?

Becker: In the superperfusion system we have shown that oestrogen enhances K$^+$-induced dopamine release, as well as amphetamine-induced dopamine release. So we do know that oestrogen affects dopamine release from exocytotic pools.

Hurd: In regard to the heterogeneity of the striatum, there are different amounts of dopamine released within the nucleus accumbens shell and core, and in the dorsal versus ventral striatal regions with, for example, sexual behaviour and drug reward. Even the anterior core region of the accumbens is different. Do you see the same effect regardless of which striatal region you examine?

Becker: I haven't looked at the shell versus the core of the accumbens, but we have conducted microdialysis on the contralateral nucleus accumbens at the same time we were conducting microdialysis in the striatum. We see the same thing in the accumbens as we do in striatum, at least during pacing of sexual behaviour. Within the striatum we have limited our studies, so far, to the dorsolateral striatum.

Gibbs: It was my understanding that sensorimotor performance in women was shown to be increased during times in the cycle when oestrogen was high. Also, thinking about the mesolimbic system and schizophrenia, there are effects of the cycle on schizophrenic symptoms. How do you see your findings fitting into those data?

Becker: In terms of the sensorimotor function changes across the menstrual cycle, they fit quite well. We see the same effect on stimulated dopamine release from the striatum with oestrogen plus progesterone that we get with oestrogen alone. During the menstrual cycle there is enhanced sensorimotor function around the period of ovulation when oestrogen is elevated and during the luteal phase when both oestrogen and progesterone are high. The data are quite consistent with this. I don't know about schizophrenia.

References

Dufy B, Vincent J-D, Fleury H, Pasquier PD, Gourdji D, Tixier-Vidal AT 1979 Membrane effects of thyrotropin-releasing hormone and estrogen shown by intracelleular recording from pituitary cells. Science 204:509–511

Österlund MK, Keller E, Hurd YL 2000 The human forebrain has discrete estrogen receptor α messenger RNA expression: high levels in the amygdaloid complex. Neuroscience 95:333–342

Powell CE, Watson CS, Gametchu B 1999 Binding partners, hormone binding affinities and post-translational modifications of the purified membrane glucocorticoid receptor. Endocrine Society Meeting, San Diego, June 1999

Saunders-Pullman R, Gordon-Elliott J, Parides M, Fahn S, Saunders HR, Bressman S 1999 The effect of estrogen replacement on early Parkinson's disease. Neurology 52:1417–1421

General discussion II

McEwen: I would like us to consider all the systems that we have looked at so far, starting with the dopaminergic system and moving to the cholinergic, serotonergic and noradrenergic. High doses of oestrogens have anti-dopaminergic actions: Jill Becker, in light of your present data, how do you think this is happening? Thinking about some of the other data we have heard, there are many highly dose-dependent and time-dependent actions of oestrogens, certainly in the cholinergic, and perhaps also in the noradrenergic and serotonergic systems. Are there other data consistent with those kinds of biphasic actions of oestrogens?

Becker: In vitro if we use 100 pg/ml oestrogen, which is the physiological dose, there is enhanced dopamine release. A higher dose, such as 1000 pg/ml, inhibits dopamine release. I believe, therefore, that high doses of oestrogen are inhibitory in the striatum. Cheryl Watson has similar data in cells from the pituitary. It is also the case that prolonged oestrogen treatment produces an inhibition of dopamine activity, with decreased dopamine turnover in the striatum.

McEwen: You mentioned γ-aminobutyric acid (GABA). One of the issues that has come up in the hippocampal system, where GABA neurons are important oestrogen targets, is that there may be biphasic effects of oestrogens on GABA production. Diane Murphy and others have shown that there seems to be a short term decrease. You said that GABA interneurons may be very important: that the inhibition of Ca^{2+} channels in those neurons would inhibit GABAergic activity which would in turn disinhibit the system and allow dopamine release. Disinhibition seems to be an important issue. It may be an issue in basal forebrain, brainstem and in a number of brain serotonergic systems. Oestrogen interactions on GABA interneurons have a key role in regulating and allowing for these releases of transmitters to occur.

Watson: In *in vitro* systems, one of the reasons we haven't been able to observe these biphasic effects in the past is that a completely defined media system, in which all serum can be eliminated, has only recently become available for culturing these cells. When a serum-free medium is used, we can begin to see picomolar effects of oestrogen which are actually stronger than the nanomolar effects in these rapid responses, and there are inhibitory doses between the picomolar and nanomolar peaks. By being able to control culture conditions more carefully for the amount of oestrogen in the system, we are revealing these different dose–response effects.

Toran-Allerand: We did a study of the distribution of oestrogen binding sites in the basal forebrain in combination with immunohistochemistry for neurotransmitters and found that there were oestrogen receptors in cholinergic neurons (Toran-Allerand et al 1992). However, there were far more oestrogen receptors in the GABAergic neurons that surrounded the cholinergic neurons, and there were immunoreactive GABAergic terminals on these cholinergic neurons. I wondered to what extent oestrogen acting on GABAergic neurons might be influencing the basal forebrain cholinergic neurons, whether or not those neurons were expressing oestrogen receptors.

Gibbs: I think that the effects on the GABAergic neurons may play an important role, and what I'm seeing are the cholinergic parameters. There is hard evidence for GABAergic synapses on those cells; there is pharmacological evidence for GABAergic effects on cholinergic transmission in the hippocampus. As we are hearing from other systems, there appear to be consistencies in the effects that we see being related to GABAergic interneurons.

On the behavioural side, there is also evidence for oestrogen being able to attenuate behavioural deficits produced by benzodiazepine or GABA agonists. This is all fitting together nicely.

Herbison: We have worked for many years on oestrogen regulation of GABAergic neurons in the medial preoptic area. I think that, as a general theme, we will find oestrogen-receptive GABA neurons as principal players mediating oestrogen effects in a number of different brain regions. However, I would be wary about thinking that oestrogen-receptive GABA neurons in the hippocampus are necessarily going to work in the same way as those in the preoptic area or in the basal forebrain. For example, oestrogen seems to have very different region-specific effects on GABA-related enzymes, such as glutamic acid decarboxylase. Oestrogen regulates the expression of this enzyme in the hippocampus but not in the preoptic area where, instead, it regulates GABA transporter and $GABA_A$ receptor subunit expression (Herbison 1997).

Luine: We have recently published a study on the effects of oestrogen and progesterone on turnover of GABA (Luine et al 1999). In the medial preoptic area and ventromedial nucleus (VMN) oestrogen did enhance turnover, and progesterone further enhanced it. In the hippocampus, oestrogen also enhanced GABA turnover, an effect which we haven't yet published. We also saw that in the VMN progesterone was associated with decreased serotonin release. We therefore think that there are $GABA_B$ receptors on serotonin neuronal endings in the VMN. By giving GABAergic drugs we can influence release of serotonin in the VMN.

Kushner: This gets back to the identity of these membrane receptors. There is a large literature suggesting that progestins can interact with the $GABA_B$ receptor. Do you think oestrogen might be doing the same?

Luine: As far as I know, it doesn't.

McEwen: One of the things I mentioned in my introduction yesterday was the late Bob Moss's work on oestrogens and the kainate currents (Wong & Moss 1991, 1992, Gu et al 1999). The oestrogen specificity does not fit the classical oestrogen receptor, in that oestrogen antagonists have no effect. For the NMDA receptor, most of the data that I am aware of point to a delayed effect of oestrogen on NMDA currents (Wong & Moss 1992, Woolley et al 1997), and this is consistent with the induction of NMDA receptors.

Woolley: There is one report from Berger's lab (Foy et al 1999) that NMDA receptor-mediated EPSPs in CA1 pyramidal cells are rapidly potentiated by oestrogen, and that this potentiation can enhance the induction and expression of LTP in the slice. But most of Bob Moss's suggested that there wasn't an effect on NMDA-mediated synaptic currents.

Gustafsson: What about 17α-oestradiol? Sometimes we hear that this metabolite has a particular effect on the CNS.

McEwen: It does produce effects that in some cases are as potent at the 17β-oestradiol. Some of this may be explained on the basis of pharmacodynamics, but this doesn't seem to be a compelling explanation.

References

Foy MR, Xu J, Xie X, Brinton RD, Thompson RF, Berger TW 1999 17β-estradiol enhances NMDA receptor-mediated EPSPs and long-term potentiation. J Neurophysiol 81:925–929

Gu Q, Korach, KS, Ross RL 1999 Rapid action of 17β-estradiol on kainate-induced currents in hippocampal neurons lacking intracellular oestrogen receptors. Endocrinology 140:660–666

Herbison AE 1997 Estrogen regulation of GABA transmission in rat preoptic area. Brain Res Bull 44:321–326

Luine VN, Wu V, Hoffman CS, Renner KJ 1999 GABAergic regulation of lordosis: influence of gonadal hormones on turnover of GABA and interaction of GABA with 5-HT. Neuroendocrinology 69:438–445

Toran-Allerand CD, Miranda RC, Bentham W et al 1992 Estrogen receptors colocalize with low-affinity NGF receptors in cholinergic neurons of the basal forebrain. Proc Natl Acad Sci USA 89:4668–4672

Wong M, Ross RL 1991 Electrophysiological evidence for a rapid membrane action of the gonadal steroid, 17β-estradiol, on CA1 pyramidal neurons of the rat hippocampus. Brain Res 543:148–152

Wong M, Moss RL 1992 Long-term and short-term electrophysiological effects of estrogen on the synaptic properties of hippocampal CA1 neurons. J Neurosci 12:3217–3225

Woolley CS, Weiland NG, McEwen BS, Schwartzkroin PA 1997 Estradiol increases the sensitivity of hippocampal CA1 pyramidal cells to NMDA receptor-mediated synaptic input: correlation with dendritic spine density. J Neurosci 17:1848–1859

Oestrogen effects in olivo-cerebellar and hippocampal circuits

Sheryl S. Smith, Fu-Chun Hsu, Xinshe Li, Cheryl A. Frye*, Donald S. Faber and Ronald S. Markowitz

*Department of Neurobiology and Anatomy, MCP-Hahnemann University, EPPI, 3200 Henry Ave., Philadelphia, PA 19129 and *Department of Psychology, SUNY Albany, Albany, NY, USA*

Abstract. 17β-oestradiol (E2) is known to exert activating effects on CNS excitability, which are in part mediated by increases in glutamate responses, as we have shown in cerebellum. In addition, this steroid is known to facilitate rapid, rhythmic limb movement. Because the inferior olive is believed to be a timer of rapid movement, we have investigated effects of E2 on patterns of discharge recorded from dorsal accessory olive (DAO) using chronically implanted microwires. E2 increases the frequency of rhythmic olivary discharge as well as the number of synchronized neurons in association with facilitation of rhythmic limb and vibrissae movement. One possible mechanism for this effect is via an increase in gap junction proteins, as olivary cells are electrotonically coupled. Levels of connexin 32 (Cx32) and the dendritic lamellar body, both markers for gap junction-associated proteins, are increased threefold after 48 h E2 exposure (2 μg, i.p.), compared to control in both ventral medulla and hippocampal neurons. Gap junction conductance has also been shown to be decreased by γ-aminobutyric acid (GABA)ergic input. For this reason, we tested effects of 48 h E2 treatment on GABA$_A$ receptor subunit proteins and GABAergic synaptic current. E2 increased levels of the α4 subunit in hippocampus via an increase in the GABA-modulatory progesterone metabolite 3α-OH-5α-pregnan-20-one. This effect was correlated with a decrease in decay time of tetrodotoxin-resistant miniature inhibitory postsynaptic currents (mIPSCs) recorded from pyramidal cells in CA1 hippocampus, an effect which would tend to reduce total GABA inhibition. In sum, these effects of E2 are consistent with the concept that E2 exerts primarily activating effects on CNS excitability.

2000 Neuronal and cognitive effects of oestrogens. Wiley, Chichester (Novartis Foundation Symposium 230) p 155–172

In addition to classic effects on reproductive function, 17β-oestradiol (E2) is known to be associated with a variety of excitatory effects in the CNS. Behaviourally, this is evidenced by activation of an array of sensorimotor parameters, including improved limb coordination, balance and sensory perception (Hampson & Kimura 1988, Smith & Chapin 1996a,b). The word 'oestrous', now used to denote the stage of the cycle following peak levels of circulating E2, was initially used to describe the sensorimotor activation observed at this time, and literally means 'frenzy'. A number of clinical studies suggest that increased levels of E2 in the circulation are also correlated with 'activating effects' on mood, memory and attentional mechanisms (Ross et al 1998). Further evidence for this activating action of E2 is provided by reports of its proconvulsant action. Cyclic elevations in circulating E2 can exacerbate ongoing convulsive activity, although this effect is dependent upon the seizure subtype (Bäckström 1976, Herzog et al 1997). E2 administration can also facilitate the acquisition of kindled seizures in the dorsal hippocampus in experimental animals (Buterbaugh & Hudson 1991). This proconvulsant action appears to be direct, as local application of E2 to the surface of the cat cerebral cortex induces focal seizures characterized by 2–3 Hz spike and slow wave discharge (Marcus et al 1966).

Excitatory amino acid receptors

These excitatory effects of E2 on the motor system may be due to effects of the steroid on neurotransmitter receptor systems of relevant sensorimotor areas, such as cerebellum and basal ganglia. Early results from this laboratory demonstrated that E2, applied locally by pressure ejection or systemically at physiological concentrations, can enhance excitatory responses of neurons within the cerebellum (Purkinje cells) to excitatory amino acid (EAA) neurotransmitters, including glutamate, within minutes (Smith et al 1987a, see Fig. 1). This effect was specific for EAA receptor subtypes quisqualate and NMDA and was long-term, persisting for at least 6–8 h after exposure of the neuron to the steroid (Smith 1989). Although enhanced quisqualate responses were seen in ovariectomized animals and were not due to classic receptor activation, potentiation of NMDA responses by E2 was only observed under conditions of E2-priming (Smith 1989). Because the β form of the E2 receptor (ERβ) has been localized to the cerebellar Purkinje cell (Shughrue et al 1997), it is possible that the effects of chronic E2 are receptor-mediated. More recent studies have demonstrated that E2 can increase kainate responses of pyramidal cells in CA1

FIG. 1.   Locally applied 17β-oestradiol (E2) augments Purkinje cell responses to glutamate. Strip chart records (left) and corresponding peri-event histograms (right) indicate changes in Purkinje cell response to glutamate before (upper records), during (middle records) and after (lower records) continuous pressure ejection of E2 (0.5 μmol/l in 0.01% propylene glycol-saline) at 1–2 p.s.i. Each histogram sums unit activity from 4–5 glutamate pulses (solid bar, 23 nA) of 10 s duration, occurring at 40 s intervals. Glutamate-induced excitation is indicated as a percent change in firing rate relative to spontaneous discharge (numbers next to bars). Purkinje cell responses to glutamate were significantly enhanced within seconds after the onset of E2 application, and did not recover to control levels of response by 30 min after termination of steroid application. (Reprinted with permission from Smith et al 1988, © Elsevier Science.)

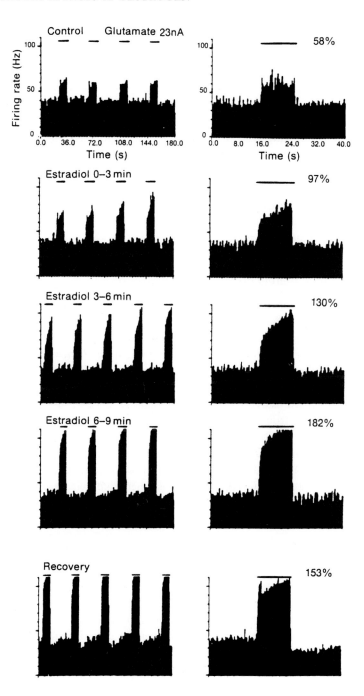

hippocampus while longer-term treatment is also able to increase NMDA responses in this CNS site (Wong & Moss 1992, Foy et al 1999), presumably as a result of increases in NMDA binding via an increase in NR1 subunit levels (Gazzaley et al 1996, Weiland 1992a). In contrast, the rapid effects of this steroid appear to be mediated by a G protein-coupled mechanism (Moss & Gu 1999). The outcome of this increase in excitatory tone is an increase in dendritic spine density in CA1 hippocampus functionally resulting in an increase in the gain of the input/ output relationship and facilitation of long-term potentiation (Woolley & McEwen 1992, Woolley et al 1997, Murphy et al 1998, Foy et al 1999). Increases in synaptogenesis in this region are seen on pro-oestrus as evidenced by the number of free postsynaptic densities (Desmond & Levy 1998). In contrast, in striatum E2 is able to increase dopamine release through novel membrane effects (Xiao & Becker 1998). Together these findings suggest a global effect of this steroid on excitatory neurotransmitter systems.

Contrast enhancement of olivo-cerebellar circuits

More recent studies in this laboratory have focused on the effects of E2 on olivo-cerebellar networks. Across the rat oestrous cycle, elevations in circulating levels of E2 are followed closely by increases in progesterone. As steroids delivered by the circulation, effects of these lipophilic molecules would be evidenced at the network level rather than at single synapses. Therefore we have examined effects of combined hormone administration on networks of neurons recorded from chronically implanted bundles of electrodes in the cerebellar Purkinje cell layer, as well as the afferents from the inferior olivary nucleus (Smith & Chapin 1996a,b). Our earlier studies (Smith et al 1987b) demonstrated that systemic injection of physiological concentrations of progesterone enhances γ-aminobutyric acid (GABA)-mediated inhibition of Purkinje cell discharge recorded extracellularly via local conversion to the GABA-modulatory metabolite, $3\alpha,5\alpha$-THP (3α-OH-5α-pregnan-20-one). In combination with the EAA responses enhanced by E2, the dual action of both hormones would increase the contrast of neuronal responses to both excitatory and inhibitory input, an effect we have demonstrated both in the cerebellum and inferior olivary nucleus (Smith & Chapin 1996b). Circuits involving the rostral dorsal accessory olive (rDAO) are believed to signal errors or motor event changes to the cerebellum (Gellman et al 1985). Acting as a selective sensory filter, this structure gates out sensory input during active movement presumably via GABAergic inhibition. During non-movement, input, reflecting motor error or event change, is gated in via direct glutamatergic input from sensory afferents. Assessment of single unit responses to peripheral stimulation indicates that E2 and progesterone enhance both the excitatory response to afferent input during non-movement, as well as the

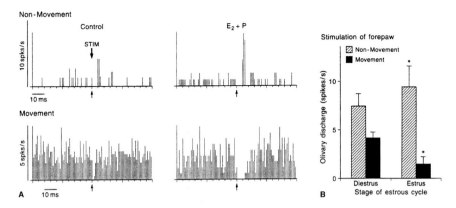

FIG. 2. Contrast enhancement of olivary responses to forepaw stimulation after administration of oestrus hormones. (A) Representative histograms from the same cell illustrate the effects of E2 and progesterone (P) on responses of rostral dorsal accessory olivary (rDAO) neurons to forepaw stimulation during movement and non-movement of this limb. Non-movement-dependent responses (upper panel) were enhanced by administration of oestrus hormones (right). During forepaw movement (lower panel), suppression of these responses lasted longer after hormone treatment (right) than in the control case (left). Hormones (30 ng E2 + 50 μg P) were administered i.p. 20 min before testing and 24 h after an initial priming dose of 2 μg E2 to replicate oestrus conditions. (B) Sensorimotor gating of the rDAO across the oestrous cycle: a summary diagram of responses to forepaw stimulation. On the night of behavioural oestrus, following elevations in circulating E2 and P, the difference or 'contrast' between responses during movement and non-movement is enhanced. ($n = 15$ neurons, 3 rats). (Reprinted with permission Smith & Chapin 1996a, © Springer-Verlag.)

inhibitory response during movement, compared to results obtained during dioestrus (Smith & Chapin 1996a,b, see Fig. 2). Similar changes were observed on the night of behavioural oestrus. Such an effect would be expected to sharpen the 'contrast' of responses elicited by expected and unexpected input. Thus, oestrous-enhanced error signalling by the rDAO might be expected to increase the resolution for detection of motor errors by this structure, an effect consistent with reports of improved motor coordination associated with the night of behavioural oestrus (Smith & Chapin 1996a,b).

Rhythmic discharge

In addition to more general effects on sensorimotor performance, E2 is known to be associated with facilitation of rapid, rhythmic movements of the limbs and digits. In human studies, finger tapping frequency increases during the midcycle peak in E2 (Becker et al 1982), as does the ability to accurately perform other repetitive rapidly alternating tasks such as repeating vowels, typing and the peg-and-board task (Broverman et al 1968, Hampson & Kimura 1988). E2-deficient girls with Turner's Syndrome experience significant improvement in the speed of both repetitive, non-guided tasks and spatially guided motor tasks following E2 therapy (Ross et al 1998). Results from the present laboratory demonstrate that following 48 h of E2 exposure (2 μg, i.p.), both limb stepping and vibrissae movement (whisking) are faster, with a more consistent frequency (Smith 1998). One potential site for E2 in producing these effects is the inferior olivary nucleus. In addition to functioning as an error signalling device, this structure is reported to act as a putative timing device for coordinated movement due to its ability to support subthreshold oscillations of membrane potential (Llinás & Yarom 1986). These oscillations may then generate rhythmic spike trains upon appropriate sensorimotor stimulation, in phase with the endogenous membrane potential fluctuations (Lang et al 1996). Membrane oscillations of individual neurons are dependent upon a low-threshold Ca^{2+} spike and can be modified by K^+ currents associated with anomalous rectification (Llinás & Yarom 1986). *In vivo*, however, oscillations are an emergent property of rhythmically firing somatotopically aligned neurons coupled via gap junctions (Lang et al 1996). For this study, we tested the hypothesis that E2 enhances synchronized, rhythmic discharge of neurons within the inferior olive in conjunction with rhythmic movement of the limbs and vibrissae. Towards this end, female rats were implanted bilaterally with two eight-microwire (25 or 50 μm) bundles into the limb and vibrissa area of the DAO (Smith 1998). In addition to determination of the sensory receptive field and characteristic waveform, cells were identified by their ability to follow high frequency antidromic stimulation from the contralateral paravermal cerebellum (Smith & Chapin 1996a). DAO discharge was recorded simultaneously from up to 48 neurons during treadmill locomotion or spontaneous whisking (rhythmic protraction and retraction of the vibrissae) before and after administration of E2, either systemically or locally applied.

For the first study, whisker-responsive DAO neurons were recorded during whisking behaviour and rhythmicity assessed across hormone state. The spatial extent of coupling was also determined by examining the diameter occupied by neurons firing synchronously. [Synchronization here is defined by the presence of a 1 ms peak at the cross-correlation node, with an $r > 0.1$, $P < 0.0005$, where $n = 10^3$–10^5 spikes, using correlogram analysis.] Characteristically, whisker-responsive DAO neurons discharge with a rhythm time-locked to whisker movement (7–9 Hz). Following treatment with E2, whisking frequency was consistent at 8.5 Hz, with no deviations (SEM, 0; variance, 0). In contrast, on dioestrus, a highly variable whisking frequency was observed (5–10 Hz), with a lower average frequency than observed on oestrus (6.9 ± 0.60) and a greater variance (3.08, dioestrus; 0, oestrus). E2 treatment produced similar effects on DAO discharge patterns, resulting in a faster, more consistent rhythmic DAO discharge frequency (mean, 8.5; variance, 1.91) versus dioestrus values (mean, 7.4; variance, 8.82; $P < 0.05$). In addition, the amplitude of the oscillation was increased 86% by E2, while trough amplitude was depressed by 80%, also suggesting that DAO rhythmicity is more consistent at this time, a result similar to that observed for the forepaw area of the DAO. Further analysis with interspike interval histograms and autocorrelation raster analysis demonstrates that the interspike interval exhibits more peaks (3.5 vs. 1.5, $P < 0.05$) of shorter duration (98 vs. 265 ms, $P < 0.05$) following treatment with E2 than observed on dioestrus.

Two days of E2 treatment also produced maximal effects on synchronization of DAO oscillations, increasing the number of synchronized neurons to 90%, which effectively increased the coupling diameter by $150 \pm 45 \mu m$ ($P < 0.001$). The correlation coefficient of coupling was also increased from 0.1 to 0.148 following E2 treatment ($n = 4.5 \times 10^5$, 2 ms bins/cell, $P < 0.05$).

These results suggest that the effect of E2 in increasing the extent of rhythmic, synchronized DAO discharge is associated with facilitation of rhythmic whisker movement. To discern more completely the relationship between DAO discharge patterns and rhythmic whisking behaviour, we also compared cross-correlation analysis of DAO discharge during periods of active whisking (rhythmic movement) versus non-whisking and across hormone state. E2 treatment resulted in rhythmic DAO discharge (8.5 Hz) during periods of both whisking and non-whisking, while dioestrus conditions were associated with rhythmic discharge dependent upon rhythmic whisker movement. These results suggest that E2 effects on whisking, *per se*, are not producing the enhanced, synchronized discharge observed from DAO recordings.

Rhythmic discharge of olivary neurons was also tightly coupled to step cycle rhythmicity during treadmill locomotion under control conditions. Increases in the number of synchronized, rhythmically discharging neurons following oestrus hormone treatment produced an increase in the spatial extent of the oscillating neuronal cluster recorded during locomotion. When recorded during treadmill locomotion at speeds of 11 cm/s under control dioestrus conditions, the average diameter of a coupled cluster was $50 \pm 10 \mu m$. Local administration of E2 increased the coupling diameter by $150 \pm 20 \mu m$ compared to dioestrus values ($P < 0.05$, see Fig. 3), suggesting that E2 produces direct effects on the DAO. In a similar manner, cyclic increases in circulating oestrus hormones on oestrus increased the diameter of the oscillating cluster by $100 \pm 11 \mu m$ (light phase, $P < 0.01$) and $151 \pm 12.5 \mu m$ (dark phase, $P < 0.01$); administration of E2 + progesterone to an E2-primed rat replicated oestrus conditions in that the spatial extent of rhythmic DAO discharge was increased by $135 \pm 5.4\%$ ($P < 0.01$).

CONTROL   **LOCAL E₂**

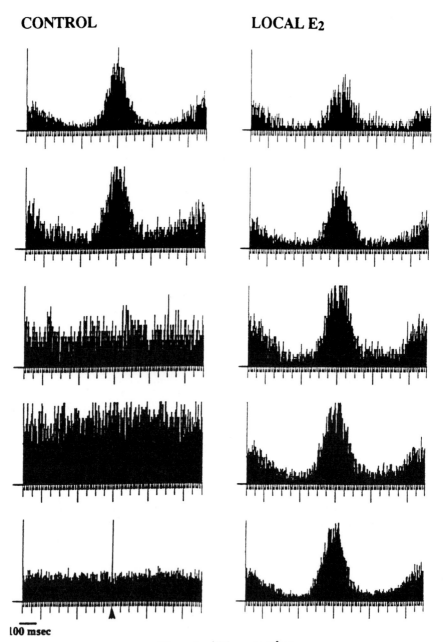

100 msec

Treadmill locomotion
(variable acceleration)

Gap junction proteins

Coupling between olivary neurons is accomplished via dendrodendritic gap junctions (Lang et al 1996). The gap junction conductance is modulated by both NMDA receptors (Pereda & Faber 1996) and GABAergic input (Lang et al 1996), systems which are in turn altered by chronic E2 exposure (Smith 1989, Murphy et al 1998, Weiland 1992b). However, more compelling evidence for E2-modulation of gap junctions is found in a number of reports demonstrating that E2 can increase formation of gap junction proteins in uterus and liver via genomic as well as non-genomic mechanisms (Grümmer et al 1999). In addition, the Cx43 gene contains an E2 receptor response element (Yu et al 1994), suggesting a transcriptional effect of E2 on this gap junction protein. In order to determine a possible mechanism for the observed E2 facilitation of rhythmic DAO discharge, we tested the hypothesis that connexin 32 (Cx32), a marker for neuronal gap junction proteins, and the dendritic lamellar body (DLB), a specific marker for dendrodendritic gap junctions, are increased by E2. As before, female rats were treated with E2 ($2\,\mu g$, i.p.) or vehicle for 2 d before testing, in some cases in conjunction with sustained periods of treadmill running. Western blot analysis of connexin band density revealed a profound effect of both hormone and locomotor activity on connexin levels. Cx32 band staining was evident as a single band at 38 kDa, as previously described. Cx32 levels were highest in rat both exposed to E2 and rhythmic treadmill running (300% increase above control; $P < 0.001$). Cx32 levels in rats only treated with E2 were threefold higher than control ($P < 0.05$). Treadmill locomotion alone produced no significant effect on Cx32 levels. Staining with the DLB antibody revealed a single band at 110 kDa. Levels of the DLB were increased 175% by E2 + treadmill running ($P < 0.05$); E2 alone increased DLB levels by 120% ($P < 0.05$). Treadmill locomotion alone increased DLB levels by 52%. In contrast, Cx43 levels were increased only by E2 treatment (200% increase above control levels, $P < 0.01$) but not by treadmill locomotion + E2. Cx26 levels were not altered by either hormonal or behavioural state. All changes in Cx band density observed were unaccompanied by corresponding changes in levels of the control protein, GAPDH.

Although less well established, gap junctions are known to exist in the adult hippocampus, where they may play a role in mediating high frequency (200 Hz)

FIG. 3.   Locally applied E2 enhances olivary oscillations. Correlograms of olivary discharge recorded during a variable acceleration paradigm before (left) and 20–40 min after (right) local infusion of 600 pM E2 to an E2-primed progesterone-treated rat. Locally applied E2 increased from two to five the number of synchronized, oscillating neurons compared to control values. Control records were obtained 20–40 min after infusion of vehicle alone. These results are representative of 15 neurons recorded in three rats. (Reprinted with permission from Smith 1998, © Elsevier Science.)

oscillations (Draguhn et al 1998). Two-day treatment with E2 also significantly increased band density for both DLB and Cx32 (by twofold and fivefold, respectively), suggesting an additional role for this hormone on electrotonic coupling and oscillation states.

In the inferior olive, E2-induced increases in levels of both Cx32 and DLB suggests an increase in dendrodendritic gap junctions. Oscillatory discharge of the DAO is an ensemble-like property which requires both an intrinsic pacemaker potential at the cellular level as well as electrotonic coupling of neuronal populations by gap junctions (Lang et al 1996). Thus, these results suggest that stimulatory effects of E2 on rhythmic DAO discharge and facilitation of rhythmic movement may be mediated by an increase in gap junction density. The results from this study not only have implications for hormonal control over the DAO circuit, but suggest that E2-induced increases in high frequency oscillations may be more of a global CNS phenomenon.

GABA$_A$ receptor subunit composition

In addition to its established activating effects via EAA receptors and gap junction formation, another possible activating effect of this steroid is via decreases in GABA inhibition. There is evidence that E2 treatment can alter levels of GAD, the GABA synthesizing enzyme in areas including CA1 hippocampus (McCarthy et al 1995, Murphy et al 1998, Weiland 1992b). Our recent results suggest that, in CA1 hippocampus, 48 h exposure to E2 decreases GABA inhibition via up-regulation of the α4 subunit of the GABA$_A$ receptor (GABA-R). Our previous findings suggest that abrupt discontinuation after chronic exposure to the GABA-modulatory metabolite 3α,5α-THP increases CNS excitability via α4 GABA-R subunit up-regulation (Smith et al 1998). This increase in excitability was evidenced by increases in seizure susceptibility and can be explained mechanistically by decreases in the decay time for GABA-gated current which would decrease total charge transfer. Both endpoints were prevented after α4 expression was suppressed using antisense technology (Smith et al 1998). Because our preliminary data suggest that 48 h exposure to 3α,5α-THP also increases GABA-R subunit levels, we tested 48 h E2 treatment on this system because this steroid is also known to enhance conversion of progesterone to 3α,5α-THP due to its facilitating effect on the 5α-reductase enzyme. E2 is able to increase levels of the α4 subunit by two- to threefold after injection of 0.4–20 μg/kg for two days (i.p.) to female adult rats, with maximal effects at 4–8 μg/kg of the steroid (Fig. 4). A dose of 8 μg/kg E2 significantly increased hippocampal levels of 3α,5α-THP (2.1 ± 0.07 ng/g) above dioestrus, control values (1.59 ± 0.19 ng/g; $P < 0.04$), assessed using radioimmunoassay procedures. These results suggest

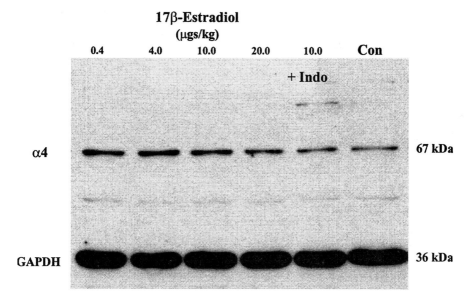

FIG. 4.   E2 increases levels of the α4 subunit of the GABA$_A$ receptor. Two-day treatment with E2 (2 μg, i.p.) increased significantly levels of the α4 subunit in hippocampus compared to control, dioestrus values. In contrast, no changes in GAPDH protein were noted. These results suggest that chronic E2 treatment alters GABA-R subunit composition ($n=3$ rats/ group, performed in triplicate).

that up-regulation of the α4 subunit in the CA1 hippocampus can be accomplished by E2.

Synaptic current

In order to assess the impact of chronic E2 exposure on GABA physiology, we recorded GABAergic synaptic current from pyramidal cells in CA1 hippocampus following 48 h E2 treatment (8 μg/kg, i.p.). Tetrodotoxin (TTX)-resistant miniature inhibitory postsynaptic currents (mIPSCs) were recorded using the whole cell approach at −70 mV at 34–35 °C and the data were analysed using software developed in the laboratory of H. Korn (Ankri et al 1994). mIPSCs represent the minimal (quantal) postsynaptic response to transmitter released from a single vesicle. Therefore, mIPSC characteristics are not complicated by factors which would influence evoked responses, such as the number or pattern of vesicle release, and would not unphysiologically activate extrasynaptic GABA-R. Frequency, peak amplitude and kinetics of mIPSCs were determined

E₂-treated

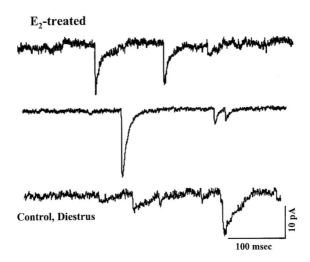

Control, Diestrus

10 pA

100 msec

FIG. 5.   E2 decreases the decay time constant of mIPSCs recorded from CA1 hippocampus. Individual traces of mIPSCs recorded from CA1 hippocampus at −60 mV at room temperature (27 °C) using whole cell recording techniques in the hippocampal slice preparation. Following two days of E2 treatment (2 μg, i.p.) mIPSCs are of shorter duration, but reflect a similar amplitude distribution as observed on the day of dioestrus. mIPSC frequency was slightly decreased after E2 treatment.

in cells from E2-treated animals and compared to those from control, vehicle-treated dioestrus rats. Decay time constants for mIPSCs were analysed as either mono- or biexponential decay time constants using non-linear curve fitting routines and least squares approximations (Fig. 5). Under control conditions, spontaneous mIPSCs were low amplitude (5–40 pA), infrequent events (6–8 Hz). A kinetic analysis revealed a bimodal population with a larger population (90%) decaying with a monoexponential decay time constant $(\tau) = 3.88 \pm 0.56$ ms. A smaller population (10%) decayed with a biexponential decay; $\tau f = 0.8 \pm 0.33$, $\tau s = 5.2 \pm 1.2$ ms. Following treatment with E2, there was a striking increase in the percentage of cells exhibiting the biexponential decay time (30–40%), and a significant decrease in the slow τ compared with control; $\tau f = 0.9 \pm 0.43$ ms, $\tau s = 3.7 \pm 0.43$ ms. A significant decrease in τ was also noted for the monoexponentially decaying time constant $(\tau = 2.77 \pm 0.32$ ms). However, mIPCs amplitude distribution was not altered by E2 treatment. The significant decrease in decay time overall would tend to decrease total GABA inhibition. Such a change is consistent overall with the concept that increases in neuronal excitability follow elevations in circulating levels of E2.

Global activation of CNS circuits by E2, via increased glutamate excitation and reduced GABA inhibition, may result in significant facilitation of processing, both for sensorimotor systems, as well as for higher cognitive events. Theoretical analysis of oscillating, excitatory systems such as would predominate under high E2 conditions suggests that faster processing of input may result. However, under conditions where both E2 and progesterone are present, contrast enhancement of excitatory versus inhibitory input may lead to finer resolution of problem-solving processing.

References

Ankri N, Legendre P, Faber DS, Korn H 1994 Automatic detection of spontaneous synaptic responses in central neurons. J Neurosci Methods 52:87–100

Bäckström T 1976 Epileptic seizures in women related to plasma estrogen and progesterone during the menstrual cycle. Acta Neurol Scand 54:321–347

Becker D, Creutzfeldt OD, Schwibbe M, Wuttke W 1982 Changes in physiological, EEG and psychological parameters in women during the spontaneous menstrual cycle and following oral contraceptives. Psychoneuroendocrinology 7:75–90

Broverman DM, Klaiber EL, Kobayashi Y, Vogel W 1968 Roles of activation and inhibition in sex differences in cognitive abilities. Psychol Rev 75:23–50

Buterbaugh GG, Hudson GM 1991 Estradiol replacement to female rats facilitates dorsal hippocampal but not ventral hippocampal kindled seizure acquisition. Exp Neurol 111:55–64

Desmond NL, Levy WB 1998 Free postsynaptic densities in the hippocampus of the female rat. Neuroreport 9:1975–1979

Draguhn A, Traub RD, Schmitz D, Jefferys JG 1998 Electrical coupling underlies high-frequency oscillations in the hippocampus *in vitro*. Nature 394:189–192

Foy MR, Xu J, Xie X, Brinton RD, Thompson RF, Berger TW 1999 17β-estradiol enhances NMDA receptor-mediated EPSPs and long-term potentiation. J Neurophysiol 81:925–929

Gazzaley AH, Weiland NG, McEwen BS, Morrison JH 1996 Differential regulation of NMDAR1 mRNA and protein by estradiol in the rat hippocampus. J Neurosci 16:6830–6838

Gellman R, Gibson AR, Houk JC 1985 Inferior olivary neurons in the awake cat: detection of contact and passive body displacement. J Neurophysiol 54:40–60

Grümmer R, Traub O, Winterhager E 1999 Gap junction connexin genes cx26 and cx43 are differentially regulated by ovarian steroid hormones in rat endometrium. Endocrinology 140:2509–2516

Hampson E, Kimura D 1988 Reciprocal effects of hormonal fluctuations on human motor and perceptual–spatial skills. Behav Neurosci 102:456–459

Herzog AG, Klein P, Ransil BJ 1997 Three patterns of catamenial epilepsy. Epilepsia 38:1082–1088

Lang E, Sugihara I, Llinás R 1996 GABAergic modulation of complex spike activity by the cerebellar nucleoolivary pathway in rat. J Neurophys 76:255–275

Llinás R, Yarom Y 1986 Oscillatory properties of guinea-pig inferior olivary neurones and their pharmacological modulation: an *in vitro* study. J Physiol (Lond) 376:163–182

Marcus EM, Watson CW, Goldman PL 1966 Effects of steroid on cerebral electrical activity. Epileptogenic effects of conjugated estrogens and related compounds in the cat and rabbit. Arch Neurol 15:521–532

McCarthy MM, Kaufman LC, Brooks PJ, Pfaff DW, Schwartz-Giblin S 1995 Estrogen modulation of mRNA levels for the two forms of glutamic acid decarboxylase (GAD) in female rat brain. J Comp Neurol 360:685–697

Moss RL, Gu Q 1999 Estrogen: mechanisms for a rapid action in CA1 hippocampal neurons. Steroids 64:14–21

Murphy DD, Cole NB, Greenberger V, Segal M 1998 Estradiol increases dendritic spine density by reducing GABA neurotransmission in hippocampal neurons. J Neurosci 18:2550–2559

Pereda AE, Faber DS 1996 Activity-dependent short-term enhancement of intercellular coupling. J Neurosci 16:983–992

Ross JL, Roeltgen D, Feuillan P, Kushner H, Cutler GB Jr 1998 Effects of estrogen on nonverbal processing speed and motor function in girls with Turner's syndrome. J Clin Endocrinol Metab 83:3198–3204

Shughrue PJ, Lane MV, Merchenthaler I 1997 Comparative distribution of estrogen receptor-α and -β mRNA in the rat central nervous system. J Comp Neurol 388:507–525

Smith SS 1989 Estrogen administration increases neuronal responses to excitatory amino acids as a long-term effect. Brain Res 503:354–357

Smith SS 1998 Estrous hormones enhance coupled, rhythmic olivary discharge in correlation with facilitated limb stepping. Neuroscience 82:83–95

Smith SS, Chapin JK 1996a The estrous cycle and the olivo-cerebellar circuit. I. Contrast enhancement of sensorimotor-correlated cerebellar discharge. Exp Brain Res 111:371–384

Smith SS, Chapin JK 1996b The estrous cycle and the olivo-cerebellar circuit. II. Enhanced selective sensory gating of responses from the rostral dorsal accessory olive. Exp Brain Res 111:385–392

Smith SS, Waterhouse BD, Woodward DJ 1987a Sex steroid effects on extrahypothalamic CNS. I. Estrogen augments neuronal responsiveness to iontophoretically applied glutamate in the cerebellum. Brain Res 422:40–51

Smith SS, Waterhouse BD, Chapin JK, Woodward DJ 1987b Progesterone alters GABA and glutamate responsiveness: a possible mechanism for its anxiolytic action. Brain Res 400:353–359

Smith SS, Waterhouse BD, Woodward DJ 1988 Locally applied estrogens potentiate glutamate-evoked excitation of cerebellar Purkinje cells. Brain Res 475:272–282

Smith SS, Gong QJ, Hsu F-C , Markowitz RS, ffrench-Mullen JMH, Li X 1998 GABA$_A$ receptor α4 subunit suppression prevents withdrawal properties of an endogenous steroid. Nature 392:926–930

Weiland NG 1992a Estradiol selectively regulates agonist binding sites on the N-methyl-D-aspartate receptor complex in the CA1 region of the hippocampus. Endocrinology 131:662–668

Weiland NG 1992b Glutamic acid decarboxylase messenger ribonucleic acid is regulated by estradiol and progesterone in the hippocampus. Endocrinology 131:2697–2702

Wong M, Moss RL 1992 Long-term and short-term electrophysiological effects of estrogen on the synaptic properties of hippocampal CA1 neurons. J Neurosci 12:3217–3225

Woolley CS, McEwen BS 1992 Estradiol mediates fluctuation in hippocampal synapse density during the estrous cycle in the adult rat. J Neurosci 12:2549–2554

Woolley CS, Weiland NG, McEwen BS, Schwartzkroin PA 1997 Estradiol increases the sensitivity of hippocampal CA1 pyramidal cells to NMDA receptor-mediated synaptic input: correlation with dendritic spine density. J Neurosci 17:1848–1859

Xiao L, Becker JB 1998 Effects of estrogen agonists on amphetamine-stimulated striatal dopamine release. Synapse 29:379–391

Yu W, Dahl G, Werner R 1994 The connexin43 gene is responsive to oestrogen. Proc R Soc Lond B Biol Sci 255:125–132

DISCUSSION

Simpkins: Have studies been done in either female athletes or musicians, looking at the stage of menstrual cycle and performance?

Smith: That's an interesting question. I have heard on one study that was fairly anecdotal, which was a study of gymnasts that showed improved performance with higher oestrogen levels. I don't know if this was ever published. The other studies I know about have to do with fine movements such as typing.

Simpkins: And do they change over the menstrual cycle?

Smith: Rapid movements such as finger tapping, walking, typing and repeating vowels are facilitated during the mid-cycle peak in oestradiol.

Herbison: In models of neural networks, the most efficient way of increasing the output of network is to increase both the excitatory and the inhibitory inputs (see Nelson & Turrigiano 1998). To alter one without the other is highly destabilizing. What you have shown here is a nice biological demonstration of this hypothesis.

Smith: That is a good point. Enhancing the contrast of both excitatory and inhibitory input as we have shown may be especially relevant for on-beam versus off-beam activity in cerebellum or centre–surround activity in the visual system. However, achieving a balance between excitatory and inhibitory events, as you suggest, would also ensure stability.

Murphy: A quick comment on your structural evidence for gap junctions with connexins, which I think is very interesting. I did some studies on hippocampal slice cultures with Lucas Pozzo-Miller. Although unpublished, we found an increase in the structural proteins N-cadherin and β-catenin in hippocampal synapses from slice cultures after oestrogen treatment. From a perspective of oestrogenic effects on synaptogenesis, it's very interesting that structural modifications occur that may enhance the function of the synapses electrically.

Smith: That is an interesting finding and certainly suggests that oestrogen produces structural changes in addition to its modulating effects.

Gustafsson: Benita Katzenellenbogen (Urbana, Illinois) has studied effects of oestrogens on MCF-7 cells. Using differential display she has picked up a couple of interesting target genes. One of them is what she calls EBP50 (ERM-binding phosphoprotein), which is a protein that seems to bind to a family of structural membrane proteins. Interestingly, she thinks that this sort of potential morphological effect of oestrogen is perhaps mechanistically related to one of the phenotypes of the ERKO mice recently described in *Nature* by Rex Hess, where it was found that there is less reabsorption of fluid in the epididymal ducts, owing to morphological changes (Hess et al 1997). This leads to a back-pressure in the testes which ultimately destroys sperm production. Those are yet other examples of interesting effects of oestrogens on structural proteins in the membrane.

I would like to point to an interesting finding that we made when we studied expression of the various nuclear receptors in the brain. Take for example PPARα, which is activated by fatty acids and prostaglandin, and LXRα and β, which are activated by oxysterols. These nuclear receptors and many others that we have studied seem invariably to be more highly expressed in two structures in the brain, namely the hippocampus and the cerebellum. I wonder about the specificity of these phenomena. If you gave ligands like WY14643, which is a PPARα agonist, do you think these parameters of locomotion would be affected?

Smith: I guess you're asking about the specificity of our effect. Initially we did see specificity in some of our initial studies investigating oestrogen effects on glutamate responses of cerebellar Purkinje cells. In this case, oestrogen exerted rapid effects (<1 hour) in potentiating glutamate responses. This effect was specific for 17β-oestradiol (not 17α), but was not blocked by tamoxifen or protein synthesis inhibitors, suggesting that it is *not* acting through a classic steroid mechanism. We are just beginning to look at our effects on gap junctions. We have started to use oestrogen blockers to see if this is really a classic effect.

Baulieu: Have you tested progesterone specifically on the gap junctions?

Smith: If you give progesterone with oestrogen it doesn't really seem to change our effect. We haven't looked at progesterone alone on the gap junctions, but we have looked at progesterone alone just on the olivary oscillations. I don't know whether it makes it worse, but it certainly doesn't do anything.

Baulieu: What mechanism do you postulate for the effect on α4?

Smith: Our initial finding was that it was an effect of chronic exposure and withdrawal. Interestingly, other GABA modulators, such as benzodiazepines and alcohol given chronically, also up-regulate α4, so it may be a kind of homeostatic mechanism to allow the $GABA_A$ receptor to be less inhibitory. It appears that the primary effect of an increase in the α4 subunit is to decrease the decay time of GABA-gated current, which would mean less Cl^- getting through the channel and less inhibition. If $GABA_A$ receptors are stimulated for a long time by one of these steroids, then a compensatory decrease in the inhibition would protect CNS circuits from becoming overly depressed.

I don't think it's going to be mediated by classic steroid receptor, because alcohol and benzodiazepines do the same thing. This makes me think that the mechanism is via the $GABA_A$ receptor.

Herbison: With regard to the effects of steroids on the hippocampus, have you looked at the effects of oestrogen and progesterone on $GABA_A$ subunits other than the α4? There was a paper a few years ago by Nancy Weiland looking at the effects of several steroids, including allopregnanolone, on several $GABA_A$ receptor subunits (Weiland & Orchinik 1995). Also, in our own experience, working on hypothalamic oxytocin neurons, we know that progesterone up-regulates the α1 subunit, and this is probably what makes that receptor sensitive to allopregnanolone. Is anything going on other than just the α4 subunit?

Smith: We have done most of this work with either progesterone or allopregnanolone; we've only recently tried oestrogen. We were actually surprised that oestrogen increased α4: we thought it would have no effect. Although we haven't looked at alterations in other GABA-R subunits using oestrogen, we have with allopregnanolone. The most striking change is an increase in the δ subunit, which is a subunit that may alter sensitivity of the GABA-R to allopregnanolone. We see other small changes. We think we see a change in $\beta2$, $\beta3$; it looks like $\alpha2$ and $\gamma2$ are not changing, and I'm not sure about $\alpha1$. We are getting conflicting results.

Gibbs: Do you know of any work that has looked at effects of oestrogen on saccadic eye movements? These are heavily influenced by GABA.

Smith: I know that Sundström et al (1997) are using this technique to look at GABAergic function, but I don't believe they have looked at oestrogen. They look at women with premenstrual syndrome (PMS), and the results are consistent with what I found. $\alpha4$ is benzodiazepine insensitive, and they find that women with PMS have saccades which are relatively insensitive to benzodiazepine, whereas in normal women benzodiazepine treatment slows the saccades.

Gibbs: Could you comment on whether your results would lead to any predictions about interactions between hormone replacement and benzodiazepine action in women?

Smith: This is hard to say, because we only give oestrogen for 48 h and then we see this change. We haven't really got into all the possible mechanisms and what their repercussions may be. It is hard to say what longer treatment with oestrogen might do. It could be that longer treatment makes this effect go away, because this is what happens with progesterone. Progesterone has an odd multiphasic effect. After 2 d there is an increase in $\alpha4$, which then gradually returns to normal. The oestrogen effect might just be a transient increase in response to a perturbation.

Murphy: We have been trying to look at the effects of progesterone on dissociated hippocampal cultures, with very heterogeneous results. Can you speculate a bit about whether you think that *in vivo* there exist subpopulations of GABAergic cells in the hippocampus that may diverge in their response to progesterone?

Smith: Absolutely. The populations of GABA receptors in the hippocampus are heterogeneous; we see different subunits. Although the subunit which is mainly responsible for altered allopregnanolone sensitivity, the δ subunit, is in low concentration, it's definitely there. This means that if you hit a cell that has that subunit, it would not respond to the same degree. I believe it is much more common in the granule cells than in the dentate gyrus, so if you're looking at everything including the dentate gyrus you should be hitting some populations which are not as sensitive.

Toran-Allerand: However, if you are making cultures of dissociated hippocampal cells, the cells used are very immature (prenatal), and many neurons have not yet been born, so one may be missing the very neurons that may be important and should be responding.

Murphy: What I meant was that for every cell that we can record a significant response from, we find one or two cells from which we get no response.

References

Hess RA, Bunick D, Lee H-K et al 1997 A role for oestrogens in the male reproductive system. Nature 390:509–512

Nelson SB, Turrigiano GG 1998 Synaptic depression: a key player in the cortical balancing act. Nat Neurosci 1:539–541

Sundström I, Ashbrook B, Bäckström T 1997 Reduced benzodiazepine sensitivity in patients with premenstrual syndrome : a pilot study. Psychoneuroendocrinology 22:25–38

Weiland NG, Orchinik M 1995 Specific subunit mRNAs of the $GABA_A$ receptor are regulated by progesterone in subfields of the hippocampus. Brain Res Mol Brain Res 32:271–278

Effects of oestradiol on hippocampal circuitry

Catherine S. Woolley

Department of Neurobiology and Physiology, Northwestern University, 2153 N. Campus Drive, Evanston, IL 60208, USA

Abstract. Oestradiol produces structural and functional changes in hippocampal circuitry of adult female rats. The density of both dendritic spines and axospinous synapses on hippocampal CA1 pyramidal cells is regulated by oestradiol. Additionally, oestradiol-induced differences in synaptic connectivity are paralleled by changes in NMDA receptor binding, immunoreactivity for NMDA receptors and sensitivity to NMDA receptor-mediated synaptic input. Curiously, while oestradiol effects are observed in CA1 pyramidal cells, most evidence indicates that these cells lack genomic oestradiol receptors. In contrast, at least some inhibitory neurons in CA1 do express oestradiol receptors. Others' *in vitro* studies suggest that oestradiol-induced increases in spine density require an initial decrease in inhibitory (GABAergic) drive onto pyramidal cells. We have used single and double label immunohistochemistry for c-Fos (as a measure of neuronal activation) and glutamic acid decarboxylase 65 (as a marker for inhibitory circuitry) to determine: (1) which hippocampal neuronal populations are activated by oestradiol and the time-course of this activation, as well as (2) whether oestradiol affects inhibitory circuitry *in vivo* as it does *in vitro*. Our findings are consistent with the suggestion that oestradiol increases dendritic spine density through a mechanism involving disinhibition of pyramidal cells.

2000 Neuronal and cognitive effects of oestrogens. Wiley, Chichester (Novartis Foundation Symposium 230) p 173–187

Oestrogen exerts a wide variety of effects on neuronal function. In addition to its well-documented role in reproductive physiology, oestrogen also regulates the connectivity and physiology of neurons in brain regions that are not typically associated with reproduction, such as the hippocampus. Oestrogen's effects on hippocampal neurons have gained recent attention because of potential links between oestrogen replacement therapy and improved cognitive function as well as neuroprotection from pathologies that involve the hippocampus such as Alzheimer's disease.

We use a multidisciplinary approach that combines light and electron microscopic analysis of synaptic connectivity with electrophysiological recording to understand how oestrogen (as oestradiol, the primary oestrogen in the rat) regulates the structure and function of hippocampal circuitry in the adult female rat. What follows is a description of: (1) anatomical studies of oestradiol's effects on hippocampal connectivity, (2) electrophysiological analyses of the consequences of oestradiol-induced structural changes, and (3) initial studies aimed at understanding the mechanism by which oestradiol regulates the structure and function of hippocampal circuitry.

Effects of oestradiol on hippocampal dendritic spines and synapses

We have studied the effects of oestradiol on dendritic spines and synapses on hippocampal CA1 pyramidal cells. Dendritic spines are small protrusions, each about 1–3 μm in length, that densely cover the dendrites of the principal neurons in the hippocampus. On hippocampal pyramidal cells, over 90% of excitatory synapses are formed on dendritic spines (Harris & Kater 1994). As such, the density (i.e. no. of spines/unit length of dendrite) or number of dendritic spines on hippocampal pyramidal cells can be taken as a light microscopic indicator of density or number of excitatory synaptic inputs to these cells. Electron microscopy has validated this approach (see below).

A series of studies has demonstrated that the density and number of dendritic spines on hippocampal CA1 pyramidal cells are sensitive to changes in circulating levels of oestradiol and progesterone in the adult female rat. Gould et al (1990) initially showed that ovariectomy for 6 d results in a decrease in the density of dendritic spines on the lateral dendritic branches of hippocampal CA1 pyramidal cells. Oestradiol treatment can either prevent (Gould et al 1990, Washburn et al 1997) or reverse (Woolley & McEwen 1993) the ovariectomy-induced decrease in spine density. The maximal changes in spine density occur in the apical dendritic tree. Reported spine density differences range from up to 45% in Golgi-impregnated cells (Gould et al 1990) to 22% in biocytin-filled cells (Woolley et al 1997). Importantly, oestradiol-induced changes in density of dendritic spines occur without changes in overall dendritic length or number of dendritic branches (Woolley & McEwen 1993, 1994), indicating that the hormone-induced differences in spine density reflect differences in spine number rather than an expansion and shrinkage of the dendritic tree. Within the hippocampus of young adult female rats, these changes occur specifically on CA1 pyramidal cells; no oestradiol-induced differences in spine density are seen on CA3 pyramidal cells or dentate gyrus granule cells.

Differences in the density of axospinous synapses parallel oestradiol-induced changes in dendritic spine density. This has been shown using both stereological as well as conventional estimation of synapse density (Woolley & McEwen 1992). Importantly, in the same tissue in which hormone-induced changes in axospinous synapse density are observed, no differences are seen in the density of synapses formed on dendritic shafts. The specificity of oestradiol's effect for axospinous synapses is further indication that these changes in synapse density are not an artefact of fluctuation in hippocampal volume, since such generalized changes would be expected to affect spine and shaft synapses in parallel.

Serial electron microscopic reconstruction of dendritic spines and presynaptic boutons in tissue from oestradiol-treated and control ovariectomized rats strongly suggests that new spines induced by oestradiol form synapses with preexisting boutons in stratum radiatum (Woolley et al 1996). Thus, the addition of new spine synapses apparently does not require axonal sprouting or new presynaptic boutons. In control tissue from ovariectomized animals, a proportion of presynaptic boutons form synapses with more than one dendritic spine (Woolley et al 1996); these boutons are termed multiple synapse boutons (Chicurel & Harris 1992). The addition of new spine synapses to preexisting boutons in oestradiol-treated animals increases the proportion of multiple to single synapse boutons as well as the average number of spine synapses formed by multiple synapse boutons (Woolley et al 1996). These changes in the configuration of excitatory synaptic input to CA1 pyramidal cells increase the number of postsynaptic spines synaptically connected to single presynaptic boutons, which may facilitate synchronization of synaptically driven CA1 pyramidal cell activity.

Electrophysiological effects of oestradiol in hippocampal CA1 pyramidal cells

The majority of spine synapses on hippocampal CA1 pyramidal cells are glutamatergic. It is generally thought that both N-methyl-D-aspartate (NMDA) and non-NMDA glutamate receptors are co-localized at glutamatergic synapses on these cells (Bekkers & Stevens 1989, Collingridge & Singer 1990, Kharazia et al 1996). However, glutamate receptor binding autoradiography has shown that oestradiol differentially regulates NMDA and non-NMDA glutamate receptors in CA1. Oestradiol increases agonist binding to NMDA receptors with no effect on non-NMDA glutamate receptor binding (Weiland 1992, Woolley et al 1997). Weiland (1992) further demonstrated that the estradiol-induced increase in NMDA receptor binding reflects a change in B_{max} rather than in receptor affinity. One explanation for a differential effect of oestradiol on NMDA and non-NMDA receptor binding is the possibility that oestradiol-induced synapses are specifically enriched in the NMDA subtype of glutamate receptor. This suggestion is supported by electrophysiological data from $in vitro$ hippocampal slices.

An initial suggestion that oestradiol regulates NMDA receptor function in CA1 pyramidal cells came from Wong & Moss (1992). In this study, CA1 pyramidal cells were recorded in hippocampal slices from oestradiol-treated and control ovariectomized animals. Approximately 20% of cells from oestradiol-treated animals showed prolonged excitatory postsynaptic potentials (EPSPs) and increased incidence of repetitive firing in response to stimulation of the major excitatory afferents (in stratum radiatum). Each of these findings could result from either increased NMDA receptor activity or reduced disynaptic inhibition. Since these experiments were done by recording cells from resting membrane potential, NMDA receptors were probably largely inactive due to a voltage dependent Mg^{2+} block of the receptor channel. Thus the ability to detect differences in NMDA receptor function was minimized and this may account for the observation that only a small proportion of cells showed signs consistent with enhanced NMDA receptor function.

We directly determined the effect of oestradiol treatment on the sensitivity of CA1 pyramidal cells to NMDA receptor-mediated synaptic input (Woolley et al 1997). This study found no effect of oestradiol on sensitivity to synaptic input under 'baseline' conditions in which non-NMDA glutamate receptors dominate the postsynaptic response. However, when NMDA receptor-mediated EPSPs were specifically analysed, it was shown that the same oestradiol treatment paradigm that increases the density of excitatory synaptic input to CA1 pyramidal cells also increases these cells' sensitivity to NMDA receptor-mediated synaptic input. Furthermore, in CA1 pyramidal cells filled with biocytin to visualize dendritic spines following recording, the sensitivity of individual cells to NMDA receptor-mediated synaptic input was well correlated with the density of dendritic spines on the same cell. These findings are consistent with the hypothesis (Warren et al 1995) that oestradiol-induced dendritic spine synapses on CA1 pyramidal cells are a functionally specialized subpopulation at which the NMDA receptor dominates the postsynaptic response.

Further evidence of oestradiol's effect on NMDA receptor function comes from immunohistochemistry. Gazzaley et al (1996) demonstrated that oestradiol treatment increases immunofluorescence for NMDAR-1, the obligatory subunit of the NMDA receptor, in the CA1 region of the hippocampus. The timing of increased NMDAR-1 immunofluorescence correlates with increases in spine synapse density. Surprisingly, this study also showed no effect of oestradiol treatment on NMDAR-1 mRNA levels, which would contradict an effect of oestradiol on production of NMDA receptors. However, NMDAR-1 mRNA levels were only analysed at the same time point as protein expression; earlier time points were not studied. Thus, it remains a possibility that oestradiol also increases NMDAR-1 mRNA levels, but at times earlier than NMDAR-1 protein.

Taken together, binding, electrophysiological and immunohistochemical data suggest that the hippocampal synapses induced by oestradiol are a specialized subpopulation at which the NMDA receptor dominates the postsynaptic response. As such, oestradiol-induced hippocampal synapses may not be the same as 'normal' (i.e. containing both NMDA and non-NMDA receptors) glutamatergic hippocampal synapses. The prevalence and plasticity of synapses containing primarily NMDA receptors is a controversial issue in hippocampal research. However, synapses mediating apparently pure NMDA receptor responses have been observed in slices from relatively young animals (Isaac et al 1995, Liao et al 1995). Additionally, CA1 synapses containing NMDA, but not non-NMDA, receptors have been reported in adult rats based on electron microscopic immunocytochemical data (He et al 1998).

The contribution of putative NMDA receptor synapses induced by oestradiol to hippocampal function is not yet clear. Such synapses may not be active under normal conditions due to the voltage-dependent Mg^{2+} block of the receptor channel. However, it is likely that NMDA receptor synapses could be recruited under conditions of substantial postsynaptic depolarization. As such, it may be relevant that *in vivo* seizure (e.g. Teresawa & Timiras 1968, Buterbaugh & Hudson 1991) and long-term potentiation (Warren et al 1995, Córdoba Montoya & Carrer 1997) studies, both of which have shown effects of oestradiol, involve substantial postsynaptic depolarization that would be likely to reveal an increase in NMDA receptor function.

Mechanism of oestrogen regulation of dendritic spines and synapses

One of the most puzzling aspects of oestradiol's effects on hippocampal circuitry is that most evidence suggests that CA1 pyramidal cells lack genomic oestradiol receptors. On the other hand, at least some inhibitory interneurons in CA1 do appear to express oestradiol receptors (Loy et al 1988, Weiland et al 1997). Importantly, a recent series of *in vitro* studies has revealed an interaction between oestradiol and γ-aminobutyric acid (GABA)ergic interneurons in regulation of hippocampal dendritic spines. Oestradiol induces new dendritic spines and synapses on cultured hippocampal neurons (Murphy & Segal 1996) as it does *in vivo*. There are several parallels between oestradiol induction of hippocampal synapses *in vivo* and *in vitro*, making it seem likely that the *in vitro* findings may apply to hormonal regulation of spines *in vivo*. In both cases, the effect of oestradiol is delayed, occurring after several days of hormone exposure and both require activation of NMDA receptors (Woolley & McEwen 1994, Murphy & Segal 1996). Additionally, inhibitory neurons in hippocampal cultures express the α form of the oestradiol receptor (Murphy et al 1998) as seen *in vivo* (Weiland et al 1997).

The *in vitro* studies have directly linked an oestradiol-induced decrease in GABAergic inhibition to spine formation on hippocampal neurons. Oestradiol treatment induces a biphasic response in GABAergic neurons *in vitro*, initially

(12–24 h) decreasing GABA and glutamic acid decarboxylase (GAD, the rate-limiting enzyme in GABA synthesis) immunoreactivity and subsequently producing an increase in both (48 h; Murphy et al 1998). Increased excitatory drive onto spiny cells parallels the initial decrease in inhibition. Pharmacological reduction of inhibition by GABA depletion mimics the effect of oestradiol on spine density. Additionally, treatment of cultures with tetrodotoxin, which would block Na$^+$ action potential discharge, also blocks the oestradiol-induced increase in spine density, suggesting an activity-dependent mechanism in spine formation.

Given the suggestion that oestradiol may act initially on interneurons to produce an activity-dependent effect on pyramidal cell connectivity, we were interested to know which hippocampal neuronal populations are activated by oestradiol, as well as the time-course of this activation. We used induction of c-Fos immunoreactivity as a measure of neuronal activation and quantified c-Fos-labelled nuclei at various time points within the treatment regimen known to regulate dendritic spines and synapses on CA1 pyramidal cells. Some tissue in this study was double labelled for c-Fos and glutamic acid decarboxylase 65 (GAD65), a marker of inhibitory neurons. Preliminary analyses have revealed that oestradiol treatment produces a phasic c-Fos response in the CA1 and CA3 pyramidal cell layers and granule cell layer of the dentate gyrus; effects in CA1 were most prominent (Rudick & Woolley 2000). Low and stable numbers of c-Fos labelled nuclei were observed in tissue from ovariectomized control animals. c-Fos immunoreactivity was initially (at 2 h) increased by oestradiol, then depressed (6–12 h), then increased again (24 h). Very few, if any, cells with c-Fos immunoreactive nuclei were also labelled for GAD65, suggesting that it is pyramidal cells rather than inhibitory neurons that are activated (as determined by c-Fos expression) by oestradiol. The oestradiol-induced c-Fos immuno-reactivity may reflect phasic activation and coupling to gene expression in pyramidal cells, which could be involved in oestradiol's effects on excitatory synaptic connectivity in the hippocampus.

Additional preliminary analyses of tissue labelled for GAD65 alone revealed an effect of oestradiol on both the density of GAD65-labelled cell bodies as well as GAD65 labelling in puncta within the CA1 cell body and apical dendritic layers (Hart et al 1999). Ovariectomy gradually decreased GAD65 immunoreactivity in CA1, whereas oestradiol treatment produced a biphasic effect, initially decreasing GAD65 further and subsequently producing an overshoot. These results are very similar to the biphasic effect of oestradiol on GAD in hippocampal cultures. Thus oestradiol may regulate hippocampal dendritic spines and synapses *in vivo* in an activity-dependent manner similar to its effects *in vitro*. Our results are consistent with a scenario in which oestradiol acts directly on GABAergic inhibitory neurons to decrease inhibitory drive onto pyramidal cells. Interestingly, the initial oestradiol-induced decrease in GAD65 (24 h) correlates precisely with a large increase in c-Fos expression in the pyramidal cell layer. It is plausible that this second wave of c-Fos expression reflects an oestradiol-induced disinhibition of CA1 pyramidal cells.

Conclusions

Our studies and those of others have demonstrated that oestradiol induces a variety of structural and functional changes in the circuitry of the adult female rat hippocampus. The density of dendritic spines and spine synapses on hippocampal CA1 pyramidal cells is regulated by oestradiol. These oestradiol-induced differences in spine/synapse density are associated with differences in NMDA, but not non-NMDA, glutamate receptor function. Our *in vivo* findings, together with others' *in vitro* work, suggest that oestradiol may act initially on GABAergic interneurons to disinhibit CA1 pyramidal cells and produce an activity-dependent increase in spine and synapse numbers. Although the functional relevance of the oestradiol-induced changes in hippocampal circuitry is not yet clear (reviewed in Woolley 1998), the hormonal sensitivity of these neurons suggests that structural and electrophysiological changes in hippocampal circuitry induced by gonadal steroids like oestradiol may impact cognitive functions associated with this brain region.

Acknowledgements

Supported by NS 37324 (NINDS) and the Alfred P. Sloan Foundation.

References

Bekkers JM, Stevens CF 1989 NMDA and non-NMDA receptors are co-localized at individual excitatory synapses in cultured rat hippocampus. Nature 341:230–233

Buterbaugh GG, Hudson GM 1991 Estradiol replacement to female rats facilitates dorsal hippocampal but not ventral hippocampal kindled seizure acquisition. Exp Neurol 111:55–64

Chicurel ME, Harris KM 1992 Three-dimensional analysis of the structure and composition of CA3 branched dendritic spines and their synaptic relationships with mossy fiber boutons in the rat hippocampus. J Comp Neurol 325:169–182

Collingridge GL, Singer W 1990 Excitatory amino acid receptors and synaptic plasticity. Trends Pharmacol 11:290–296

Córdoba Montoya DA, Carrer HF 1997 Estrogen facilitates induction of long term potentiation in the hippocampus of awake rats. Brain Res 778:430–438

Gazzaley AH, Weiland NG, McEwen BS, Morrision JH 1996 Differential regulation of NMDAR1 mRNA and protein by estradiol in the rat hippocampus. J Neurosci 16:6830–6838

Gould E, Woolley CS, Frankfurt M, McEwen BS 1990 Gonadal steroids regulate dendritic spine density in hippocampal pyramidal cells in adulthood. J Neurosci 10:1286–1291

Harris KM, Kater SB 1994 Dendritic spines: cellular specializations imparting both stability and flexibility to synaptic function. Annu Rev Neurosci 17:341–371

Hart SA, Rudick CN, Spruston N, Woolley CS 1999 Estradiol regulates GAD immunoreactivity in the CA1 region of the adult female rat hippocampus. Soc Neurosci Abstr, in press

He Y, Janssen WG, Morrison JH 1998 Synaptic coexistence of AMPA and NMDA receptors in the rat hippocampus: a postembedding immunogold study. J Neurosci Res 54: 444–449

Isaac JT, Nicoll RA, Malenka RC 1995 Evidence for silent synapses: implications for the expression of LTP. Neuron 15:427–434

Kharazia VN, Phend KD, Rustioni A, Weinberg RJ 1996 EM colocalization of AMPA and NMDA receptor subunits at synapses in rat cerebral cortex. Neurosci Lett 210: 37–40

Liao D, Hessler NA, Malinow R 1995 Activation of postsynaptically silent synapses during pairing-induced LTP in CA1 region of the hippocampal slice. Nature 375:400–404

Loy R, Gerlach JL, McEwen BS 1988 Autoradiographic localization of estradiol-binding neurons in the hippocampal formation and entorhinal cortex. Dev Brain Res 39: 245–251

Murphy DD, Segal M 1996 Regulation of dendritic spine density in cultured rat hippocampal neurons by steroid hormones. J Neurosci 16:4059–4068

Murphy DD, Cole NB, Greenberger V, Segal M 1998 Estradiol increases dendritic spine density by reducing GABA neurotransmission in hippocampal neurons. J Neurosci 18:2550–2559

Rudick CN, Woolley CS 2000 Estrogen induces a phasic Fos response in the hippocampal CA1 region of adult female rats. Hippocampus, in press

Terasawa E, Timiras PS 1968 Electrical activity during the estrous cycle of the rat: cyclical changes in limbic structures. Endocrinology 83:207–216

Warren SG, Humphreys AG, Juraska JM, Greenough WT 1995 LTP varies across the estrous cycle: enhanced synaptic plasticity in proestrus rats. Brain Res 703:26–40

Washburn SA, Lewis CE, Johnson JE, Voytko ML, Shively CA 1997 17α-dihydroequilenin increases hippocampal dendritic spine density of ovariectomized rats. Brain Res 758:241–244

Weiland NG 1992 Estradiol selectively regulates agonist binding sites on the N-methyl-D-aspartate receptor complex in the CA1 region of the hippocampus. Endocrinology 131: 662–668

Weiland NG, Orikasa C, Hayashim S, McEwen BS 1997 Distribution and hormone regulation of estrogen receptor immunoreactive cells in the hippocampus of male and female rats. J Comp Neurol 388:603–612

Wong M, Moss RL 1992 Long-term and short-term electrophysiological effects of estrogen on the synaptic properties of hippocampal CA1 neurons. J Neurosci 12:3217–3225

Woolley CS 1998 Estrogen-mediated structural and functional synaptic plasticity in the female rat hippocampus. Horm Behav 34:140–148

Woolley CS, McEwen BS 1992 Estradiol mediates fluctuation in hippocampal synapse density during the estrous cycle in the adult rat. J Neurosci 12:2549–2554

Woolley CS, McEwen BS 1993 Roles of estradiol and progesterone in regulation of hippocampal dendritic spine density during the estrous cycle in the rat. J Comp Neurol 336:293–306

Woolley CS, McEwen BS 1994 Estradiol regulates hippocampal dendritic spine density via an N-methyl-D-aspartate receptor dependent mechanism. J Neurosci 14:7680–7687

Woolley CS, Wenzel HJ, Schwartzkroin PA 1996 Estradiol increases the frequency of multiple synapse boutons in the hippocampal CA1 region of the adult female rat. J Comp Neurol 373:108–117

Woolley CS, Weiland NG, McEwen BS, Schwartzkroin PA 1997 Estradiol increases the sensitivity of hippocampal CA1 pyramidal cells to NMDA receptor-mediated synaptic input: correlation with dendritic spine density. J Neurosci 17:1848–1859

DISCUSSION

Murphy: It is always nice to see good, quantified morphology in the real brain. Greenough's lab has shown that input/output curves are enhanced on the afternoon of pro-oestrus. Are you seeing anything similar to that with oestrogen replacement?

Woolley: If you make slices from control animals and oestrogen-treated animals and look at long-term potentiation (LTP) with nothing special in the media, there is no effect of oestrogen treatment on LTP. This is not consistent with the *in vivo* data either during oestrous cycle from Juraska's lab (Warren et al 1995), or other data looking at oestrogen treatment. In both of these cases, oestrogen enhances LTP. However, if you induce LTP *in vitro* and do all subsequent recording in low levels of bicuculline, a $GABA_A$ receptor blocker, the degree of initial potentiation that one sees following tetanus is no different between oestrogen-treated and control, but the time course of decay is different. In the slices from the oestrogen-treated animals the decay of LTP is slower and LTP stabilizes at a higher value than in slices from the control-treated animals. The decay of LTP is a function of NMDA receptor activation. One might expect, then, that more NMDA activity during the tetanus would be reflected in a change in decay. And since we now have evidence that there really is quite a dramatic difference in GABAergic innervation with oestrogen, bicuculline might be needed to reduce inhibition and reveal this difference in NMDA receptor function.

Murphy: This leads to another comment. You didn't mention this in your paper, but you have shown that the increase in spine density can be blocked with an NMDA antagonist, as we see in culture. The point needs to be made that effects of oestrogen specifically on NMDA, at least in the hippocampus, are very important. We have heard in this meeting about L-type Ca^{2+} channels, but if this morphological change can be blocked from the onset with a specific NMDA receptor blocker, then there's something very specific about Ca^{2+} coming in through an NMDA channel that is needed for changes in morphology. Finally, in terms of your Fos immunoreactivity, it is interesting that you haven't seen this in interneurons. When we did the CREB study, looking at phosphorylation of CREB in cultures, about 80% of the cells were positive for phosphorylated CREB after oestrogen treatment. There were about 20% that weren't. Coincidentally, in these particular cultures, about 20% of cells are interneurons. We never actually looked to see which cell types were expressing CREB, so it might be worth going back to look to see whether the 20% or so of cells that don't express CREB after oestrogen are interneurons.

Woolley: That would be very interesting.

Pfaff: I would like to reinforce the neuroanatomical robustness of the kind of oestrogen-stimulated growth responses that we have talked about (Cohen & Pfaff

1981, 1992). They are not just present in the hippocampus and hypothalamus, but also in other more surprising places. VanderHorst & Holstege (1997) published a surprising finding. They were looking at projections from the nucleus retroambiguus to the spinal cord, and in oestrous cats compared to anoestrous cats, they saw a marked degree of axonal sprouting at lumbar 6.

Woolley: That is a large effect.

Pfaff: I have a question with respect to ultimate mechanisms. To what extent do you think your results are dependent upon oestrogen-stimulated biosynthetic events (e.g. Jones et al 1990) in the cell body?

Woolley: I certainly wouldn't rule that out. My suspicion is that these effects are the result of an interaction between rapid effects of oestrogen on excitability and transcription-dependent effects in the pyramidal cells themselves, perhaps not mediated by genomic oestrogen receptors, but dependent upon transcription, nevertheless. Certainly to do all that you have to do as a neuron to make new spines and synapses, there has to be quite a bit of biosynthesis. There is no reason to think that this might not occur, at least in part, in the cell body.

Pfaff: Even the first phase of your triphasic response?

Woolley: I suppose the initial Fos induction is a genomic effect, but it is not mediated by genomic oestrogen receptors. My thinking is that the first phase is probably the result of non-genomic potentiation of glutamatergic synaptic transmission in CA1, but that the second phase (the 24 h phase) does involve genomic oestrogen receptors, but perhaps in GABAergic neurons. Thus what we're looking at is disinhibition, which is producing the second wave of Fos.

Murphy: I agree that, at least in culture, it looks like there has to be some initial rapid response mediated through the NMDA receptor which could somehow begin a cascade of signalling events in the dendritic tree. Then the latter effects of genomic activity happen in the cell body. There is quite a large body of literature as to how rapid dendritic changes are able to occur at the level of the membrane. If a cell generates new dendritic spines, it has to put all the proteins into the membrane that are necessary for this. It is hard to imagine that all of this is occurring genomically. There may be some local processes taking place in the dendrites. There are ribosomes at the base of dendritic spines, so there could be local translation of proteins which can rapidly put things like NMDA and other receptors into the membrane. There is probably a concert of events between rapid effects, local dendritic effects and effects going on in the cell body which are mediated by the oestrogen receptor.

Woolley: We know that the effects of oestrogen both in culture and *in vivo* require NMDA receptor activation, but we don't know when that receptor activation is required. We don't know that there is an initial NMDA receptor-dependent event that's subsequently followed by something else. It would be very interesting to know if that were the case, but we don't know yet at what point NMDA receptors are involved.

Gibbs: Did you look at the numbers of symmetric synapses which, as you commented, are inhibitory?

Woolley: In our original synapse study we did look at symmetric synapses in stratum radiatum and we did not see any changes in the density of these synapses with oestrogen. We were relieved by this because we were worried that changes in the synapse density could be an artefact of volume changes in the hippocampus, and the fact that we saw effects of oestrogen specifically on one type of synapse and not another one in the same place was reassuring. This suggests that in the dendrites, at 48 h following the second injection, there is no difference in overall numbers of inhibitory synapses in the dendrites. We don't know yet about the cell body layer, however, where the GAD changes are actually greater. The other thing to make clear is that although it is tempting to imagine that each of those GAD-positive puncta is a presynaptic bouton, we don't really know this.

Gibbs: Are the changes in spine density and synapse numbers maintained with prolonged oestrogen treatment?

Woolley: I don't know. Spine density decreases rather gradually following ovariectomy and then stays low. If you leave animals ovariectomized for a long time, spine density stabilizes and stays low. The effect of long-term oestrogen treatment, though, is not known.

Gibbs: Do you think some of these effects are initiated by changes in growth factor production? We have reported that oestrogen and progesterone affect brain-derived neurotrophic factor (BDNF) mRNA levels in the hippocampus (Gibbs 1998).

Woolley: The most convincing evidence for the role of growth factors in oestrogen-induced spine density changes comes from Diane Murphy's work. She has shown that a decrease in BDNF level is critical for the oestrogen-induced increase in spine density *in vitro* (Murphy et al 1998). As you say, BDNF levels are regulated by oestrogen *in vivo* in the hippocampus, so there's a good chance that BDNF is in fact very important in this system. However, we have yet to address this in our experiments.

Fillit: Does the level of dendritic spines remain low for long periods? Is there a point that the changes become irreversible, a point at which if you were to give oestrogen you still couldn't reverse the decline in dendritic spine density?

Woolley: That is a good question. I have only done a few experiments on this. When we ovariectomize an animal, treat on the third and fourth days following ovariectomy, and then look 48 h later, we only see a partial reversal and partial blockade effect. Spine density is decreasing gradually, we come in with oestrogen when spine density is half-way down, and we see this difference at the end. It was important originally to show that if you let spine density go down as far it's going to go, wait a little while and then give oestrogen, you can in fact reverse it. We did show this, but we didn't go out much beyond a couple of weeks. This is not really a very long time; it would be interesting to know whether this circuitry can be locked in a different state. There are obvious clinical implications to this.

Gustafsson: Is there any indication that there is plasticity with reference to oestrogen modulation, in that the dendritic spines would vary with the age of the animals with more dramatic effects with young animals? Also, what are the sex differences?

Woolley: That is a good question. I have only used young adult females in my own studies. An intriguing paper was published by Miranda et al (1999) looking at the effects of oestrogen on granule cells in aged females. They found that chronic treatment with oestradiol had no effect on the density of dendritic spines in aged females, but acute oestradiol did produce an increase. There was also a sex difference in the response to oestradiol: males responded differently to oestradiol than females. I have never done experiments with male rats and oestrogen treatment, although in initial studies we looked at the density of spines in male animals compared to female animals. Male animals have spine density that is comparable to females at their high levels of spine density.

Gustafsson: How would you explain that hypothetically? What turns the spines on in males—androgens?

Woolley: I don't think that spines are the sole consequence of oestrogen exposure. Oestrogen is a modulator of one of many structural components of dendrites; spines happen to be sensitive to oestrogen.

Toran-Allerand: Some years ago there was a paper on sex differences in hippocampal dendritic spine formation occurring at puberty in mice that was dependent on androgens (Meyer et al 1978).

Woolley: That's right, there is an androgen-dependent increase in spine density on CA1 pyramidal cells in male mice that occurs at puberty.

Smith: One of the endpoints of an increase in GAD synaptically might be to increase the frequency of miniature inhibitory postsynaptic currents (mIPSCs). Although we didn't see that with our 48 h oestrogen treatment, as we have mentioned previously we are using a different paradigm to your lab.

Although I didn't present any data, we do see a big increase in the frequency of mIPSCs on pro-oestrus. Have you looked across the oestrous cycle to see whether GAD levels increase on pro-oestrus? Alternatively, have you looked to see whether oestrogen plus progesterone modulates GAD levels?

Woolley: Across the oestrous cycle spine density does change, but we have not done any oestrous cycle studies or progesterone studies in which we looked at GAD.

Smith: If you see an increase on pro-oestrus it would be very interesting.

Woolley: Again, I am really interested in potential distinctions that can be made between what one sees during the natural cycle, and what one sees by treating ovariectomized animals with oestrogen plus progesterone. There may be a predisposition to relate everything that happens during the oestrous cycle solely to oestrogen and progesterone levels. This might not be the case.

Becker: The finding that there are these rapid changes in the structure and synapses of neurons in the hippocampus is something that gets everybody really excited. What do you suppose this has to do with the cognitive effects of these hormones and the role of the hippocampus?

Woolley: That's a very important question. There has been keen interest in the last few years, among people studying learning and memory in rats, to try to associate some of these oestrogen-induced changes in hippocampal structure with differences in learning. The results have been equivocal at best. Many people find no differences either across the oestrous cycle or with oestrogen treatments that are similar in time-course to what we use here, although there are, as we discussed this morning, more rapid effects of oestrogen. Some people find a small impairment of behaviour either with elevated oestradiol during the oestrous cycle or following oestrogen treatment. So it's been a question up in the air. However, I was at the meeting of the Society for Behavioural Neuroendrocrinology recently, and Tina Williams presented some very interesting data. She had designed a working memory version of the Morris water maze, and used precisely the same hormone treatment paradigm we have used in anatomical studies and tested spatial working memory in young adult female animals. She presented data showing that there was a dramatic improvement in spatial working memory associated with the points at which we are seeing increased density of dendritic spines and synapses. This wasn't only in the case of oestrogen, but also in the case of oestrogen plus progesterone; where we see a biphasic effect on spines, they saw a biphasic effect on this behaviour. I have also discussed with you before some of my ideas about the potential relationship of these changes to more natural behaviours, particularly to maternal behaviour. One of my ideas is that changes in the hippocampus that occur on the time-course of the oestrous cycle may not actually be the most relevant, because in the wild there aren't very many free oestrous cycles. Female rats are either pregnant or caring for pups, so there aren't many free oestrous cycles for natural selection to have worked upon. The oestrogen sensitivity of the hippocampus may actually have something to do with pregnancy and maternal behaviour. One avenue we would like to pursue in trying to understand the behavioural consequences of these changes is to determine whether they have something to do with setting up the appropriate circuitry for efficient maternal behaviour.

Herbison: You showed dramatic changes in the GAD65 activity. What is the physiological significance of this? When we consider that there are two GADs and that GAD67 is widely thought to be the active enzyme, I am not sure how important GAD65 is in regulating the amount of GABA available at the synapse.

Woolley: We are now looking at GAD67. We will soon be able to put those two pieces of the puzzle together. We first looked at GAD65 because we were most interested in puncta, and GAD65 labels puncta more strongly than GAD67. GAD67 labels cell bodies better. You are right that we need to put the two GADs together to make a case for functional changes in inhibition. The next step is to study these hormonal states electrophysiologically and determine whether or not there is a functional change in the inhibition of pyramidal cells associated with these anatomical changes. The immunofluorescence is really an initial step to providing some basis for further functional studies.

Murphy: We have used an antibody that recognizes both GAD65 and GAD67. Although the more dramatic decrease is seen in the 65 kDa band it also decreases in the 67 kDa band.

Levin: What about the developmental plasticity? If you could make a fetal animal relatively oestrogen-deficient, would this affect hippocampal or other neuronal plasticity?

Woolley: Oestrogen is certainly very important in the development of a brain. It is a major factor.

Levin: What about the male brain?

Woolley: Oestrogen is very important for the development of the male phenotype of the brain.

Gustafsson: The aromatase knockout mice are oestrogen-deficient. They don't go crazy. There is no gross behavioural deficit.

References

Cohen RS, Pfaff DW 1981 Ultrastructure of neurons in the ventromedial nucleus of the hypothalamus in ovariectomized rats with or without estrogen treatment. Cell Tissue Res 217:451–470

Cohen RS, Pfaff DW 1992 Ventromedial hypothalamic neurons in the mediation of long-lasting effects of estrogen on lordosis behavior. Prog Neurobiol 38:423–453

Gibbs RB 1998 Levels of trkA and BDNF mRNA, but not NGF mRNA, fluctuate across the estrous cycle and increase in response to acute hormone replacement. Brain Res 787:259–268

Jones K, Harrington C, Chikaraishi D, Pfaff DW 1990 Steroid hormone regulation of ribosomal RNA in rat hypothalamus: early detection using *in situ* hybridization and precursor-product ribosomal DNA probes. J Neurosci 10:1513–1521

Meyer G, Ferres-Torres R, Mas M 1978 The effects of puberty and castration on hippocampal dendritic spines of mice. A Golgi study. Brain Res 155:108–112

Miranda P, Williams CL, Einstein G 1999 Granule cells in aging rats are sexually dimorphic in their response to estradiol. J Neurosci 19:3316–3325

Murphy DD, Cole NB, Segal M 1998 Brain-derived neurotrophic factor mediates estradiol-induced dendritic spine formation in hippocampal neurons. Proc Natl Acad Sci USA 95:11412–11417

VanderHorst V, Holstege G 1997 Estrogen induces axonal outgrowth in the nucleus retroambiguus-lumbosacral motoneuronal pathway in the adult female cat. J Neurosci 17:1122–1136

Warren SG, Humphreys AG, Juraska JM, Greenough WT 1995 LTP varies across the estrus cycle: enhanced synaptic plasticity in proestrus rats. Brain Res 703:26–30

Oestrogen and cognitive function throughout the female lifespan

Barbara B. Sherwin

Department of Psychology & Department of Obstetrics and Gynecology, McGill University, 1205 Dr. Penfield Avenue, Montreal, Quebec, Canada H3A 1B1

Abstract. Evidence that oestrogen helps to maintain verbal memory in women comes from several sources. Studies that have tested cognitive functioning at different phases of the menstrual cycle have found few differences, perhaps because oestrogen levels are sufficiently high, albeit variable, during all cycle phases. Experimental studies in postmenopausal women have generally found a protective effect of oestrogen, specifically on verbal memory. Results of several large, longitudinal studies that have become available recently have also demonstrated that women who were oestrogen users performed better on certain tests of cognitive function than non-users of similar age. On the basis of this body of evidence, it is possible to conclude that oestrogen may attenuate or prevent the decline in aspects of memory that occur with normal ageing in women.

2000 Neuronal and cognitive effects of oestrogens. Wiley, Chichester (Novartis Foundation Symposium 230) p 188–201

The past two decades have witnessed an abundance of evidence from basic neuroscience that oestrogen influences the chemistry and morphology of brain areas known to be involved in cognition. In turn, the elucidation of these mechanisms has provided biological plausibility for investigations of oestrogen and cognitive functions in women. The clinical relevance of this line of research pertains to increasing life expectancy, particularly in women. Although the age of spontaneous menopause has remained stable at 51.8 years throughout recorded history, women now live an average of 82 years in industrialized countries (compared with an average of 52 years in 1900). Unfortunately, this increase in female life expectancy has not been paralleled by a decrease in the length of disability before death. Thus, the purported ability of oestrogen to attenuate or even prevent deterioration in some aspects of cognitive functions that occur with normal ageing as well as with pathological states would considerably enhance the quality of life for older women.

Several research paradigms have been used to investigate oestrogenic effects on cognition. The most common of them take advantage of the naturally-occurring hormonal fluctuations that occur during a variety of reproductive events during the female lifespan. Before reviewing studies on the menstrual cycle and the postmenopause, it is worthwhile to address, briefly, what is known about sex differences in cognitive functions because it may inform the findings from studies on oestrogen and cognition.

Sex differences in cognitive functions

Sex differences in brain structure and function are thought to occur as a result of differential exposure of men and women to sex hormones during fetal life (Gorski 1996). These differences between men and women are not evident on full-scale IQ scores but rather in specific cognitive abilities. The best established of these findings indicate that, on average, men excel in spatial and quantitative abilities and in gross motor strength, whereas women excel in verbal abilities, in perceptual speed and accuracy, and in fine motor skills (Jarvik 1975). Although the magnitude of the sex difference is modest (between 0.25 and 1 standard deviation from normative scores), it has been detected reliably across studies. Confirmatory evidence that these sex differences in cognitive skills are due to differential prenatal exposure to sex hormones comes from studies of individuals with genetic disorders that caused them to be exposed to abnormal levels of sex hormones during fetal life. For example, individuals with congenital adrenal hyperplasia (CAH) have an adrenal 21-hydroxylase deficiency that results in decreased glucocorticoid production and increased adrenal androgen production because of their high adrenocorticotropic hormone (ACTH) levels .Therefore, female fetuses with CAH are exposed to abnormally high levels of adrenal androgens (Nass & Baker 1991). If it is the case that prenatal exposure to androgens serves to enhance spatial skills, then one would predict that these girls would perform better on tests of spatial abilities compared to their unaffected sisters, which is precisely what has been demonstrated (Resnick et al 1986). If it is true that in adulthood, circulating levels of sex hormones serve to activate neural pathways organized during fetal life, then we would expect specific sex hormones to affect specific cognitive functions differentially. A corollary of this hypothesis is that no hormone would exert a global effect on all cognitive functions.

Menstrual cycle studies

Because of the reliable cyclicity of ovarian hormone secretion during various phases of the menstrual cycle, investigators have sought to determine whether levels of oestrogen and progesterone were associated with scores on tests of cognitive function during different menstrual cycle phases. Several studies have found that women perform better on tests of spatial abilities during phases of the cycle that are characterized by low levels of oestrogen (Hampson 1990, Hampson & Kimura 1988, Komenich et al 1978). Bearing in mind that the effect size was small and that not all studies used hormone assays to confirm menstrual cycle phase, these findings suggest that oestrogen has a negative influence on spatial skills. On the other hand, verbal articulatory skills were better during cycle phases characterized by high levels of oestrogen (Anderson 1972, Hampson 1990, Wickham 1958) suggesting that oestrogen facilitates simple verbal tasks. We studied 35 healthy, regularly cycling women not taking oral contraceptives during the luteal and during the menstrual phase of their cycles (Phillips & Sherwin 1992a). Cycle phases were confirmed via radioimmunoassays of oestrogen and progesterone. Women performed better on a test of long-term visual memory during the luteal phase of the cycle and their scores on this test were significantly and positively correlated with plasma levels of progesterone but not with oestrogen at that time. This tentatively suggested that progesterone may enhance visual memory in women. However, we failed to find between-phase differences on tests of verbal memory that we had predicted a priori. On the basis of the available evidence, it is possible to conclude that although there is some fluctuation in test scores of some cognitive functions that occur simultaneously with fluctuations in ovarian hormones during the menstrual cycle, at all times levels of sex hormones are sufficient to maintain cognitive functions in women of reproductive age.

Postmenopausal women

The menopause, a universal event in the female life cycle, occurs at a mean age of 50.8 years in industrialized countries and marks the transition from the reproductive to the non-reproductive phase of life. In premenopausal women, the ovary secretes 95% of the oestradiol that enters the circulation (Lipsett 1986). After the menopause, the ovaries atrophy and virtually cease their production of oestradiol. The drastic alterations in the hormonal milieu that occur at the time of menopause, coupled with the increasing practice of prescribing hormone replacement therapy provides a unique opportunity for investigating the possible effects of the sex hormones on cognitive functions.

Findings from several experimental studies of postmenopausal women have provided interesting information. In a prospective study of premenopausal women who had their uterus and ovaries surgically removed for benign disease, those who randomly received treatment with oestrogen maintained their preoperative scores on verbal memory whereas scores of those given placebo decreased significantly compared to preoperative baseline when their ovaries were producing oestrogen (Sherwin 1988). A subsequent prospective controlled study of women who were about to undergo a surgical menopause employed a more comprehensive battery of neuropsychological tests to investigate whether the oestrogenic enhancement of memory we had found earlier affected cognitive functions globally or whether it was specific to verbal memory (Phillips & Sherwin 1992b). Women were tested several weeks before surgery and again following three months of treatment with either oestrogen or placebo after their ovaries were removed. Although none of the mean scores on the cognitive tests differed at the preoperative test session, by the third postoperative treatment month, scores had decreased significantly on the immediate and delayed recall of the paired-associate test only in the placebo group coincident with the fairly drastic decreases in their plasma oestrogen levels. Although these findings demonstrated that oestrogen maintained verbal memory in women, it was without effect on visual memory. Importantly, since mood scores remained stable in both groups during the course of the study, we were able to conclude that oestrogen exerted its effect directly on cognitive functioning and not secondarily via an effect on mood.

Recently, two other experimental studies failed to document an effect of oestrogen on cognition in menopausal women. In one study, women who had undergone a surgical menopause several years earlier were treated with either 0.625 mg or 1.25 mg of conjugated equine oestrogen daily (Ditkoff et al 1991). The Digit Span and Digit Symbol tests were the only tests of cognitive function administered and are both primarily measures of attention. In a crossover study of postmenopausal women, oestrogen replacement therapy was not superior to placebo on any test of cognitive performance (Polo-Kantola et al 1998). The fact that cognitive testing was carried out by means of a computer software program instead of conventional administration by a psychologist and that two different transdermal oestrogen preparations were used (gel and patch) instead of the oral drugs used in most other studies makes these findings difficult to compare with those of previous investigations.

Numerous observational studies of postmenopausal oestrogen users and non-users have also provided information on the putative oestrogen–cognition relationship. When 65 year old women were matched for age, socioeconomic status, years of education and marital status, those who had been taking oestrogen since their natural menopause approximately 15 years earlier scored significantly higher on tests of short- and long-term memory compared to the never-users (Kampen & Sherwin 1994). Similar studies also found an advantage in oestrogen-users on a test of verbal fluency (Robinson et al 1994) and on tests of verbal and visual memory (Kimura 1995) compared to non-users. In a large, longitudinal cohort of 800 women living in California, oestrogen-users had better scores on The Mini-Mental Status Examination and verbal fluency but no differences were found on tests of explicit memory between oestrogen-users and non-users (Barrett-Connor & Kritz-Silverstein 1993). However, their analyses failed to exclude women who had become demented (3.6% of the sample). On the other hand, two recent longitudinal, observational studies of women in the USA have reported positive findings. In the Baltimore Longitudinal Study of Aging, oestrogen-users made significantly fewer errors on the Benton Visual Retention Test which measures short-term visual memory, visual perception and constructional skills, and which was the only cognitive test administered (Resnick et al 1997). In a similar large, longitudinal, community-based study in Manhattan, New York, women who had used oestrogen-replacement therapy scored significantly higher on tests of verbal memory and verbal abstract reasoning than non-users and their performance improved slightly with increasing age (Jacobs et al 1998).

In a longitudinal investigation, we matched three groups of healthy elderly men and women on age and level of education (Carlson & Sherwin 1998). Our 72 year old men and female oestrogen-users both had higher levels of plasma oestradiol and higher scores on a test of verbal memory than the age-matched oestrogen non-users, although they did not perform better on other tests of explicit memory compared to the non-users.

Recently, we tested the oestrogen–memory relationship using a different experimental paradigm. Women between the ages of 32 and 36 years with uterine myomas (benign, oestrogen-dependent growths in the uterus) were treated with a gonadotropin-releasing hormone agonist (GnRH-a) for 12 weeks, which caused complete suppression of ovarian function (Sherwin & Tulandi 1996). Then, one-half randomly received the GnRH-a with the addition of conjugated equine oestrogen 0.625 mg daily or with the addition of a placebo daily for two additional months. Scores on tests of verbal memory decreased significantly in these women after 12 weeks of GnRH-a therapy compared to pre-treatment baseline when their ovaries were functioning normally. The memory deficit was reversed in the group given 'add-back' oestrogen, but scores remained depressed in women who received the GnRH-a plus 'add-back' placebo. These findings in young, healthy, well-functioning women provide further support for the idea that oestrogen helps to maintain verbal memory in women. They also suggest that, at least in the short-term, cognitive changes due to oestrogen deprivation are reversible with replacement therapy.

Conclusion

Overall, the findings from both experimental and observational studies that have investigated the association between oestrogen and cognitive functioning in postmenopausal women provide increasing support for the hypothesis that oestrogen helps to maintain aspects of short- and long-term memory in women. However, it is also clear that there is some inconsistency in findings which may be due to experimental confounds in the extant studies. For example, several investigations used only one or two neuropsychological tests of only one single domain of cognition and then generalized their findings to the entire realm of cognitive functions (Ditkoff et al 1991, Resnick et al 1997). In other cases, positive effects of oestrogen on specific tests were trivialized because the hormonal effect was not evident in every test in the neuropsychological battery that measured a large number of cognitive domains (Barrett-Connor & Kritz-Silverstein 1993). In fact, we have consistently found a specificity of the oestrogenic effect on neuropsychological tests that measure explicit memory and no influence of this sex hormone on other aspects of cognition such as visual or spatial memory. Indeed, the literature on the organizational effects of sex hormones on cognitive functions discussed earlier suggests that oestrogens enhance verbal skills while androgens enhance visual and spatial skills. On that basis, it would be hypothesized that, in adult life, circulating levels of oestrogen would activate the same neural pathways that are organized by that hormone prenatally. If that is true, then oestrogen would have a specific effect on verbal memory, which is what we have found in numerous studies.

Another potential confound in these studies is related to the fact that treatment of postmenopausal women who have an intact uterus requires that a progestin be added to the oestrogen replacement therapy regimen in order to protect the endometrium against the stimulatory effects of oestrogen. Although there are no studies on effects of progestins on cognition in women, these synthetic forms of progesterone are known to be neuroactive and cause involution of neuronal dendritic spines (Woolley & McEwen 1993). Several observational studies on oestrogen and cognition reviewed above failed to ensure that oestrogen-users were tested while receiving oestrogen alone.

Third, in these studies of the oestrogen–cognition relationship, many different oestrogen preparations were given to subjects in different doses and via different routes of administration. Almost no information is available regarding dose–response relationships with respect to cognitive functioning or to possible differential rates of diffusion into the brain of the large number of commercially available oestrogen preparations. Neither is it known whether oral hormone drugs which are heavily metabolized by the liver reach the brain in similar concentrations as parenterally administered preparations which bypass the so-called 'first-pass hepatic effect'. Moreover, in the majority of the studies in this area, conjugated equine oestrogen, an oral preparation with numerous metabolites, was the drug being consumed and it is not known whether some of these metabolites not contained in other oestrogen preparations may be neuroactive. Thus, it is likely that the human studies on oestrogen and cognition reviewed above are not comparable, albeit in unknown ways.

It is also important to consider that oestrogen helps to maintain mood and can even alleviate dysphoric symptoms in postmenopausal women (Sherwin 1988) It is likely that oestrogen accomplishes this enhancement of mood by up-regulating the serotonergic system (Sherwin & Suranyi-Cadotte 1990). Therefore, in any investigation of oestrogen and cognition, it would be important to also measure mood so that mood scores can be used as a covariate in analysis of the cognitive data in order to ensure that any hormonal effect on cognition did not occur secondary to an enhancement of mood. Only a small percentage of the above studies concurrently measured mood in their subjects at the time of neuropsychological testing.

Of course, observational and cross-sectional studies are inherently biased since medication is not randomly assigned to the oestrogen and to the no-treatment groups. This is a significant concern in studies involving oestrogen replacement therapy since it has been well established that women who take hormone replacement therapy after the menopause constitute approximately 20% of the eligible population and tend to be more highly educated women from higher socioeconomic groups. Since more education and higher socioeconomic status are themselves protective against cognitive decline with ageing, this is a serious confound in studies that are attempting to attribute causality to the oestrogenic prevention of cognitive deterioration with ageing. Ensuring that treated and untreated subjects are matched on these relevant variables helps to mitigate these confounds but it is unlikely that this strategy negates them entirely.

Obviously, one way to control for many of the experimental confounds in the existing literature is to conduct a large, randomized, clinical trial of oestrogen in older women. The Women's Health Initiative Memory Study (WHIMS) currently underway in the USA is a double-blind, randomized, placebo-controlled, long-term clinical trial of oestrogen or placebo in women aged 65 or older at the time of recruitment (Shumaker et al 1998). In this multicentre trial, women will be tested and followed for four to six years in order to test the hypothesis that oestrogen replacement therapy reduces the incidence of Alzheimer's disease, whose classical symptom is memory deficits. Although not a stated goal, findings from this study will likely also shed light on differential deterioration in memory that may occur between the oestrogen and the placebo groups in this ageing population across time. Ultimately, the clinical implications of this line of research are to prevent deterioration of memory in older women so as to protect their quality of life during the latter one-third of their lifespan.

Acknowledgements

The preparation of this manuscript was supported by a grant from The Medical Research Council of Canada (No. MT-11623) awarded to B. B. Sherwin.

References

Anderson ET 1972 Cognitive performance and mood change as they relate to menstrual cycle and estrogen level. Dissert Abstr Int 33:1758B

Barrett-Connor E, Kritz-Silverstein D 1993 Estrogen replacement therapy and cognitive function in older women. JAMA 269:2637–2641

Carlson LE, Sherwin BB 1998 Steroid hormones, memory and mood in a healthy elderly population. Psychoneuroendocrinology 23:583–603

Ditkoff EC, Crary WG, Cristo M, Lobo RA 1991 Estrogen improves psychological function in asymptomatic postmenopausal women. Obstet Gynecol 78:991–995

Gorski RA 1996 Androgens and sexual differentiation of the brain. In: Bhasin S, Gabelnick HL, Spieler JM, Swerdloff RS, Wang E (eds) Pharmacology, biology and clinical applications of androgens. Wiley-Liss, New York, p 159–168

Hampson E 1990 Variations in sex-related cognitive abilities across the menstrual cycle. Brain Cogn 14:26–43

Hampson E, Kimura D 1988 Reciprocal effects of hormonal fluctuations on human motor and perceptual–spatial skills. Behav Neurosci 102:456–459

Jacobs DM, Tang MX, Stern Y et al 1998 Cognitive function in nondemented older women who took estrogen after menopause. Neurology 50:368–373

Jarvik LF 1975 Human intelligence: sex differences. Acta Genet Med Gamellol (Roma) 24:189–211

Kampen DL, Sherwin BB 1994 Estrogen use and verbal memory in healthy postmenopausal women. Obstet Gynecol 83:979–983

Kimura D 1995 Estrogen replacement therapy may protect against intellectual decline in postmenopausal women. Horm Behav 29:312–321

Komenich P, Lane DM, Dickey RP, Stone SC 1978 Gonadal hormones and cognitive performance. Physiol Psych 6:115–120

Lipsett MB 1986 Steroid hormones. In: Yen SSC, Jaffe RB (eds) Reproductive endocrinology, physiology, pathophysiology and clinical management, 2nd edn. Saunders, Philadelphia, PA, p 140–153

Nass R, Baker S 1991 Androgen effects on cognition: congential adrenal hyperplasia. Psychoneuroendocrinology 16:189–201

Phillips SM, Sherwin BB 1992a Variations in memory function and sex steroid hormones across the menstrual cycle. Psychoneuroendocrinology 17:497–506

Phillips SM, Sherwin BB 1992b Effects of estrogen on memory function in surgically menopausal women. Psychoneuroendocrinology 17:485–495

Polo-Kantola P, Portin R, Polo I, Helenius H, Irjala K, Erkkola R 1998 The effect of short-term estrogen replacement therapy on cognition: a randomized, double-blind, cross-over trial in postmenopausal women. Obstet Gynecol 91:459–466

Resnick SM, Berenbaum S, Gottesman I, Bouchard T 1986 Early hormonal influences on cognitive functioning in congenital adrenal hyperplasia. Dev Psychol 22:191–198

Resnick SM, Metter J, Zonderman AB 1997 Estrogen replacement therapy and longitudinal decline in visual memory. A possible protective effect? Neurology 49:1491–1497

Robinson D, Friedman L, Marcus R, Tinklenberg J, Yesavage J 1994 Estrogen replacement therapy and memory in older women. J Am Geriatr Soc 42:919–922

Sherwin BB 1988 Affective changes with estrogen and androgen replacement therapy in surgically menopausal women. J Affect Disord 14:177–187

Sherwin BB, Suranyi-Cadotte B 1990 Up-regulatory effect of estrogen on platelet ^{3}H-imipramine binding sites in surgically menopausal women. Biol Psychiatry 28:339–348

Sherwin BB, Tulandi T 1996 'Add-back' estrogen reverses cognitive deficits induced by a
    gonadotropin releasing-hormone agonist in women with leiomyamata uteri. J Clin
    Endocrinol Metab 81:2545–2549
Shumaker SA, Reboussin BA, Espeland MA et al 1998 The Women's Health Initiative Memory
    Study (WHIMS): a trial of the effect of estrogen therapy in preventing and slowing the
    progression of dementia. Control Clin Trials 19:604–621
Wickham M 1958 Effects of the menstrual cycle on test performance. Brit J Psychol 49:34–41
Woolley CS, McEwen BS 1993 Roles of estradiol and progesterone in regulation of hippocampal
    dendritic spine density during the estrous cycle in the rat. J Comp Neurol 336:293–306

DISCUSSION

Levin: What is the neuroanatomical correlate with verbal memory?

Sherwin: It is hippocampally dependent memory, involving the medial temporal lobe.

Levin: Is there one specific cell type or region involved?

Resnick: It really depends on which aspect of memory you are looking at.

Hurd: Going back to yesterday's discussion about mood and depression, I have a question: did the young women that you studied show mood disorders before and after the oestrogen treatment?

Sherwin: We did measure mood in that study: I would never do a study on cognition and not measure mood, because one could argue, justifiably, that changes in cognition may have occurred secondary to changes in mood. Their mood dipped somewhat during the GnRH treatment. It wasn't clinically significant, but their mood scores were somewhat higher during the ovarian suppression phase of the study. I used those scores as a covariate when I was doing analysis of variance and looking at the cognitive data, and found that mood did not influence cognitive scores significantly.

Gustafsson: With reference to the effects on memory, it will be interesting to consider what happens when oestrogen antagonists, such as raloxifene and tamoxifen, are given. This is becoming increasingly common. These are given because they are supposed to prevent breast cancer and are good for bones. Raloxifene does not act on the uterus, but effects on the CNS are much less well considered. For instance, raloxifene gives rise to increased frequency of hot flushes, which are supposed to be a CNS-type of phenomenon. If anything, you would expect that these drugs would act on the CNS as an anti-oestrogen. In this case, your data are of great interest, because raloxifene treatment may cause memory loss.

McEwen: The anti-oestrogen CI-628 does block oestrogen-induced synapse formation (McEwen et al 1999).

Sherwin: I think it's becoming very much part of the public debate, and researchers are going to have to catch up and provide data. There is not much evidence one way or another right now on whether or not raloxifene acts as an oestrogen agonist or antagonist on cognitive functioning in women. My own guess would be that raloxifene might be acting as an anti-oestrogen, and thus dampening some cognitive functions.

Toran-A llerand: You showed that women who had an hysterectomy without the bilateral oopherectomy performed quite normally compared with women who had had hysterectomy with or without ovariectomy. I am aware that gynaecologists like to combine removal of the uterus with removal of the ovaries, but wouldn't this be an argument for saying that unless there was a medical reason to remove the ovaries, they should be left?

Sherwin: Most gynaecologists I know would say that it's a non-issue, because after surgery the women will receive oestrogen. But I don't think it's a non-issue at all, and I'm happy to say that surgical removal of healthy ovaries is being done much less routinely these days. Retained ovaries in premenopausal women who have had a hysterectomy are functional and our findings indicate that the oestrogen they continue to produce protects memory.

Baulieu: I'm interested in the relative contribution of peripheral oestrogens and oestrogens made in the brain. You alluded to this when you spoke about men. Clearly, testosterone of testicular origin is an important contribution. But there are also other C19 steroids such as dehydroepiandrosterone (DHEA) sulfate circulating in the blood in both sexes and thus in women including after menopause. What is your opinion of the relative contribution of the two sources at different ages in the two sexes? Do you have overlapping data comparing Δ5-androstenediol $3\beta,17\beta$ and DHEA upon cognition in women?

Sherwin: Yes, in that older cohort (the 72 and 74 year olds) we also measured cortisol and DHEA sulfate. High levels of cortisol cause hippocampal atrophy and cognitive deficits. We found that the women whose cortisol levels increased over time did less well on tests of cognitive functioning, whereas those whose levels remained stable didn't have any change in performance over time. DHEA sulfate levels were, of course, higher in the men, but they were the same in both groups of women whether they were oestrogen users or not, and they didn't correlate with any scores on tests of cognitive function in our own study.

Baulieu: What about in the female non-users?

Sherwin: DHEA sulfate levels were not different between our healthy elderly oestrogen users and non-users. I didn't measure androstenediol. What prompted me to even measure those steroids was a report in the Alzheimer's literature that women with higher levels of DHEA sulfate seem to perform better on some tests of cognition than women with lower levels of DHEA (Sunderland et al 1989). But we failed to confirm that finding.

Gibbs: People are always asking me why don't I give testosterone to my female rats. What is the role of testosterone on cognitive functioning?

Sherwin: We really do not know the potential role of testosterone on cognitive functioning. There are no studies at all that address the possible roles of testosterone on cognition in women, and this needs to be done. All the studies that I have described also need to be done in men. The effects of oestradiols and/or testosterone on cognition in ageing men are largely unknown.

Gibbs: Have you looked at attentional processes, since this will affect performance on these behaviours? I'm interested in that question because the cholinergic system is also involved in attention.

Sherwin: Attention is definitely affected by oestrogen. Digit span is a good test of attention, and this is influenced in some studies by oestrogen. This is a test when subjects are asked to recall a progressively longer series of numbers. Forward digit span, which measures attention, wasn't affected in many of my own studies. Reverse digit span, which really does require more working memory, is positively influenced by oestrogen in my lab studies.

Resnick: With reference to DHEA, we have a paper in press in which we examined the association between plasma levels of DHEA sulfate and cognition in participants in the Baltimore Longitudinal Study of Aging (BLSA). Since DHEA sulfate levels had been assayed through a BLSA study of prostate disease, we were able to examine associations between longitudinal change in DHEA sulfate and longitudinal change in cognition, as well as mean levels. Controlling for a number of factors, we found no relationship between DHEA sulfate and a battery of cognitive tests in men.

Baulieu: How old were your men?

Resnick: They were aged 50–90 for most of the measures. We analysed the data in several ways, and we found absolutely no evidence for any beneficial effects of DHEA sulfate on cognition.

As one of the people who has found some differences between oestrogen replacement therapy (ERT) users and non-users on visual memory tests, I wanted to make the point that these effects are small and are only seen with fairly large samples. Pauline Maki, a National Research Council Fellow in our group, and I have recently concluded a study of verbal memory, using the California Verbal Learning Test, which assesses memory for a 16-item shopping list. On that test, enhanced performance in active ERT users compared to never-treated women was much greater than the types of effects we typically observe on the visual memory test.

The ERT effects on memory that we observed in this study involved both the initial encoding and learning of the items and retrieval from memory after a delay. However, there was no difference between groups in the amount forgotten between a short and longer delay. The specific types of memory affected by ERT may have some relevance to the question that was asked before, because it is the initial encoding phase which is more hippocampally mediated, whereas many of the more recent imaging studies have suggested that the right frontal lobe shows the greatest activation during retrieval of information.

Luine: Some of our recent data on oestrogen treatment of ovariectomized rats mirrors this idea about working memory. If we use either a spatial or a non-spatial memory task, oestrogen enhances performance to a greater degree when delays are given within the task (i.e. more working memory required). Interestingly, oestrogen appears more effective in the non-spatial (frontal cortex) than in spatial tasks (hippocampal). In object recognition (non-spatial), oestrogen enhances performance following a 24 h delay, but in a radial arm maze task (spatial) it enhances performance only up to a 5 h delay. I don't know whether this difference has to do with encoding or the different parts of the brain where oestrogen may be acting.

Resnick: In the studies Pauline Maki and I have done using the California Verbal Learning Test, we found that oestrogen appeared to offer some benefit in terms of development of a clustering strategy. This is one way in which information can be encoded, and oestrogen may be affecting the type of processing that's going on.

Sherwin: One of the problems in this area is that human researchers are making the assumption that an oestrogen is an oestrogen. There are many commercial oestrogens being used. Premarin is the most frequently used oestrogen in the world, and has 26 metabolites. Added to this, there are many different routes of administration, including transdermal, oral and injectable. No one really knows what different effects on the brain these may be having.

Watson: Does anyone know about the cognitive effects of dietary phytooestrogens that differ between cultures?

Sherwin: I think there are some studies underway, but we don't yet have any data.

Henderson: The Honolulu Heart Study included an analysis of Japanese–American men who had diets with varying amounts of tofu (White et al 1996). The investigators reported an association between greater amounts of tofu and an elevated risk of cognitive decline and dementia. A second observation is from a different Japanese–American cohort in the state of Washington. Here, oestrogen therapy was thought to protect against Alzheimer's disease, except among a subgroup of women who consumed large amounts of tofu, suggesting that there might have been a blocking effect (Rice et al 1995). On the other hand, Dr Tom Clarkson's recent data in non-human primates suggests soy phytooestrogens have a beneficial effect on cognitive skills.

Gibbs: You mentioned that the verbal memory task is a hippocampal task, and spatial memory is also a hippocampal task, but you get very different effects of oestrogens on one versus another. I'm not used to thinking about two hippocampal tasks being affected differently. Do you know of a neurobiological substrate or have structural information that might help to differentiate between the tasks and could underlie the different effects?

Murphy: There may be different cell types to regulate those activities. There is a theoretical hippocampal 'place cell' which actually fires a particular 'place field' as an animal learns its spatial environment. If it is indeed a different kind of cell within the hippocampus, this could be one explanation.

Gibbs: In the animal literature, people talk about two memory systems, hippocampal memories and non-hippocampal memories, but I never hear people talk about memories that are associated with specific cell types in the hippocampus.

Becker: The role of the hippocampus in humans is clearly different from the role of the hippocampus in rats. In rats, some of the neurons in the hippocampus fire when an animal moves through a particular location in space, and these cells have been called 'place cells'. In humans and primates, neurons in the hippocampus do not seem to serve this same function. There is PET evidence of increased blood flow in the right hippocampus during object location or navigation tasks (Johnstrude et al 1999). However, the hippocampus in humans is not dedicated to solving spatial tasks. The current view of hippocampal function in learning and memory in humans, as I understand it, is that the hippocampus is necessary for the formation of long-term declarative memories (which may or may not be spatial), but not for the retrieval of these memories (Teng & Squire 1999).

Resnick: The other thing to consider is that the types of tests that have typically been used in studies of oestrogen effects are not spatial memory tests in the way you think about them in the rat. They are figural memory tests, or memory for designs. The types of tests used so far don't require women to do any navigation.

Henderson: Particular types of 'new learning' involve hippocampus, parahippocampal regions and cerebral neocortex. In the neocortex, there are well-known left–right differences, with the left hemisphere, for example, being primarily concerned with certain verbal tasks. To the extent that memory encoding, storage, and retrieval involve reciprocal connects with other parts of the brain, a task that depends on hippocampus may show lateralized effects that are not really hippocampally dependent.

McEwen: There was an fMRI study in London cab drivers recalling the route from one place to another, showing that one hippocampus becomes activated during the recall (Maguire et al 1997).

Wolf: That was the right hippocampus, and there's actually some evidence that the left hippocampus might be more important for verbal memory and the right one for spatial memory. It might be that oestradiol effects are more on the left hippocampus.

Resnick: I don't think any of us have yet used truly parallel tests to compare figural and verbal memory. The tests used often involve different types of input (e.g. visual versus auditory) and different types of responses. They require different types of processing. We have some tests that we have developed which are parallel but we haven't used them in a large-scale study yet.

Woolley: The hippocampus is a large and complicated structure. I don't see any intrinsic reason to think that all hippocampus-dependent functions need to change in parallel.

McEwen: If it is true that with oestrogen treatment there is an impairment of some aspects of spatial ability and an improvement of some aspects of declarative memory, it may be that it is serving as a switch to favour a certain kind of memory.

References

Johnsrude IS, Owen AM, Crane J, Milner B, Evans AC 1999 A cognitive activation study of memory for spatial relationships. Neuropsychologia 37:829–841

Maguire EA, Frackowiak RSJ, Frith CD 1997 Recalling routes around London: activation of the right hippocampus in taxi drivers. J Neurosci 17:7103–7110

McEwen BS, Tanapat P, Weiland NG 1999 Inhibition of dendritic spine induction on hippocampal CA1 pyramidal neurons by a nonsteroidal estrogen antagonist in female rats. Endocrinology 140:1044–1047

Rice MM, Graves AB, Larson EB 1995 Estrogen replacement therapy and cognition: role of phytoestrogens. Gerontologist (suppl 1) 35:169(abstr)

Sunderland T, Merril CR, Harrington MG et al 1989 Reduced plasma DHEA concentrations in Alzheimer's disease. Lancet 2:570

Teng E, Squire LR 1999 Memory for places learned long ago is intact after hippocampal damage. Nature 400:675–677

White L, Petrovitch H, Ross GW, Masaki K 1996 Association of mid-life consumption of tofu with late life cognitive impairment and dementia: the Honolulu–Asia Study. Neurobiol Aging (suppl) 17:S121(abstr)

Neuroprotective effects of phenolic A ring oestrogens

Pattie S. Green*, Shao-Hua Yang*‡, and James W. Simpkins*†[1]

*Center for the Neurobiology of Aging, †Department of Pharmacodynamics, and ‡Department of Neurosurgery, PO Box 100487, University of Florida, Gainesville, FL 32610, USA

Abstract. We have formulated the 'phenolic A ring hypothesis' for the neuroprotective effects of oestrogens based upon several observations: (i) structure–activity relationships show that a phenolic A ring and at least two additional rings are required for neuroprotection while oestrogenicity requirements are more stringent; (ii) neuroprotection with phenolic A ring compounds occurs in cells that lack oestrogen receptors and are not antagonized by anti-oestrogens; (iii) phenolic A ring compounds rapidly activate a variety of signal transduction pathways that are known to be involved in cell homeostasis; and (iv) *in vivo*, treatment with oestrogens results in a neuronal type-independent neuronal protection from ischaemic insult. Potential mechanisms of actions that may be involved in the neuroprotective effects of phenolic A ring compounds are: (i) oestrogen redox cycling that potently inhibits oxidative stress; (ii) interactions with signal transduction pathways including the transcription factor cAMP response element binding protein; and (iii) induction of anti-apoptotic proteins. These signalling pathways may individually or collectively contribute to the plethora of neuronal cell types that are protected from a variety of insults by oestrogen-like compounds.

2000 Neuronal and cognitive effects of oestrogens. Wiley, Chichester (Novartis Foundation Symposium 230) p 202–220

Postmenopausal oestrogen replacement therapy is associated with several neurological benefits including a reduction in the risk of Alzheimer's disease, improved cognitive performance of Alzheimer's patients and a reduction in stroke-related mortality. These effects of oestrogens may be mediated, in part, by direct enhancement of neuronal survival.

We first described the direct neuroprotective effects of the potent oestrogen, 17β-oestradiol (βE2), on human SK-N-SH neuroblastoma cells under conditions of serum-deprivation (Bishop & Simpkins 1994). In this model, serum-deprivation reduces neuronal viability by approximately 80% and concurrent

---

[1]This chapter was presented at the symposium by James W. Simpkins, to whom correspondence should be addressed.

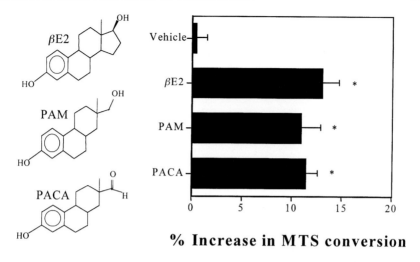

% Increase in MTS conversion

FIG. 1.   The potent oestrogen, 17β-oestradiol (βE2), and both PAM and PACA, which represent the A-B-C rings of oestradiol and oestrone structure, respectively, protect SK-N-SH cells from the toxic effects of serum-deprivation. The compounds (2 nM) were added concurrent with onset of serum-deprivation and MTS conversion was determined 48 h later. Data are expressed as mean ± SEM for 14–22 wells per group. * = P < 0.05 versus vehicle.

treatment with βE2 (2 nM) significantly increases neuronal viability (Fig. 1; Bishop & Simpkins 1994, Green et al 1997a,b). This is not a mitogenic effect of the steroid as this concentration of βE2 does not increase [³H]-thymidine uptake in these cells (Bishop & Simpkins 1994). Since this first report, more than 25 reports have similarly described the neuroprotective effects of βE2 in a variety of neuronal cell types (Table 1). βE2 can attenuate the toxicity associated with serum-deprivation (Fig. 1), exposure to amyloid β peptide (Aβ) (Fig. 2; Behl et al 1995, 1997, Goodman et al 1996, Green et al 1996, 1998a, Mook-Jung et al 1997, Bonnefont et al 1998, Roth et al 1999), excitatory amino acid exposure (Singer et al 1996, 1999, Weaver et al 1997, Zaulynov et al 1999), and oxidative stress (Behl et al 1995, 1997, Blum-Degen et al 1998, Singer et al 1998). The concentrations of βE2 required for neuroprotection range from a low near 10 pM (Bishop & Simpkins 1994, Green et al 1996, 1997a, Brinton et al 1997) to a high of 50 μM (Weaver et al 1997, Sagara 1998). This variation in neuroprotective concentrations may relate to the neuronal type used, the presence of glutathione in the culture media (Green et al 1998a), duration of oestrogen pretreatment (Mook-Jung et al 1997), or the severity and type of insult. Interestingly, evidence is accumulating to suggest that the neuroprotective effects of oestrogens do not require the participation of a classic nuclear oestrogen receptor (ER) mechanism.

TABLE 1 Reported neuroprotective effects of β-oestradiol in neuronal cultures

Toxicity	Neuronal type	References
Serum deprivation	SK-N-SH	Bishop & Simpkins 1994, Green et al 1997a,b
Aβ peptide	HT-22	Behl et al 1995, 1997, Green et al 1998a
	SK-N-SH	Green et al 1996
	Primary rat hippocampal	Goodman et al 1996, Pike 1999
	PC-12	Mattson et al 1997, Bonnefont et al 1998, Roth et al 1999
	B103	Mook-Jung et al 1997
	NT-2a	Bonnefont et al 1998
Glutamate	Primary rat neocortical	Singer et al 1996, 1999, Zaulynov et al 1999
	Primary rat hippocampal	Goodman et al 1996, Weaver et al 1997
	Primary murine hippocampal	Regan & Guo 1997
	Primary mesencephalic	Sawada et al 1998
	PC-12	Bonnefont et al 1998
	NT-2	Singer et al 1998
Anoxia	Primary rat neocortical	Zaulynov et al 1999
	Primary murine hippocampal	Regan & Guo 1997
Hypoglycaemia	Primary rat hippocampal	Goodman et al 1996
Age in culture	Primary rat neocortical	Brinton et al 1997
GP120	Primary rat neocortical	Brooke et al 1997
	Primary rat hippocampal	Brooke et al 1997
Haloperidol	HT-22	Sagara 1998
	Primary rat neocortical	Sagara 1998
Hydrogen peroxide	HT-22	Behl et al 1995, 1997
	Primary rat neocortical	Behl et al 1997
	IMR-32	Blum-Degan et al 1998
	PC-12	Bonnefont et al 1998
	NT-2	Singer et al 1998
	SK-N-MC	Moosmann & Behl 1999

Oestrogen structure–neuroprotective activity relationship

We (Green et al 1996, 1997a,b) found that the weak oestrogen, 17α-oestradiol (αE2), was similar in neuroprotective efficacy and potency to βE2 (Table 2). This led us to evaluate the structure–activity relationship (SAR) for oestrogen-mediated neuroprotection in serum-deprived SK-N-SH neuroblastoma. Several oestrogens with D-ring modifications were similar to βE2 in neuroprotective activity at a 2 nM concentration including αE2, 1,3,5(10)-oestratriene-3-ol (E-3-ol), oestrone,

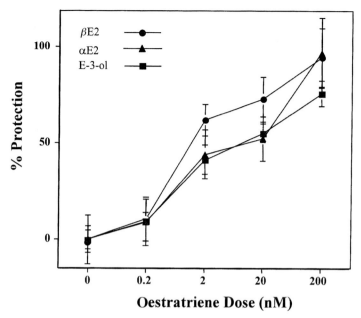

Oestratriene Dose (nM)

FIG. 2.   17β-oestradiol (βE2), 17α-oestradiol (αE2), and oestratriene-3-ol (E-3-ol) attenuate Aβ (25–35)-induced toxicity in HT-22 cells. Cells were exposed to 20 μM Aβ in the presence of the indicated dose of steroid for 48 h prior to viability assessment. Data were normalized to the Aβ-free control group as 100% and the Aβ-alone group as 0% protection. Data are presented as mean ± SEM for 3–5 wells per group.

TABLE 2   Comparison of relative neuroprotective effect and relative potencies in ER binding, uterotrophic growth stimulation and MCF-7 cell proliferation

| Steroid | % of 17β-oestradiol (βE2) activity | | | |
	Neuroprotection[a]	RBA	Uterotrophic[b]	MCF-7 Proliferation[c]
βE2	100	100	100	100
αE2	89	49[b]	1	<1
E-3-ol	102	79[c]	ND	1
Oestrone	88	66[b]	29	2
Oestriol	75	16[b]	1	3

% of βE2 neuroprotection refers to percentage protection of serum-deprived SK-N-SH cells with a 2 nM concentration. RBA, uterotrophic growth stimulation and MCF-7 cell proliferation are reported as relative potencies. RBA, relative binding affinity; ND, not determined.
[a]Green et al 1997b; [b]Korenman 1969; [c]Wiese et al 1997.

oestriol and ethynyl oestradiol (Green et al 1997a,b). However, abolition of the phenolic nature of the A ring by replacement of the 3-positioned hydroxyl function with a methyl ether function renders the βE2 inactive in this assay even at concentrations as high as 20 nM. Similarly all 3-O-methyl ether-substituted oestrogens tested were inactive as was cholesterol which possesses a hydroxy function in the 3 position but has a fully saturated A ring. All steroids without phenolic A rings, including testosterone, dihydrotestosterone, progesterone, corticosterone, aldosterone, prednisolone and 6-methylprednisone, lacked neuroprotective activity at the low nanomolar concentrations tested. The importance of the phenolic A ring is further illustrated by the diphenolic oestrogen mimic, diethylstilbesterol (DES), which was a potent neuroprotective agent in this assay. DES retained nearly full neuroprotective activity when one, but not both, of the phenolic hydroxy functions were replaced with a methyl ether group.

To assess the requirement for the steroid ring structure, we further evaluated phenol and 5,6,7,8-tetrahydronapthol which represent the A and A-B rings of the oestratriene structure, respectively. Phenol was toxic, resulting in an enhancement of the toxicity and 5,6,7,8-tetrahydronapthol had no effect on live cell number. The two lipophilic phenols, butylated hydroxytoluene and butlyated hydroxyanisol, were inactive in this assay system. However, octahydro-7-hydroxy-2-methyl-2-phenanthrenementhanol (PAM) and octahydro-7-hydroxy-2-methyl-2-phenanthrenecarboxaldehyde (PACA) which represent the A-B-C rings of oestradiol and oestrone, respectively, both attenuate serum-deprivation toxicity in SK-N-SH cells (Fig. 1).

From these data, we formulated the 'phenolic A ring hypothesis' of oestrogen neuroprotection; a phenolic A ring and at least three rings of cyclophenanthrene structure are required for low dose neuroprotective effects of oestrogens. Since, we and others have found that phenolic A ring oestrogens with weak ER agonist activity, such as 17α-oestradiol (αE2), are nevertheless similar in neuroprotective potency to βE2 (Table 2) in a variety of neurotoxicity paradigms including not only serum-deprivation (Green et al 1997a,b), but Aβ toxicity (Fig. 3, Green et al 1996, 1998a, Behl et al 1997, Roth et al 1999), excitotoxicity (Behl et al 1997, Sawada et al 1998, Zaulynov et al 1999), anoxia (Zaulynov et al 1999) and oxidative stress (Behl et al 1997, Blum-Degen et al 1998, Sawada et al 1998).

Oestrogen protection in absence of nuclear oestrogen receptors

The SAR for oestrogen-mediated neuroprotection differs markedly from the SAR for ER activation suggesting that an ER is not required for the neuroprotective effects of oestrogens. This is supported by the observation that ER antagonists does not attenuate oestrogen-mediated neuroprotection in SK-N-SH cells (Green et al

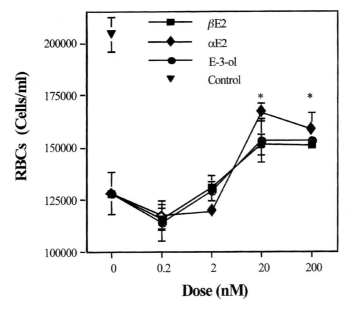

FIG. 3. Oestratrienes attenuate FeCl₃-induced lysis in erythrocytes (RBCs). RBCs were exposed to 200 μM FeCl₃ in the presence of the indicated dose of either βE2, αE2 or E-3-ol. Data are presented as mean $\pm$ SEM for 3–6 samples. $* = P < 0.05$ versus no steroid.

1997a). Further, neither SK-N-SH or HT-22 cells, both of which are protected by oestrogens, demonstrate specific [³H]-βE2 binding (Green et al 1997b, 1998a).

To further evaluate the ER/genomic independence of the protection provided by oestrogens, we evaluated the cytoprotective effects of βE2 and two non-feminizing, phenolic A ring oestrogens, αE2 and E-3-ol, in erythrocytes (RBCs). A 4 h exposure to 200 μM FeCl₃ resulted in lysis of 35–44% of RBCs from three different subjects (Fig. 3). Co-treatment with 20 nM βE2, αE2 and E-3-ol significantly attenuated the FeCl₃-induced decrease in RBC number in each experiment, protecting an average of 44%, 45% and 44% of the cells, respectively. Higher concentrations of oestrogens did not increase the magnitude of the protection. Oestrogen-mediated protection of RBCs required 10- to 100-fold higher doses that oestrogen-mediated protection of neuronal (Bishop & Simpkins 1994, Green et al 1996, 1997a, Singer et al 1996, Brinton et al 1997, Mook-Jung et al 1997), glial (Bishop & Simpkins 1994) and endothelial cells (Shi et al 1997). As RBCs are anuclear, these results portend that a nucleus is not required for a cytoprotective effect of oestrogens. However, the higher concentrations of oestrogens required to protect RBCs suggest that nuclear events may contribute to the cytoprotective effects of the steroid.

Oestrogen neuroprotection in animal models of focal ischaemia

In culture models, phenolic A ring oestrogens attenuate the toxic effects of excitotoxicity, anoxia/reoxygenation and oxidative stress, suggesting a role for circulation oestrogens in reducing neuronal death associated with cerebral ischaemia. In rats, middle cerebral artery (MCA) occlusion causes a clinically relevant cerebral ischaemia with neuronal death focused in the parietal cortex and basal ganglia. In a transient MCA occlusion model, ovariectomy increases post-operative mortality to 24% from 12% in intact female rats (Zhang et al 1998). Pretreatment with either βE2 or the non-feminizing oestrogen, αE2, reduces MCA occlusion-related mortality by 46 and 36%, respectively (Simpkins et al 1997).

A reduction in the ischaemic lesion volume may underlie this remarkable reduction in mortality (see Table 3). Pretreatment (1 week to 2 hours) with βE2 reduces ischaemic lesion size by 45–66% (Simpkins et al 1997, Shi et al 1997, Zhang et al 1998) in ovariectomized female rats and by 53% in intact male rats (Hawk et al 1998). Similar reductions in lesion area with βE2 pretreatment in ovariectomized female rats have been reported by Dubal et al (1998) and Toung et al (1998). Treatment with αE2 similarly reduces lesion area by an average of 55% in two separate studies (Simpkins et al 1997, Stubley et al 1998). Further, *ent*-oestradiol,

TABLE 3   Reported neuroprotective effects of oestrogens in focal ischaemia model

Steroid and administration	Model type	Percentage lesion reduction	References
βE2 prior to ischaemia	Temporal MCAO in OVX females	45–66%	Simpkins et al 1997 Shi et al 1997 Zhang et al 1998
	Permanent MCAO in OVX females	60% cortical	Dubal et al 1998
	Temporal MCAO in males	53% 51–69% cortical	Hawk et al 1998 Toung et al 1998
βE2 30 min to 3 h post-ischaemia	Permanent MCAO in OVX females	38–71%	Yang et al 1999
αE2 prior to ischaemia	Temporal MCAO in OVX females	55%	Simpkins et al 1997 Stubley et al 1998
Ent-oestradiol prior to ischaemia	Temporal MCAO in OVX females	60%	P. S. Green, S.-H. Yang, K. R. Nilsson, A. S. Kumar, D. F. Covey & J. W. Simpkins, unpublished observations

MCAO, middle cerebral artery occlusion; OVX, ovariectomized.

the non-feminizing enantiomer of βE2, results in a 60% reduction in ischaemic lesion area (P. S. Green, S.-H. Yang, K. R. Nilsson, A. S. Kumar, D. F. Covey & J. W. Simpkins, unpublished observations). The protective effects of these non-feminizing, phenolic A ring oestrogens suggest that the SAR for neuroprotection in culture models can be extended to animal models of cerebral ischaemia.

Physiological circulating concentrations of βE2 (10–75 pg/ml) are sufficient to confer neuronal protection from focal ischaemia (Simpkins et al 1997, Dubal et al 1998, Yang et al 1999). Supraphysiological plasma concentrations of βE2 do not increase the extent of neuronal protection observed, as the same degree of lesion reduction was seen using supraphysiological concentrations (greater than 1000 pg/ml) as physiological concentrations (about 50 pg/ml) of βE2 (Simpkins et al 1997). In fact, Dubal et al (1998) reports near maximal lesion reduction with 20 pg/ml plasma βE2 concentration.

Oestrogens do not appear to abate ischaemic damage by attenuating the occlusion-induced decrease in cerebral blood flow (CBF). Neither βE2 nor αE2 diminish the extent of decline or the extent of recovery in CBF due to temporary MCA occlusion (Shi et al 1998, Stubley et al 1998) although oestrogen treatment did improve the rate of CBF recovery (Shi et al 1998). This is consistent with MRI studies revealing that oestrogens may be exerting their neuroprotective effect during reperfusion rather than during the ischaemic period (J. Shi, J. D. Bui, S.-H. Yang, D. L. Buckley, S. J. Blackband, M. A. King, A. L. Day & J. W. Simpkins, unpublished observations).

βE2 is effective at reducing lesion size when administered either prior to (Simpkins et al 1997, Shi et al 1997, 1998, Toung et al 1998) or after (Simpkins et al 1997, Zhang et al 1998) onset of transient MCA occlusion. βE2-CDS is effective at reducing lesion area when administered up to 90 minutes after onset of occlusion. In models of permanent MCA occlusion, βE2 is protective at time points as much as 3 h post occlusion onset (Yang et al 1999). The effectiveness of βE2 post-treatment in these models adumbrates a non-genomic mechanism of action.

Potential mechanisms of the neuroprotective activity of phenolic A ring oestrogens

As outlined earlier, a classic nuclear ER mechanism of action is not required for the neuroprotective effects of oestrogens. Although ER-mediated gene transcription may contribute to βE2-mediated neuroprotection, it is unlikely to contribute to the neuroprotection conferred by non-feminizing oestrogens.

Interestingly, the SAR for oestrogen-mediated neuroprotection is identical to the SAR for antioxidant activity of the steroid; however, although phenolic A

100 nM βE2

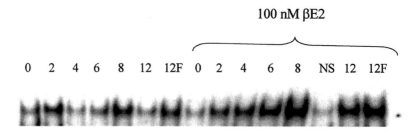

0   2   4   6   8   12   12F   0   2   4   6   8   NS   12   12F

FIG. 4.   Effects of serum deprivation and 17β-oestradiol treatment on CRE binding activity in SK-N-SH neuroblastoma cells. SK-N-SH cells were serum-deprived for the indicated length of time and the nuclear protein analysed by EMSA analysis. 12F represents SK-N-SH cells which were retained in the presence of 10% FBS for the entire incubation period, and NS represents non-specific binding as determined by an excess of unlabelled CRE oligonucleotides. Results shown are representative of four individual experiments.

ring oestrogens are potent antioxidants, these effects require micromolar concentrations of the steroid (Sugioka et al 1987). We have found that a sub-protective concentration of glutathione (around $3 \mu M$) in culture media interacts synergistically with phenolic A ring oestrogens to increase the neuroprotective potency by approximately 400-fold (Green et al 1998a). We hypothesize that this synergistic interaction is due to redox cycling of phenolic A ring oestrogens with glutathione which would increase the antioxidant potency of the oestrogens.

Several studies indicated that βE2 induces phosphorylation of cAMP response element binding protein (CREB) in neuronal cells (Zhou et al 1996, Green et al 1998b). Similarly, we have found that both low concentrations (10 nM) of both βE2 and αE2 increases CREB phosphorylation by about 50% within one hour of exposure in SK-N-SH neuroblastoma cells. While the importance of CREB phosphorylation in oestrogen-mediated neuroprotection remains to be elucidated, studies indicates that a decline in CREB levels precedes neuronal death in ischaemic injury (Walton et al 1997) and hypoglycaemic seizure (Panickar et al 1997). Similarly, we have recently examined the effect of serum-deprivation of both CREB and phosphorylated CREB levels in SK-N-SH cells. At 4–6 h of serum-deprivation, these cells showed a decline in both phosphorylated CREB immunoreactivity and CRE binding capacity (Fig. 4). No loss of cell viability as determined by propidium iodide exclusion was detected at these time points. Interestingly, a 100 nM dose of βE2 attenuated the serum-deprivation induced decline in CRE binding.

Non-feminizing, phenolic A ring oestrogens have many other cellular effects which may contribute to neuroprotective efficacy. We have found that αE2 attenuates hydrogen peroxide-induced NF-κB activation in a manner similar to βE2 (P. S. Green & J. W. Simpkins, unpublished observations). Further, phosphorylation of the MAP kinase, ERK, is increased in cortical explants by treatment with both βE2 (Singh et al 1999) and αE2 (C. D. Toran-Allerand & M. Singh, personal communication). In chicken granulosa cells, αE2 is similar to βE2 in modulation of intracellular Ca^{2+} concentrations (Morley et al 1992).

Summary

Oestrogens are potent neuroprotective agents in a wide variety of neurotoxicity models. We and others have demonstrated that an ER is not required for the neuroprotective effects of oestrogens either in culture models or in rodent models of focal ischaemia. Further, the neuroprotective effects of oestrogens are dependent upon the presence of a phenolic A ring in the oestrogen structure and can be disassociated from the classical oestrogenic effects. The neuroprotective mechanism of action of these non-feminizing oestrogens remains to be elucidated; however, non-feminizing, phenolic A ring oestrogens have many cellular actions which may contribute to their observed protective activity. These include antioxidant effects (Sugioka et al 1987) and activation of signal transduction pathways. In summary, these data suggest that phenolic A ring oestrogens without feminizing effects may be effective in treatment of neuronal degeneration and acute forms of brain injury such as ischaemia. Such a treatment would be useful for men as well as for women in which conventional oestrogen therapy is contraindicated.

Acknowledgements

Supported by grants NIH AG 10485, Apollo BioPharmaceutics, Inc. and the US Army. Ent-oestradiol, PAM and PACA were generously supplied by Douglas Covey, Washington University, St. Louis, MO.

References

Behl C, Widmann M, Trapp T, Holsboer F 1995 17β-estradiol protects neurons from oxidative stress-induced cell death *in vitro*. Biochem Biophys Res Commun 216:473–482

Behl C, Skutella T, Lezoualc'h F et al 1997 Neuroprotection against oxidative stress by estrogens: structure–activity relationship. Mol Pharmacol 51:535–541

Bishop J, Simpkins JW 1994 Estradiol treatment increases viability of glioma and neuroblastoma cells *in vitro*. Mol Cell Neurosci 5:303–308

Blum-Degen D, Haas M, Pohli S et al 1998 Scavestrogens protect IMR 32 cells from oxidative stress-induced cell death. Toxicol Appl Pharmacol 152:49–55

Bonnefont AB, Muñoz FJ, Inestrosa NC 1998 Estrogen protects neuronal cells from the cytotoxicity induced by acetylcholinesterase–amyloid complexes. FEBS Lett 441:220–224

Brinton RD, Tran J, Proffitt P, Montoya M 1997 17β-estradiol enhances the outgrowth and survival of neocortical neurons in culture. Neurochem Res 22:1339–1351

Brooke S, Chan R, Howard S, Sapolsky R 1997 Endocrine modulation of the neurotoxicity of gp120: implications for AIDS-related dementia complex. Proc Natl Acad Sci USA 94: 9457–9462

Dubal DB, Kashon ML, Pettigrew LC et al 1998 Estradiol protects against ischemic injury. J Cereb Blood Flow Metab 18:1253–1258

Goodman Y, Bruce AJ, Cheng B, Mattson MP 1996 Estrogens attenuate and corticosterone exacerbates excitotoxicity, oxidative injury, and amyloid β-peptide toxicity in hippocampal neurons. J Neurochem 66:1836–1844

Green PS, Gridley KE, Simpkins JW 1996 Estradiol protects against β-amyloid (25-35)-induced toxicity in SK-N-SH human neuroblastoma cells. Neurosci Lett 218:165–168

Green PS, Bishop J, Simpkins JW 1997a 17α-estradiol exerts neuroprotective effects on SK-N-SH cells. J Neurosci 17:511–515

Green PS, Gordon K, Simpkins JW 1997b Phenolic A ring requirement for the neuroprotective effects of steroids. J Steroid Biochem Mol Biol 63:229–235

Green PS, Gridley KE, Simpkins JW 1998a Nuclear estrogen receptor-independent neuroprotection by estratrienes: a novel interaction with glutathione. Neuroscience 84:7–10

Green PS, Monck EK, Panickar KS, Simpkins JW 1998b Rapid phosphorylation of cyclic AMP response element binding protein by estradiol in SK-N-SH neuroblastoma cells. Soc Neurosci Abstr 24:269

Hawk T, Zhang YQ, Rajakumar G, Day AL, Simpkins JW 1998 Testosterone increases and estradiol decreases middle cerebral artery occlusion lesion size in male rats. Brain Res 796:296–298

Korenman SG 1969 Comparative binding affinity of estrogens and its relation to estrogen potency. Steroids 13:163–177

Mattson MP, Robinson N, Guo Q 1997 Estrogens stabilize mitochondrial function and protect neural cells against the pro-apoptotic action of mutant presenilin 1. Neuroreport 8:3817–3821

Mook-Jung I, Joo I, Sohn S, Kwon HJ, Huh K, Jung MW 1997 Estrogen blocks neurotoxic effects of β-amyloid (1-42) and induces neurite extension on B103 cells. Neurosci Lett 235:101–104

Moosmann B, Behl C 1999 The antioxidant neuroprotective effects of estrogens and phenolic compounds are independent from their estrogenic properties. Proc Natl Acad Sci USA 96:8867–8872

Morley P, Whitfield JF, Vanderhyden BC, Tsang BJ, Schwartz JL 1992 A new, nongenomic estrogen action: the rapid release of intracellular calcium. Endocrinology 131:1305–1312

Panickar KS, Guan G, King MA, Rajakumar G, Simpkins JW 1997 17β-estradiol attenuates CREB decline in the rat hippocampus following seizure. J Neurobiol 33:961–967

Pike CJ 1999 Estrogen modulates neuronal Bcl-x_L expression and β-amyloid-induced apoptosis: relevance to Alzheimer's disease. J Neurochem 72:1552–1563

Regan RF, Guo Y 1997 Estrogens attenuate neuronal injury due to hemoglobin, chemical hypoxia, and excitatory amino acids in murine cortical cultures. Brain Res 764:133–140

Roth A, Schaffner W, Hertel C 1999 Phytoestrogen kaempferol (3,4′,5,7-tetrahydroxyflavone) protects PC12 and T47D cells from beta-amyloid-induced toxicity. J Neurosci Res 57:399–404

Sagara Y 1998 Induction of reactive oxygen species in neurons by haloperidol. J Neurochem 71:1002–1012

Sawada H, Ibi M, Kihara T, Urushitani M, Akaike A, Shimohama S 1998 Estradiol protects mesencephalic dopaminergic neurons from oxidative stress-induced neuronal death. J Neurosci Res 54:707–719

Shi J, Zhang YQ, Simpkins JW 1997 Effects of 17β-estradiol on glucotransporter 1 expression and endothelial cell survival following focal ischemia in rats. Exp Brain Res 117:200–206

Shi J, Panickar KS, Yang SH, Rabbini O, Day AL, Simpkins JW 1998 Estrogen attenuates overexpression of β-amyloid precursor protein messenger RNA in an animal model of focal ischemia. Brain Res 810:87–92

Simpkins JW, Rajakumar G, Zhang YQ et al 1997 Estrogens may reduce mortality and ischemic damage caused by middle cerebral artery occlusion in the female rat. J Neurosurg 87:724–730

Singer CA, Rogers KL, Strickland TM, Dorsa DM 1996 Estrogen protects primary cortical neurons from glutamate toxicity. Neurosci Lett 212:13–16

Singer CA, Rogers KL, Dorsa DM 1998 Modulation of Bcl-2 expression: a potential component of estrogen protection in NT2 neurons. Neuroreport 9:2565–2568

Singer CA, Figueroa-Masot XA, Batchelor RH, Dorsa DM 1999 The mitogen-activated protein kinase pathway mediates estrogen neuroprotection after glutamate toxicity in primary cortical neurons. J Neurosci 19:2455–2463

Singh M, Sétáló G Jr, Guan Xi, Warren M, Toran-Allerand D 1999 Estrogen-induced activation of mitogen-activated protein kinase in cerebral cortical explants: convergence of estrogens and neurotrophin signaling pathways. J Neurosci 19:1179–1188

Stubley, L, Yang SH, Shi J, Day AL, Simpkins JW 1998 The effects of 17α-estradiol of cerebral blood flow and cortical protection during middle cerebral artery occlusion in rats. Soc Neursci Abstr 24:216

Sugioka K, Shimosegawa Y, Nakano M 1987 Estrogens as natural antioxidants of membrane phospholipid peroxidation. FEBS Lett 210:37–39

Toung TJK, Traystman RJ, Hurn PD 1998 Estrogen-mediated neuroprotection after experimental stroke in male rats. Stroke 29:1666–1670

Walton M, Sirimanne E, Williams C, Gluckman P, Dragnow M 1997 The role of cAMP response element binding protein in hypoxic–ischemic brain damage and repair. Mol Brain Res 43:21–29

Weaver CE Jr, Park-Chung M, Gibbs TT, Farb DH 1997 17β-estradiol protects against NMDA-induced excitotoxicity by direct inhibition of NMDA receptors. Brain Res 761:338–341

Wiese TE, Polin LA, Palomino E, Brooks SC 1997 Induction of the estrogen specific mitogenic effect of MCF7 cells by selected analogs of 17β-estradiol: a 3D QSAR study. J Med Chem 40:3659–3669

Yang SH, Shi J, Day AL, Simpkins JW 1999 Estradiol exerts neuroprotective effect when administered after ischemic insult. Stroke 31:745–750

Zaulynov L, Green PS, Simpkins JW 1999 Glutamate receptor requirement for neuronal death from anoxia-reoxygenation: an *in vitro* model for the neuroprotective effects of estrogens. Cell Mol Neurobiol 19:705–715

Zhang YQ, Shi J, Rajakumar G, Day AL, Simpkins JW 1998 Effects of gender and estradiol treatment of focal brain ischemia. Brain Res 784:321–324

Zhou Y, Watters JJ, Dorsa DM 1996 Estrogen rapidly induces the phosphorylation of the cAMP response element binding protein in rat brain. Endocrinology 137:2163–2166

DISCUSSION

McEwen: What are the advantages of oestrogens over other kinds of antioxidants?

Simpkins: One advantage is in the dose that needs to be administered. If the *in vitro* cellular environment is correct, we can get protective effects in the picomolar concentrations of oestrogens. If you take away certain constituents from the media then it requires much higher concentrations that are in the order of what's seen with other antioxidants. This system is regenerating, so I think the oestrogen effects may last a good deal longer than the effects of other antioxidants.

McEwen: Phyllis Wise was a co-author of a recent paper (Dubal et al 1999) that looked at ischaemic damage and talked about the involvement of ER. Could you comment on that in relation to your model?

Simpkins: It is true that she sees induction of anti-apoptotic proteins or prevention of decline, as well as induction of ERs when she induces ischaemia. This could explain part of what is being seen. However, those data are difficult for me to reconcile with ours. If one needs to have an ER involved in this kind of protection, then ERα should work considerably less potently than ERβ, and other compounds that we put into the system and have seen effects with should not work at all. She does see induction of the ER, but I don't know how that could explain the observed protection throughout the cortex.

McEwen: If you see only partial protection, and she sees only partial protection, is it possible that both mechanisms are working?

Simpkins: It is certainly possible. We both see very good cortical protection: where we don't see good protection is in the basal ganglia. Our interpretation of that has been that the lentate arteries, which are terminal arteries, don't form anastomoses supplying the basal ganglia. So we think the blood flow reduction that we're inducing is much more severe in the striatum. We are not sure that anything is going to be able to protect the striatum in this paradigm. In the cortex, there are anastomoses. The best you can do even with the largest thread you can get in, is reduce blood flow by 80%. In the few times that we attempted to measure striatal blood flow during and after these occlusions we had much more severe decline. So I don't think our data differ in brain areas protected; where we differ is in how we are interpreting the mechanism.

Levin: What happens when you give oestradiol and a variety of oestrogen receptor antagonists? Does this change anything?

Simpkins: We have not done this in the animal studies, because it is very difficult to do. *In vitro,* tamoxifen does not affect the neural protection that we see. We have used the ICI compound, but the problem we run into is that because it is a phenolic A ring oestrogen that has a chain off the 6-carbon, in our assays it is neuroprotective, so we haven't been able to use it to test the receptor antagonism.

Levin: In the coronary vasculature, it has been found in various animal models that endothelin antagonists dramatically limit the extent of infarct size. Mark Fisher believes that endothelin is very much involved in the response to acute hypoxia. It is an intermediate early gene that is rapidly produced and secreted, and extends the extent of the infarct and the damaging. Oestrogen, as you mentioned earlier, can shut this off. There might be multiple mechanisms involved here.

Gustafsson: I would also like to emphasize the possibility of multiple mechanisms. Certainly one shouldn't discount a role for the classical ERs. Peter Kushner and I collaborated some years ago on the AP-1 element (Paech et al

1997): tamoxifen or raloxifen can via ERβ turn on genes under the control of an AP-1 element. A physiological correlate of this was independently discovered by Benita Katzenellenbogen who studied the effects of tamoxifen in MCF-7 cells (Montano et al 1998). She could see using differential display that one of the genes turned on was quinone reductase, which is supposed to be an enzyme that protects against free radicals and reactive oxygen species. She also found that she could define the response element, which proved to be an antioxidant response element, which is similar but not identical to the AP-1 element. Subsequently, we have used micro arrays to look at various ERβ regulated genes, and one that has popped up in the prostate is glutathione peroxidase. One possibility is that phytooestrogens and other compounds could act on ERβ which would then turn on antioxidant response elements, perhaps resulting in transcription of a cascade of genes protecting against various potentially harmful molecular species. This could be one mechanism that is worth considering. We believe that this could be quite important to protect proliferating cells from going wrong. We do see that ERβ is increased in many proliferating cells.

Toran-Allerand: I would like to return to the classical ER, ERα. The cerebral cortex is a major target of oestrogen during development and has high levels of ERα, primarily during this developmental period which corresponds to the major period of cortical differentiation. This is the stage after most of the neuronal migration has taken place, and the layers have been formed. One of the things that has always intrigued me is that the levels of this cortical ER fall dramatically during the second postnatal week. There is, therefore, a relatively short period when there are high levels of ERα in the cerebral cortex. The number of receptors per cell is not particularly high, which is why it took a long time to identify oestrogen binding, but the number of cells that express the binding is very high. Following injury might there be a re-expression of this developmentally related ERα which could play a role in repair and make that part of the brain again responsive to oestrogen? The striatum doesn't have similarly developmentally regulated high levels of ER which then disappear, although it is no doubt an oestrogen target, perhaps mediated at the membrane. I refer here to the classical nuclear receptor, ERα. It may be that this regional difference, and the fact that the cortex is more protected than the basal ganglia, is not solely because of differences in the respective anatomy of the vasculature (which admittedly makes things much worse), but because the cerebral cortex is a region of the brain which was a major target of oestrogen during development. And while it no longer responds in the same manner during normal adult life, ischaemic injury reactivates the developmental pattern (Dubal et al 1999).

Simpkins: That is a good point: that is exactly what Phyllis Wise's paper is attempting to demonstrate (Dubal et al 1999). With injury there is increased expression of ERα. If one is going to contend however, that ERs — either nuclear or membrane — are major mediators of these responses, then we have to think about the ER as a phenolic A ring-recognizing molecule, rather than the oestrogen receptors that have been described over the last few days. Because we just cannot reconcile the high diversity of structures so long as the phenolic A ring that is retained that have these protective effects, with stereospecific interactions with proteins.

Gibbs: You gave a large number of examples in which oestrogen was shown to have neuroprotective effects. I think all of these were *in vitro* examples. Conversely, we've done studies recently to look at the ability of oestrogen to be neuroprotective for cholinergic neurons following either ibotanate injections into the basalis or mechanical lesions of the fimbria fornix. In neither case is there any evidence at all of a neuroprotective effect. Could you comment on this? Do you think that all the protective effects you described involve protection against oxidative damage, and do you think that the ibotanate and mechanical lesions are inducing neuron death in a completely different way that oestrogen doesn't protect against?

Simpkins: Both the severity and the type of insult can affect the ability of oestrogen to provide protection. Certainly, Aβ has a strong oxidative component, and glutamate toxicity has the Ca^{2+} influx and subsequently reactive oxygen species generation. Ion-stimulated cell death involves the Fenton reaction and production of hydrogen peroxide. I don't know the mechanism of the cell death with the GP120 protein. As for *in vivo* protection, what is clear is that oestrogen's ability to protect *in vitro* is very much dependent on the severity of the insult. We had to be very careful to titre kill so that we're killing about half the cells. When we get to 80–90% kills we either lose the oestrogen protection or we push that dose–response curve way out to the right, and it takes much more to give us any protection. So severity of the insult is an issue.

Gibbs: We are using concentrations of ibotanate that have been used in the past to selectively destroy cell bodies, and have been shown to have behavioural effects. We haven't done a dose–response curve to look at the interaction between oestrogen and different doses of ibotanate. With the mechanical lesion there's been debate in the past of whether the cells die or not, and the current thinking is that they don't die but just stop expressing choline acetyltranferase (ChAT). Our data suggest that the death response is less following the mechanical lesion than following the ibotanate injection, but in neither case do we see a protective effect.

Henderson: Different oestrogen compounds have different degrees of antioxidant activity, as measured, for example, in different non-biological systems. One way to see whether protective effects are mediated primarily through antioxidant activity would be to see whether the degree of protection in your model system parallels the degree of antioxidant activity of the various oestrogenic compounds.

Simpkins: My reading of the antioxidant literature is that it's fairly confused, with some folks showing that some oestratrienes have higher activities than others, but this seems to depend upon the oxidative stress that's applied. We use malandialdehyde production as an indicator, and we are now developing isoprostane assays which are more stable and should be more reliable. Christian Behl has done lots of that structure–activity work, and will mention this in his paper (Behl et al 2000, this volume).

Behl: I don't agree that these extra ring systems are necessary. We are more reductionist, and think that all that is necessary for the antioxidant activity is the phenol group and some additional inputs of electrons, which may come from additional alkyl groups at the ring. This is enough for antioxidant activity as shown by prevention of lipid peroxidation and low density lipoprotein oxidation. I want to add that vitamin E is also a phenolic compound, and it works by exactly the same mechanism of activity as an antioxidant. Vitamin E becomes a radical during its antioxidant activity, and this is recycled by vitamin C. This may interfere with your hypothesis of thione cycling and one may add the ascorbate as an additional recycling system for oestrogens to this hypothesis.

Becker: It has been reported that females are less vulnerable to quinolinic acid lesions of the striatum. Have you have looked at sex differences in vulnerability to hypoxia or whether there are effects of oestrogen on the size of lesion produced?

Simpkins: Yes, there are sex differences in infarct size. When lesions are made in intact males and intact females, the male lesions are larger. The size of the infarct in ovariectomized females is similar to that in the males, but when they are treated with oestrogen the infarct is smaller. If you treat males with oestrogens the infarct size decreases, but this is probably not due to an oestrogen effect but more likely a very prompt suppression in testosterone. You can demonstrate that by putting the androgens back in, then giving the oestrogens, and by doing this you can largely override the oestrogen protection in males. This led to another strategy of reducing infarcts in males: it turns out that whatever way you reduce androgens in males, there is a better infarct outcome. We think that in males the androgens are bad players more than oestrogens being good.

Levin: Most of the paradigms of cell death you are reducing by oestrogen involve apoptotic cell death. Perhaps this explains some of the differences in your findings: it is just this type of cell death that is protected against by oestrogen and not mechanical cell death.

Looking more globally at oestrogen as an antioxidant, not only does it prevent lipid peroxidation, it also increases NO formation and may suppress free radical formation as well as the action of those free radicals. So it seems to be having a comprehensive effect.

Simpkins: There is evidence for all of those functions. We have used NF-κB, which is a very sensitive marker of cellular oxidative state. Very rapidly after administration of oestradiol, oxidative stress-induced production of NF-κB is suppressed. Oestrogens are reducing reactive oxygen species in the cell, and many researchers have shown in various models that oestrogens prevent membrane peroxidation. I think oestrogens are at least initially reducing formation of these reactive oxygen species in response to stresses on the cell. There are certainly other mechanisms that are activated: antioxidant enzyme systems are well known to be activated by oestrogens in a variety of tissues.

Gustafsson: One thing that worries me a little is how well you have defined your assay system. The implication is that another type of ER is involved here. When you look at the relative potencies of drugs, is what you have told us here true, not only for one cell type, but for five or 10 different cell types? And when you look at neuroprotection *in vivo*, is this also true? That is, is there a lot of variation in the relative potency of these antioxidants?

Simpkins: That is a difficult question. We develop our insults so that we produce a 40–60% kill of cells. With this defined insult we do direct comparisons in the same assay of the various oestrogens, so we can get some indication of potency differences. When we change cell types or change insults, we re-optimize the insult to get the same level of kill. However, the oestrogens have performed relatively similarly in various systems. There is some variation, but it's not remarkable. However, when we look at these compounds in any system, we test them at the same time under the same conditions. We are not looking at day-to-day variations in how the assays perform: the neuroprotection assays are run the same for all compounds.

Gustafsson: Does this appear to be a relatively specific mechanism?

Simpkins: There are lots of signalling pathways that are activated to the same extent by βE2 and αE2. We haven't looked at these other compounds for activation of other signalling pathways. We do know that the time-course and the extent of increase in phosphorylation of CREB for example, appears the same for βE2 and αE2. Because of the labour involved in those assays, we simply haven't used them to screen other compounds.

Hurd: Are there any differences clinically in postmenopausal women in the incidence of stroke with and without oestrogen replacement?

Simpkins: There are five published reports in which stroke outcome is measured by mortality. There's little change in incidence of stroke in postmenopausal women on or off oestrogens, but there is a small reduction in death from stroke in those taking oestrogen. The data are mixed.

Kushner: You described an experiment in which you used *ent*-oestradiol in cell culture, and it was as equally protective as oestradiol. Have you tried this in animals?

Simpkins: Ent-oestradiol has been evaluated in animals in two paradigms. One was to assess uterus response and it simply had no agonist activity that we could pick up using that endpoint. The other was in the stroke protection. In the latter study we did a pre-treatment 2 h prior to the occlusion, and the extent of protection was the same as with βE2.

Toran-Allerand: I found a paper that suggested that αE2 was an extremely potent oestrogen on the uterus during development (Hajek et al 1997). Might the fact that you got such a good effect with αE2 in your cortical lesions be a re-expression of a developmental type of response?

Simpkins: It's certainly possible. αE2 binds very well to both ERα and ERβ. It doesn't seem to stick on the receptor and the complex doesn't seem to bind well to the oestrogen response element. For membrane receptors we know nothing about interactions. It is possible that αE2 is interacting with a lesion-induced expression of ERα and that can explain part of it's activity, but we haven't studied that.

Bethea: Do you remember what the structure of phenol red looks like? You said that you didn't get any effect with just the phenol ring.

Simpkins: We didn't get any effect with just the phenol ring. The design for the survey of all those compounds was to administer at 10 nM concentrations. This was because this was about two orders of magnitude higher than we saw significant protection with βE2, so if it didn't show up with 10 nM we assume we didn't have activity. We also used some branch chain phenols at these low concentrations and didn't see protection, although Christian Behl will tell us about some other compounds which at higher doses are antioxidants and are neuroprotectants. We've done studies with and without phenol red in the media, and we don't see any differences either in our ability to kill cells or on the potency of βE2 in protecting cells, so our assumption was that phenol red was not interacting with the mechanism that were looking at.

Levin: When we first got into this field, in our naïvety we used phenol red to culture cells in. We did many experiments before we found out we weren't supposed to be using phenol red. We then went back and repeated all of those experiments and it didn't make a whole lot of difference.

Toran-Allerand: In our hands it makes a tremendous difference, and it does make a difference in MCF-7 cells (Berthois et al 1986, Welshons et al 1988). Moo-Ming Poo, who was at Yale and is now at UCSD, has shown electrical changes in hippocampal neurons by the addition of phenol red. One has to be careful about using phenol red, especially in studies where one is interested in certain types of oestrogen action.

Simpkins: We use charcoal-stripped serum. We started with phenol red, did our dose responses and then went back and re-did them in phenol red-free media and we didn't see any difference.

Toran-Allerand: MCF-7 cells behave differently with respect to survival, whether or not phenol red is in the medium.

Kushner: The situation is a little more complicated. It was traced down not to the phenol red itself but to some other minor contaminant.

Levin: So you're suggesting that the new phenol red-containing media are OK.

Kushner: Yes.

References

Behl C, Moosmann B, Manthey D, Heck S 2000 The female sex hormone oestrogen as neuroprotectant: activities at various levels. In: Neuronal and cognitive effects of oestrogens. Wiley, Chichester (Novartis Found Symp 230) p 221–238

Berthois Y, Katzenellenbogen JA, Katzenellenbogen BS 1986 Phenol red in tissue culture media is a weak estrogen: implications concerning the study of estrogen-responsive cells in culture. Proc Natl Acad Sci USA 83:2496–2500

Dubal DB, Shughrue PJ, Wilson ME, Merchenthaler I, Wise PM 1999 Estradiol modulates Bcl-2 in cerebral ischemia: a potential role for estrogen receptors. J Neurosci 19:6385–6393

Hajek RA, Roberston AD, Johnston DA et al 1997 During development, 17α-estradiol is a potent estrogen and carcinogen. Environ Health Perspect (suppl 3) 105:577–581

Montano MM, Jaiswal AK, Katzenellenbogen BS 1998 Transcriptional regulation of the human quinone reductase gene by antiestrogen-liganded estrogen receptor-α and estrogen receptor-β. J Biol Chem 273:25443–25449

Paech K, Webb P, Kuiper GGJM et al 1997 Differential ligand activation of estrogen receptors ERα and ERβ at AP1 sites. Science 277:1508–1510

Welshons WV, Wolf MF, Murphy CS, Jordan VC 1988 Estrogenic activity of phenol red. Mol Cell Endocrinol 57:169–178

The female sex hormone oestrogen as neuroprotectant: activities at various levels

Christian Behl, Bernd Moosmann, Dieter Manthey and Stefanie Heck

Max-Planck Institute of Psychiatry, Kraepelinstrasse 2-16, D-80804 Munich, Germany

Abstract. The female sex hormone oestradiol (oestrogen) is a steroidal compound that binds to specific intracellular receptors which act as transcription factors. Oestrogen displays many of its effects by the classical mode of action through receptor binding, transactivation and binding to consensus oestrogen response elements on DNA. Although the primary role of oestrogen as an ovarian steroid was thought to be the regulation of sex differentiation and maturation, since oestrogen receptors are expressed in a variety of other tissues besides sex organs, oestrogen is believed to exert multiple activites in several target sites throughout the body, including the nervous system. In the brain oestrogens have multiple activities. Potential neuroprotective functions of oestrogens are being intensively studied and it is becoming increasingly clear that oestrogens are (1) neuroprotective hormones acting via oestrogen receptor-dependent pathways at the genomic level and (2) neuroprotective steroidal structures acting independently of the activation of specific oestrogen receptors. One striking activity of the molecule oestradiol is its intrinsic antioxidant activity which makes it a potential chemical shield for neurons. Nerve cells frequently encounter oxidative challenges during the normal physiology, but also under pathophysiological conditions. Oxidative stress has been implicated in a variety of neurodegenerative disorders including amyotrophic lateral sclerosis, Parkinson's disease and Alzheimer's disease. It is important to stress that the antioxidant neuroprotective activity of oestrogens is independent of oestrogen receptor activation, since oestrogen derivatives and aromatic alcohols that do not bind to oestrogen receptors share the same antioxidant neuroprotective activity. Although this effect of oestrogens can clearly be separated from oestrogen receptor binding, oestrogens may interact with intracellular signalling pathways, such as the mitogen activated protein kinase, cyclic AMP pathways, and with the activity of the redox-sensitive transcription factor NF-κB.

2000 Neuronal and cognitive effects of oestrogens. Wiley, Chichester (Novartis Foundation Symposium 230) p 221–238

Oxygen metabolism is essential for life under anaerobic conditions. Nevertheless, oxygen is also a threat to cells because of its potential for forming free oxygen radicals. By definition, a radical is a molecule with an unpaired and therefore highly reactive electron. The most prominent reactive oxygen species (ROS) are the superoxide radical ($O_2^{\cdot-}$), hydrogen peroxide (H_2O_2) and the hydroxyl radical (OH·). Although H_2O_2 is not a free radical in itself, it is the substrate for the formation of highly reactive hydroxyl radicals which are formed during the Fenton reaction in the presence of iron ions. As a consequence of the adaptation of anaerobic organisms to the constant generation of free radicals during oxygen metabolism, there is a normal cellular homeostasis and a delicate balance between the rate and magnitude of oxidant formation and the rate of oxidant elimination. On one hand there are various mechanisms that can lead to the formation of ROS, such as the physiological oxidative phosphorylation in the respiratory chain of the mitochondria, and the enzymatic and non-enzymatic oxidation of catecholamines through the activity of lipooxygenases and cyclooxygenases. On the other hand there are various antioxidant defence systems which can be divided into enzymatic and non-enzymatic forms. The most important enzymes for the detoxification of ROS are:

(a) the superoxide dismutase (SOD), which dismutates generated superoxide to H_2O_2;
(b) catalase, which breaks down the H_2O_2; and
(c) glutathione peroxidase, which also dismutates hydrogen peroxide.

The essential non-enzymatic antioxidants are:

(1) vitamin C, as most important hydrophilic antioxidant;
(2) vitamin E, as most important lipophilic antioxidant; and
(3) glutathione, as potent intracellular reducing compound.

While the enzymatic antioxidants are regularly formed during physiological metabolism throughout the body, vitamins C and E have to be supplemented mainly through nutrition. Of course, there are several additional mechanisms with an antioxidant activity that can be found in the organism in a region-specific manner. Nevertheless, the above-mentioned antioxidants form the central defence against ROS. Whenever the delicate balance between ROS formation and ROS detoxification is impaired either by an overproduction of ROS or by a weakened antioxidant defence activity, the phenomenon of oxidative stress occurs (Halliwell & Gutteridge 1989).

Oxidative stress during neurodegeneration

Neuronal cells of the CNS are particularly vulnerable to oxidative stress. Firstly, the brain has a high throughput of oxygen and therefore increased opportunity for the build up of ROS. Secondly, in the metabolism of nerve cells there are various mechanisms that can directly form ROS (e.g. through catecholamine metabolism, flavinooxidases, lipooxygenases and cyclooxygenases). Thirdly, this increased vulnerability is also due to the specific composition of the membranes of neuronal cells. These membranes are highly enriched by polyunsaturated fatty acids which are ideal substrates for oxidation and peroxidation through hydroxyl radicals (Halliwell & Gutteridge 1989). Moreover, oxidative stress is particularly detrimental to nerve cells, since neuronal cells lack the potential to grow, and once a nerve cell is damaged and dysfunctional, degenerative events may take place and the nerve cells may die.

Oxidative stress can endanger virtually all cellular macromolecules, including enzymes or cellular membranes. Oxidation of these molecules may lead to a complete loss of function or impaired function with respect to enzymatic activity. In general, oxidative stress is believed to be associated with the pathogenesis and the progression of various neuronal and non-neuronal disorders including arteriosclerosis, cancer, inflammation, stroke, amyotrophic lateral sclerosis (ALS), Parkinson's disease (PD) and Alzheimer's disease (AD) (Behl 1999).

Oxidations in neurodegenerative disorders may come from the lack of or gain of function of antioxidant enzymes, as reported for ALS, where SOD appears to be functionally altered. Interestingly, in familial cases of ALS a mutation in the enzyme SOD1 has been found (Rosen et al 1993). During stroke, which is linked to ischaemia and reperfusion processes, oxidation may occur as a secondary damaging effect. In PD the metabolism of the neurotransmitter dopamine and its chemical breakdown by monoamine oxidases may lead to the formation of free radicals that could damage dopaminergic neurons (for review see Ebadi et al 1996). Frequently, in neurodegenerative disorders there is a regional specificity of vulnerability of neurons against oxidative stress. In the last decade, evidence has accumulated that in the pathogenesis of AD, ROS may also be involved (Behl et al 1994, for review see Behl 1999).

Oxidative stress in Alzheimer's disease

There are many lines of evidence supporting the view that oxidative stress is involved in AD:

(a) Post mortem evidence for an oxidative burden in AD: increased levels of oxidation endproducts are found in Alzheimer's tissue compared with age-matched controls; increased oxidation of DNA and of membrane lipids have been reported.

(b) Inflammatory processes (arthritis of the brain): since AD has a large inflammatory component, it is evident that ROS formed by inflammatory cells may also challenge the surrounding nerve cells.
(c) Global age-associated decline of the antioxidant defence.
(d) Particular vulnerability of the brain to oxidation.

Therefore, it appears that oxidative stress is an eminent problem of AD pathogenesis and some triggers of ROS formation have already been identified (for review see Behl 1999).

Amyloid β protein and glutamate as oxidative stressors in the pathogenesis of Alzheimer's disease

The major hallmark of AD pathology is the occurrence of extracellular deposits of amyloid β protein (Aβ) and the intracellular formation of so-called neurofibrillar tangles. Aβ is a 42–43 amino acid polypeptide that is derived from a larger precursor, the Aβ precursor protein (APP) (Glenner & Wong 1984, Masters et al 1985). Aβ is a peptide that can easily build up β-pleated sheets that ultimately form non-soluble fibrils, thus resulting in deposits of Aβ. Since the first observation that Aβ can act directly as a neurotoxin (for review see Yankner 1996), many laboratories have investigated the potential toxic effect of Aβ and also secondary effects of Aβ deposition that endanger neurons during the AD-associated events. In 1992 we found that upon addition of Aβ to nerve cells in culture, very rapid membrane changes occur that are indicative of ROS reactions. Therefore, we have been able to show that the free radical scavenger α-tocopherol (vitamin E) can protect neuronal cells in culture against toxic events induced by Aβ aggregates (Behl et al 1992). Shortly after this first indirect evidence that oxidative stress may occur during Aβ toxicity, we have extended these data and found that H_2O_2 is in fact a potent mediator of Aβ toxicity (Behl et al 1994). Indeed, the accumulation of H_2O_2 in the nerve cells caused by Aβ is a first step in Aβ toxicity. In a next step the membrane lipids are peroxidized, membranes are damaged, and ultimately the nerve cells are breaking up. Cell lysis induced by Aβ can be measured in cell culture.

The current understanding, nevertheless, is that not only can Aβ directly challenge nerve cells in an oxidative manner, but also Aβ deposits may attract other oxidative players that challenge nerve cells. Such secondary players include inflammatory cells, mainly activated microglia (Rogers et al 1992), and so-called advanced glycation endproducts (AGEs) (Smith et al 1995) so that a concert of oxidative reactions may challenge nerve cells over time. All these oxidative challenges happen in the framework of an age-related disease. And AD is, indeed, an age-associated disorder, since more than 85% of all AD cases are

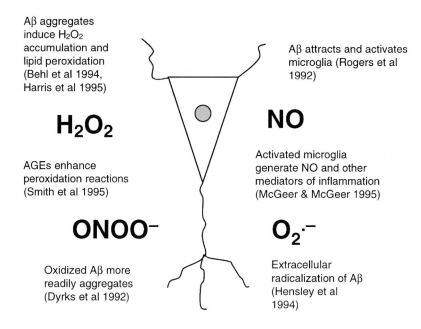

Aβ aggregates induce H$_2$O$_2$ accumulation and lipid peroxidation (Behl et al 1994, Harris et al 1995)

Aβ attracts and activates microglia (Rogers et al 1992)

H$_2$O$_2$

NO

AGEs enhance peroxidation reactions (Smith et al 1995)

Activated microglia generate NO and other mediators of inflammation (McGeer & McGeer 1995)

ONOO$^-$

O$_2^{\cdot-}$

Oxidized Aβ more readily aggregates (Dyrks et al 1992)

Extracellular radicalization of Aβ (Hensley et al 1994)

- Age-associated decrease in neuronal antioxidant activity
- Genetic predisposition (e.g. presenilin mutations, ApoE genotype)
- Additional oxidative insults (e.g. glutamate)
- Tau pathology

FIG. 1.   Oxidative stress in neurons may be induced by Aβ aggregates directly and indirectly. Aβ aggregates in the Alzheimer's disease brain induces oxdiative stress via direct interaction with the neurons or via the attraction and activation of microglial cells. The most common reactive oxygen species are superoxide (O$_2^{\cdot-}$), hydroxyl radicals (OH$^{\cdot}$), nitric oxide (NO), and peroxy nitrite (ONOO$^-$), which is formed by the O$_2^{\cdot-}$ and NO. Aβ more easily forms aggregates in an oxidative environment. The concert of oxidations and peroxidation reactions is enhanced by the occurrence of advanced glycation endproducts (AGEs). The oxidative challenges occur in association with age-related pathophysiological alterations or on top of a genetic predisposition.

strictly age-associated (Evans et al 1989). Of course various genetic predisposition factors may render the nerve cells vulnerable to oxidation. Some of these factors are already identified (e.g. ApoE allelic variations) (Fig. 1). From today's standpoint and summarizing the current knowledge of the role of Aβ deposition, an altered Aβ metabolism is, indeed, the most convincing approach to understand the pathogenesis of AD (for review see Hardy 1997). But with respect to the direct neurodegenerative mechanisms that occur in the tissue, other players in addition to Aβ — including the previously mentioned inflammatory cells — have to be taken into account (for review see McGeer & McGeer 1995). Moreover, the excitatory amino acid L-glutamate may also come into play, since an increased concentration of glutamate in the tissue may lead to excitotoxicity (Choi 1992).

In conclusion, with respect to AD pathology, a scenario is given that toxic events may be directly and indirectly driven by the deposition of Aβ. One major downstream effect of these challenges is oxidative stress. This observation combined with the evidence from histopathology is forming the so-called oxidative stress hypothesis of the pathogenesis of AD (for review see Behl 1999). In the search for vulnerability factors, disease-associated and disease-accompanying events it has become clear that there is an interesting link between the female sex hormone oestrogen and AD:

(1) Women are twice as likely to develop AD than men (Aronson et al 1990, Rocca et al 1991).
(2) The loss of oestrogen during menopause might play a role in age-associated cognitive decline (Simpkins et al 1994).
(3) There is increased incidence of AD in older women that may be due to oestrogen deficits (Paganini-Hill & Henderson 1994, Henderson et al 1994, Tang et al 1996).

A recent data search identifies 29 publications that can be found when searching *Current Contents* from 1995 to date with the keyword combination 'AD' and 'oestrogen' (July 1999). Many of these publications focus on the potential protective role of oestrogens in nerve tissue with respect to oestrogen replacement therapy in postmenopausal women. Our approach to the role of oestrogens during nerve cell death and neuroprotection started at a more molecular and chemical level focusing at the structure of the steroidal compound oestrogen and its direct interaction with neuronal cells.

Oestradiol and oestrogen derivatives as neuroprotective antioxidants

Since AD is an age-associated disorder, one major focus of our investigations of the toxicity of Aβ and glutamate in nerve cell culture was to study the effects of hormones in particular toxicity paradigms. During these studies we found that the activation of glucocorticoid receptors, for instance, in hippocampal neurons increases the toxicity of Aβ, glutamate, and other oxidative stressors (Behl et al 1997a). These data are consistent with the view that age-related hypercortisolism endangers the CNS (Sapolsky 1992). On the other hand, we observed a potent protective effect of 17β-oestradiol in the same toxicity paradigm using the identical neurotoxins (Behl et al 1995). It became evident that micromolar concentrations of 17β-oestradiol block oxidative stress caused by Aβ, glutamate and H_2O_2. In a following study we identified the structural prerequisites of this

TABLE 1 Protection of clonal hippocampal cells against oxidative stress-induced cell death by different oestrogens

Reagents	Concentration	MTT	Cell count	Peroxide
Control		100	100	100
Glutamate (or H_2O_2)	1 mM (60 μM)	9±7 (15±4)	16±5 (19±4)	168±4
Oestrone	10 μM	91±6* (74±2)*	96±3* (91±6)*	107±4**
17α-oestradiol	10 μM	86±3* (63±6)*	92±5* (89±5)*	105±3**
17β-oestradiol	10 μM	91±6* (66±5)*	90±6* (93±5)*	108±4**
Oestriol	10 μM	80±6* (80±2)*	95±9* (96±2)*	103±2**
Ethinyl oestradiol	10 μM	85±2* (65±4)*	83±5* (90±5)*	110±4**
Mestranol	10 μM	5±3 (21±5)	6±7 (14±4)	164±2
Quinoestrol	10 μM	3±5 (10±3)	5±4 (17±5)	165±2
2-OH oestradiol	10 μM	78±6* (85±5)*	74±6* (69±4)*	105±4**
4-OH oestradiol	10 μM	83±5* (46±4)*	82±6* (73±4)*	107±4**

HT22 cells were incubated with various oestrogens for 20 h at 10 μM before the indicated toxins were added. After 6 h the intracellular formation of H_2O_2 and related peroxides were determined using dichloroflourescein (DCF) fluorescence. Parallel cultures were treated for an additional 14 h. At that time, either the reduction of 3-(4,5-dimethylthiazol-2-yl)-2,5 diphenyl tetrazolium bromide (MTT) as a measure of cellular viability or trypan blue exclusion and cell counts as a measure of cell lysis were performed. The data in parentheses refer to the treatment of the cells with H_2O_2. All results were normalized to control values (no addition of reagent) as 100%. Peroxide formation as detected with DCF flourescence is expressed as percent fluorescent cells. Data are the means ± SEM of five independent experiments. The P-values of *$P < 0.01$ and **$P < 0.05$ were considered significant. Compounds in bold represent oestrogen-derivatives with an intact phenolic structure. MTT, 3,[4,5-dimethylthiazol-2-yl]-2,5-diphenyltetrazolium bromide. (Table derived from Behl et al 1997b.)

potent antioxidant activity of 17β-oestradiol. We found that the central feature of antioxidant activity of oestradiol is the intact phenolic group at the steroidal molecule (Behl et al 1997b). Neuronal cells and organotypic hippocampal slice cultures are protected against oxidative stress by compounds including oestrone, 17α-oestradiol, 17β-oestradiol, oestriol, ethinyl oestradiol, 2-hydroxyoestradiol and 4-hydroxyoestradiol (Table 1) (Behl et al 1997b, Green et al 1997). Whenever the phenolic group of oestradiol was altered, such as in mestranol or in quinestrol (ether-modified forms of oestradiol), the protective effect was prevented. Moreover, in these studies it became clear that an structurally intact oestradiol can also protect nerve cells that do not express classical oestrogen receptors. And, indeed, it turned out that the phenolic structure of oestrogen is the structure responsible for oestradiol's antioxidant activity. This is consistent with the fact that the highly neuroprotective lipophilic antioxidant vitamin E is also a phenolic compound (Fig. 2). In a more recent study, the concept of receptor-independence of the neuroprotective antioxidant effect of oestradiol has been

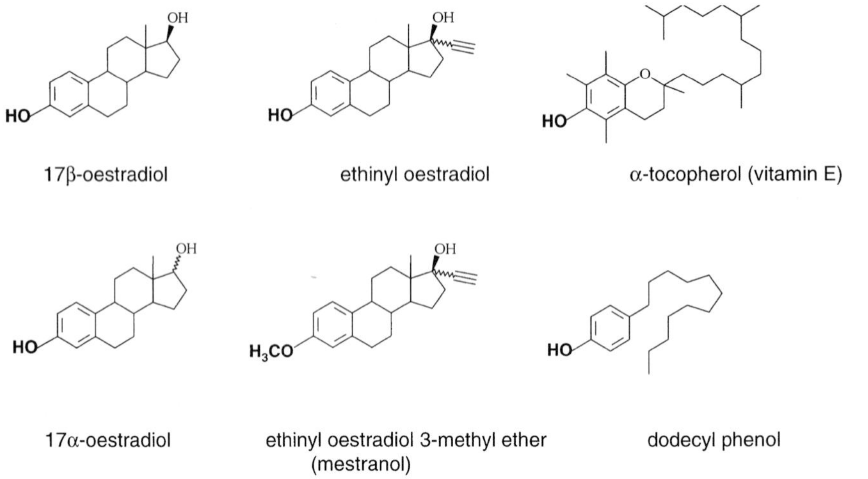

FIG. 2.   Oestrogen and selected oestrogen-derivatives with intact aromatic alcohol structure. 17β-oestradiol, 17α-oestradiol and ethinyl oestradiol as oestrogen-derivatives with an intact phenolic group at the steroid structure exert antioxidant neuroprotective activities in oxidative stress paradigms *in vitro*. Following modification of the phenolic group such as in mestranol, the neuroprotective effect is lost. Aromatic alcohols such as dodecyl phenol or vitamin E (α-tocopherol), lacking the oestrogen receptor-activating properties but carry a phenolic group and a lipohpilic tail, are also effective neuroprotective antioxidants.

intensively studied (Moosmann & Behl 1999). There, a variety of phenolic compounds including 4-dodecylphenol, 2-naphthol, resveratrol, 2-, 4-, 6-trimethylphenol, serotonin and others have been compared with 17β-oestradiol with respect to their ability for oestrogen receptor binding, oestrogen receptor activation and antioxidant effect. And, indeed, 17β-oestradiol had the highest affinity for the oestrogen receptor as prepared from rat uterus compared to, for instance, dodecylphenol, biphenol or trimethylphenol. The same picture arose when we investigated the ability of these compounds to activate oestrogen receptor-dependent gene transcription by using oestrogen response element reporter plasmids. Again, as expected, 17β-oestradiol led to a maximum activation of oestrogen receptor-dependent transcriptional activation, while compounds such as trimethylphenol had no activating abilities any more. So, there was a clear distinction between oestradiol and other phenols with respect to oestrogen receptor binding and oestrogen receptor activation. On the other hand, all these phenolic compounds exerted a similar profile of antioxidant activity in various paradigms used. Trimethylphenol did prevent the oxidation of low-density lipoprotein or the peroxidation of freshly prepared brain membranes to a similar extent as 17β-oestradiol. Moreover, the cytoprotective activity of trimethylphenol and oestradiol was similar when human neuroblastoma

SK-N-MC cells or mouse clonal hippocampal HT22 cells were challenged with high concentrations of H_2O_2. With respect to the neuroprotective antioxidant effect of these phenols there was basically no significant difference between 17β-oestradiol and dodecylphenol, biphenol and trimethylphenol in the assays used. These data are summarized and discussed in a recent publication from our lab (Moosmann & Behl 1999).

In conclusion, in focusing on antioxidant activities of oestradiol and oestradiol derivatives there is an intrinsic activity provided by the intact phenolic group in the steroidal structure that can be mimicked by mere aromatic phenols. The identification of such a chemistry-dependent antioxidant and neuroprotective effect in phenols may lead to the design of antioxidants with an improved pharmacokinetic profile. It is intriguing that oestrogen's protective effect may be separated from its hormonal oestrogen receptor-dependent effects, since it could be envisaged that such non-hormonal oestrogens may also be applied to male organisms without the problem of hormonal feminizing side effects.

Of course, oestrogen is a physiological molecule that is present in various levels at various sites throughout the body. Therefore, when aiming at a neuroprotective effect of oestrogens in women (e.g. hormone replacement therapy in postmenopausal women), one has to take into account that oestrogen is also neuroactive. Moreover, with the identification of oestrogen receptor (ER)β (Shughrue et al 1999, for review see Gustafsson 1997, Kuiper et al 1998) a whole network of receptor interactions between ERα and ERβ has arisen that may introduce a wide range of ER-dependent neuroactivities. Therefore, when oestrogens are used as hormones, they don't have to be considered as mere antioxidants but rather as female sex hormones with intrinsic antioxidant activities.

Oestrogen as a pleiotropic neuroactive compound

The window that was opened by the structural research on oestrogen antioxidant activities is, of course, rather small compared to the whole range of activities a hormone may develop. Oestrogen as the female sex hormone has so-called 'classical' effects during sex differentiation and sex maturation. These activities have been attributed to oestrogen's genomic activities (for review see McEwen 1991). Since it has been established that ERs are expressed in a variety of tissues throughout the body, including the brain, attention has focused on the hormonal effects of oestrogen in other target sites as well. With respect to its activity in the brain, the term 'neuroactive steroid' has been coined for oestrogen. These neuroactivities are mediated by both genomic and non-genomic effects (Paul & Purdy 1992, for review see Rupprecht 1997). Very recently, many groups have focused on oestrogen's potential as a neuroprotective steroid. Again, it turned out that these effects may be both of genomic nature and, as in the case of the identification of an antioxidant activity of oestrogens, also of non-genomic nature. Very important protective effects of oestrogens are the modulation of:

(a)   basal forebrain cholinergic activity and integrity (Luine 1994);
(b)   dendritic plasticity (Woolley & McEwen 1993);
(c)   NMDA receptor density (Woolley et al 1997);
(d)   neurotrophin expression and neurotrophin signalling (Toran-Allerand 1996); and
(e)   APP processing (Xu et al 1998).

Specifically, the recent discovery that oestrogen can also influence the processing of the APP molecule is highly intriguing and attractive given the evidence supporting the pathogenetic role of the protein processing of APP in neurons during AD development (Xu et al 1998).

Additional effects may be caused by oestrogens. With the increasing knowledge about the signalling events occurring during regular cell survival and also cell death, this complex network has been proposed to be an additional target for oestrogen activity. It is already shown that oestrogen can affect signalling via the cAMP pathway (Moss & Gu 1999, Kelly et al 1999).

Also, the interaction of oestrogen with the MAP kinase pathway has been recently introduced (Improta-Brears et al 1999, Watters & Dorsa 1998, Singer et al 1999). Another potential target of the ER is the activity of the transcription factor NF-κB, which is a prominent redox-sensitive transcriptional regulator (Baeuerle & Henkel 1994). Interestingly, NF-κB has been proposed to play a role during nerve cell death and neuroprotection, and may be one of the key factors in the decision whether a nerve cell is driven into the apoptosis or is kept alive (Lipton 1997, Lezoualc'h & Behl 1998). A detailed analysis of the interaction between oestradiol and NF-κB signalling in various toxicity paradigms is currently under way and will probably introduce another level of complexity when looking at potential activities of oestrogens in nerve cells (Fig. 3).

Outlook

Oestrogen is a pleiotropic neuroactive modulator, with a range of activities on neuronal cell survival and cognition. Since these two main conditions are directly associated with pathogenic mechanisms and with neurodegenerative disorders, oestrogens are, of course, of primary interest with respect to the prevention and therapy of neurodegenerative disorders. Much has been learned about oestrogen — initially thought of just as a sex hormone — over the last few years. And it seems that the more insight that is gained into the complexity of intracellular signalling and ER function, the more potential target sites of oestrogens occur.

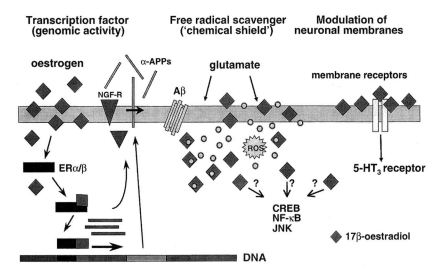

| Transcription factor (genomic activity) | Free radical scavenger ('chemical shield') | Modulation of neuronal membranes |

FIG. 3. Oestrogen serves neuromodulatory and neuroprotective functions at several cellular levels. 17β-oestradiol may exert multiple neuromodulatory and neuroprotective functions. (1) Transcription factor activation: following the activation of cognate oestrogen receptors (ERs), oestrogens may induce the transcription of modulators of neuroprotection, e.g. the activation of nerve growth factor receptors (NGF-R) and the induction of non-amyloidogenic processing of APP. (2) Free radical scavenger (chemical shield): oxidative nerve cell death as induced, e.g. by Aβ or glutamate may be prevented by the free radical scavenging of induced reactive oxygen species (ROS). (3) The modulation of membrane receptors by oestrogens may change signal transduction by these receptors. Moreover oestrogens may interact with various intracellular signalling pathways (e.g. NF-κB, CREB, JNK), ultimately modulating transcriptional programs. (Modified after Behl & Holsboer 1999.)

Moreover, oestrogen can be seen as a pure antioxidant structure and may act as neuromodulator in cells completely lacking ER activities. Therefore, when studying potential protective effects of oestrogens, one has to take into account that oestrogen is, indeed, a player at many levels and may therefore represent a model hormone that can be active in various ways. An increasing understanding of the expression and function of the ER, and perhaps the identification of novel oestrogen binding sites or ERs, may yield an even better understanding of the activities of oestrogens in neuronal tissues.

References

Aronson MK, Ooi WL, Morgenstern H et al 1990 Women, myocardial infarction, and dementia in the very old. Neurology 40:1102–1106

Baeuerle PA, Henkel T 1994 Function and activation of NF-kappa B in the immune system. Annu Rev Immunol 12:141–179

Behl C 1999 Alzheimer's disease and oxidative stress: implications for novel therapeutic approaches. Prog Neurobiol 57:301–323

Behl C, Holsboer F 1999 The female sex hormone oestrogen as a neuroprotectant. Trends Pharmacol Sci 20:441–444

Behl C, Davis J, Cole GM, Schubert D 1992 Vitamin E protects nerve cells from amyloid β protein toxicity. Biochem Biophys Res Comm 186:944–952

Behl C, Davis JB, Lesley R, Schubert D 1994 Hydrogen peroxide mediates amyloid β protein toxicity. Cell 77:817–822

Behl C, Widmann M, Trapp T, Holsboer F 1995 17-β estradiol protects neurons from oxidative stress-induced cell death in vitro. Biochem Biophys Res Comm 216:473–482

Behl C, Lezoualc'h F, Trapp T, Widmann M, Skutella T, Holsboer F 1997a Glucocorticoids enhance oxidative stress-induced cell death in hippocampal neurons in vitro. Endocrinology 138:101–106

Behl C, Skutella T, Lezoualc'h F et al 1997b Neuroprotection against oxidative stress by estrogens: structure–activity relationship. Mol Pharmacol 51:535–541

Choi DW 1992 Excitotoxic cell death. J Neurobiol 23:1261–1276

Dyrks T, Dyrks E, Hartmann T, Masters C, Beyreuther K 1992 Amyloidogenicity of β/A4 and β/A4-bearing amyloid protein precursor fragments by metal-catalyzed oxidation. J Biol Chem 267:18210–18217

Ebadi M, Srinivasan SK, Baxi MD 1996 Oxidative stress and antioxidant therapy in Parkinson's disease. Prog Neurobiol 48:1–19

Evans DA, Funkenstein HH, Albert MS et al 1989 Prevalence of Alzheimer's disease in a community population of older persons. JAMA 262:2551–2556

Glenner GG, Wong CW 1984 Alzheimer's disease: initial report of the purification and characterization of a novel cerebrovascular amyloid protein. Biochem Biophys Res Comm 120:885–890

Green PS, Gordon K, Simpkins JW 1997 Phenolic A ring for the neuroprotective effects of steroids. J Steroid Biochem Mol Biol 63:229–235

Gustafsson JÅ 1997 The estrogen receptor beta — getting in on the action? Nat Med 3: 493–494

Halliwell B, Gutteridge JMC 1989 Free radicals in biology and medicine, 2nd edn. Oxford University Press, Oxford

Hardy J 1997 Amyloid, the presenilins and Alzheimer's disease. Trends Neurosci 20: 154–159

Harris ME, Hensley K, Butterfield A, Leedle RA, Carney JM 1995 Direct evidence of oxidative injury produced by the Alzheimer's β-amyloid (1-40) in cultured hippocampal neurons. Exp Neurol 131:193–202

Henderson VW, Paganini-Hill A, Emanuel CK, Dunn ME, Buckwalter JG 1994 Estrogen replacement therapy in older women. Comparisons between Alzheimer's disease cases and nondemented control subjects. Arch Neurol 51:896–900

Hensley K, Carney JM, Mattson MP et al 1994 A model for β-amyloid aggregation and neurotoxicity based on free radical generation by the peptide: relevance to Alzheimer's disease. Proc Natl Acad Sci USA 91:3270–3274

Improta-Brears T, Whorton AR, Codazzi F, York JD, Meyer T, McDonnell DP 1999 Estrogen-induced activation of mitogen-activated protein kinase requires mobilization of intracellular calcium. Proc Natl Acad Sci USA 96:4686–4691

Kelly MJ, Lagrange AH, Wagner EJ, Rønnekleiv OK 1999 Rapid effects of estrogen to modulate G protein-coupled receptors via activation of protein kinase A and protein kinase C pathways. Steroids 64:64–75

Kuiper GG JM, Shughrue P J, Merchenthaler I, Gustafsson JÅ 1998 The estrogen receptor beta subtype: a novel mediator of estrogen action in neuroendocrine systems. Front Neuroendocrinol 19:253–286

Lezoualc'h F, Behl C 1998 Transcription factor NF-κB: friend or foe of neurons? Mol Psychiatry 3:15–20

Lipton SA 1997 Janus faces of NF-κB: neurodestruction versus neuroprotection. Nat Med 3: 20–22

Luine VN 1994 Steroid hormone influences on spatial memory. Ann NY Acad Sci 743: 201–211

Masters CL, Simms G, Weinman NA, Multhaup G, McDonald BL, Beyreuther K 1985 Amyloid plaque core protein in Alzheimer disease and Down syndrome. Proc Natl Acad Sci USA 82:4245–4249

McEwen BS 1991 Non-genomic and genomic effects of steroids on neural activity. Trends Pharmacol Sci 12:141–147

McGeer PL, McGeer EG 1995 The inflammatory response system of brain: implications for therapy of Alzheimer and other neurodegenerative diseases. Brain Res Brain Res Rev 21:195–218

Moosmann B, Behl C 1999 The antioxidant neuroprotective effects of estrogens and phenolic compounds are independent from their estrogenic properties. Proc Natl Acad Sci USA 96:8867–8872

Moss RL, Gu Q 1999 Estrogen: mechanisms for a rapid action in CA1 hippocampal neurons. Steroids 64:14–21

Paganini-Hill A, Henderson VW 1994 Estrogen deficiency and risk of Alzheimer's disease in women. Am J Epidemiol 140:256–261

Paul SM, Purdy RH 1992 Neuroactive steroids. FASEB J 6:2311–2322

Rocca W, Hofman A, Brayne C et al 1991 Frequency and distribution of Alzheimer's disease in Europe: a collaborative study of 1980–1990 prevalence findings. Ann Neurol 3:381–390

Rogers J, Cooper NR, Webster S et al 1992 Complement activation by β-amyloid in Alzheimer disease. Proc Natl Acad Sci USA 89:10016–10020

Rosen DR, Siddique T, Patterson D et al 1993 Mutations in Cu/Zn superoxide dismutase gene are associated with familial amyotrophic lateral sclerosis. Nature 362:59–62 (erratum: 1993 Nature 364:362)

Rupprecht R 1997 The neuropsychophramacological potential of neuroactive steroids. J Psychiatr Res 31:297–314

Sapolsky RM 1992 Stress, the aging brain, and the mechanisms of neuron death. MIT Press, Cambridge, MA

Shughrue P J, Lane MV, Merchenthaler I 1999 Biologically active estrogen receptor-beta: evidence from in vivo autoradiographic studies with estrogen receptor alpha-knockout mice. Endocrinology 140:2613–2620

Simpkins JW, Singh M, Bishop J 1994 The potential role for estrogen replacement therapy in the treatment of the cognitive decline and neurodegeneration associated with Alzheimer's disease. Neurobiol Aging (suppl) 15:S195–S197

Singer CA, Figueroa-Masot XA, Batchelor RH, Dorsa DM 1999 The mitogen-activated protein kinase pathway mediates estrogen neuroprotection after glutamate toxicity in primary cortical neurons. J Neurosci 19:2455–2463

Smith MA, Sayre LM, Monnier VM, Perry G 1995 Radical AGEing in Alzheimer's disease. Trends Neurosci 18:172–176

Tang MX, Jacobs D, Stern Y et al 1996 Effect of oestrogen during menopause on risk and age of onset of Alzheimer's disease. Lancet 348:429–432

Toran-Allerand CD 1996 The estrogen/neurotrophin connection during neural development: is co-localization of estrogen receptors with the neurotrophins and their receptors biologically relevant? Dev Neurosci 18:36–48

Watters JJ, Dorsa DM 1998 Transcriptional effects of estrogen on neuronal neurotensin gene expression involve cAMP/protein kinase A-dependent signaling mechanisms. J Neurosci 18:6672–6680

Woolley CS, McEwen BS 1993 Roles of estradiol and progesterone in regulation of hippocampal dendritic spine density during the estrous cycle in the rat. J Comp Neurol 336:293–306

Woolley CS, Weiland NG, McEwen BS, Schwartzkroin PA 1997 Estradiol increases the sensitivity of hippocampal CA1 pyramidal cells to NMDA receptor-mediated synaptic input: correlation with dendritic spine density. J Neurosci 17:1848–1859

Xu H, Gouras GK, Greenfield JP et al 1998 Estrogen reduces neuronal generation of Alzheimer β-amyloid peptides. Nat Med 4:447–451

Yankner BA 1996 Mechanism of neuronal degeneration in Alzheimer's disease. Neuron 16: 921–932

DISCUSSION

Gustafsson: You see a neuroprotective effect of oestradiol in a nanomolar range. But when you talked about neuroprotective effects of the other artificial compounds you are talking about higher concentrations, even in the micromolar range. Is it likely that at the physiological concentrations of oestradiol there will be these non-genomic neuroprotective effects?

Behl: This is the crucial question: are these effects in a physiological range? We see antioxidant effects of oestradiol in the low micromolar range. We have to use a lot of oxidative pressure on these cells in order to induce cell death, and we therefore need correspondingly high levels of antioxidants to protect them. But compared with vitamin E, for example, oestrogen is working much better. Rather high concentrations of vitamin E are needed to protect cells in cell culture against oxidation. In addition, with respect to disease prevention or therapy, one has to look at pharmacological concentrations rather than just physiological concentrations.

Gustafsson: But it's a big jump. Oestradiol usually exists at 0.1–1 nM, and if I understand correctly, at these concentrations you don't see an effect in your *in vitro* system. You begin to see the effect when the concentration is raised by three orders of magnitude. Even in very diseased conditions I fail to see how these high levels could be reached *in vivo*.

Behl: That is correct. However, it is not possible to tell how much of these oestrogens might accumulate in cell membranes, specifically, and in the brain, in general. Also, oestrogen is not the only phenol in the body fluid, so there might be a lot of additive effects. In our cell culture systems we can get protection with lower concentrations by using certain culture conditions and cell densities. We wanted to demonstrate a defined biochemical effect of this molecule. It is safe to say that we have shown an intrinsic antioxidative activity.

Baulieu: I'm also concerned with the concentration issue. What is special about oestradiol? Is there a nerve receptor — a 'receiving' protein — with which oestradiol interacts specifically to promote the antioxidant effect? Have you tested other steroidal phenols, for instance, oestriol?

Behl: In the biochemical assays oestriol works just like 17β-oestradiol. With respect to the concentrations, we are currently using our favoured antioxidant molecule, the trimethyl phenol, in an *in vivo* experiment which is performed in the Mongolian gerbil. We get a global ischaemia and hope that we can relieve some amount of oxidation-related cell death. In that *in vivo* experiment, we used concentrations that are usually used with 17β-oestradiol and that are in a physiological range or a little bit higher. I don't know exactly what the outcome is because this experiment is still ongoing. We will see if the big advantage of these molecules would be that it's easier to get into the brain. I don't want to propose that everybody should have daily injections of oestrogens or other phenols in order to get the antioxidant levels high enough, but I want to make a point that for intervention we may use small molecules that carry phenols and may have antioxidant activities. I'm sure that in the libraries of the pharmaceutical companies, there are a lot of phenolic compounds with perfect pharmacology and also available information on *in vivo* toxicity, and the companies might check these compounds for additional antioxidant activities.

Baulieu: If a phenol ring system is attached to a small lipophilic molecule, could this be in principle an ideal drug? Do you believe that it is possible to replace the steroidal skeleton and obtain a better antioxidant effect with molecules which would not bind to the same membrane sites for oestrogens? So could you find a specific chemical structure replacing the steroid molecule but able to carry the antioxidant effect?

Behl: I agree. If you have a phenol group and additional groups, for instance lipophilic groups, then you can modulate and target either the accumulation in membranes, but oxidation also takes place in the aqueous environment of the cell. Melatonin, for example, is also an antioxidant and mainly protects DNA from oxidation in the nucleus. It does not accumulate in membranes, but instead scavenges radicals in hydrophilic environments. There are many body components that carry these intrinsic phenols, and I think it's only natural that these molecules protect themselves from oxidation and degradation, since they evolved over billions of years under this oxidative pressure.

Toran-Allerand: I would urge you not to abandon oestrogen entirely. At the beginning of your talk, you showed a list of factors that have been suggested as possibly involved in Alzheimer's disease in addition to oxidative stress, such as neurotrophins, inflammatory processes, Aβ and cholinergic neuron dysfunction. Oestrogen has been shown by numerous people to be involved in absolutely every one of these processes, including the regulation of inflammatory genes like complement. Oestrogen doesn't involve just one aspect of brain function but multiple aspects, and if one were to focus only on that portion of the oestrogen molecule that may be important for oxidative stress, one might lose out on other portions of the molecule that would be important for regulation of APP or neurotrophin activity, for instance.

Behl: That is true. I hope you did not get the impression I was saying we don't need oestrogen but just phenol. If I were a woman, I would count on oestrogen, but what would you say to male patients?

Toran-Allerand: I would hope that their aromatase activity is strong.

Gibbs: Yesterday, Jim Simpkins was telling us about how he has to optimize his kill conditions in order to be able to show neuroprotective effects of the low concentrations of oestrogen that he uses. From your pictures it looked to me as though you were getting more than a 50% kill. Is that contributing to the high concentrations that you need?

Behl: Yes, that is likely. When we initiated these projects, we were not looking at lowering concentrations to the physiological range; we were more interested in the biochemistry, in identifying the structure which has the antioxidant activity. With this culture system there is always the problem of cell densities and serum conditions — there are many variables. I would not focus on the concentration issue too much.

Gibbs: Did you see the same sort of interaction with glutathione that Jim Simpkins described?

Behl: No.

Gibbs: I think this interaction with glutathione is very important. I'm curious as to what is different between your systems.

Behl: When we add pure glutathione, we don't get it into our cells under these conditions. We have to use glutathione ester that more readily enters the cells. One also has to look at interactions not only with glutathione but also with vitamin C in the culture system as well as with other recycling systems. The perfect experiment would be to work in a clean culture environment: you also have to leave out the phenol red and other serum additives, all these little tricky things. It is very dangerous if you point to too low concentrations since these culture systems are very variable.

Watson: There is a theory that some of the 'dangerousness' of oestrogen has to do with going through these chemical reactions, producing reactive species that can then in a damaging way bind to proteins and DNA. This can occur in many other tissues where we are worried about oestrogen exposure being harmful instead of protective. When thinking of oestrogens as therapeutics, you have to keep in mind that there are other tissues that may be at risk from the same reaction. Might these other phenolic compounds participate in that?

Behl: That's exactly why we started to look at these compounds and to compare their pharmacology. We want to find out whether there is metabolism of these small phenolic compounds. You are right in that there are recycling mechanisms, and this is why we are looking for alternatives to oestrogen.

Fillit: We think of Alzheimer's disease as a multifactorial illness. We are getting into an era when we will see a lot of combination therapy, so we may be giving oestrogen, plus cholinesterase inhibitors, plus better antioxidants.

Behl: In collaboration with our memory clinic at the Institute, we are currently launching a pilot study with a combination of anti-inflammatory and antioxidant approaches, compared with antioxidants alone.

Bethea: At the beginning of your paper you mentioned serotonin and *N*-acetyl serotonin: are these antioxidants in the same manner as oestrogen? What do we know about serotonin and Alzheimer's?

Behl: In Alzheimer's there is a drop in certain neurotransmitter systems, including serotonin. Indeed, serotonin does have antioxidant activity. It is a central modulator, a very important neuroactive molecule, and from the biochemical standpoint it may also have some structural protective activities at its site of action besides the serotonin receptor-dependent effects.

Simpkins: We have seen the glutathione synergy with Aβ killing and with serum deprivation, but we haven't seen it with glutamate-induced toxicity. However, we believe that's complicated because glutamate depletes intracellular glutathione.

The logP water partition coefficient for oestradiol is about 4, which means, if you give oestradiol a choice the vast majority is going to go into a lipid environment. Christian Behl's comment is well taken: we don't know when we put a nanomolar concentration of oestradiol in the media, what concentration ends up in the plasma membrane. Related to the potential toxicity question, we've been concerned about the same thing. We think what may be going on with oestradiol is that the lipophilicity indicates that most of the molecules are going to be in the membrane, and we think they're going to be oriented in the membrane so that quinone can be formed in a fairly innocuous environment and then be reduced to oestradiol while still in the plasma membrane.

Henderson: Dietary oestrogens, particularly phytooestrogens, are ubiquitous and we are exposed to huge quantities of these compounds. Some phytooestrogens have phenolic structures. If they do have comparable antioxidant effects, might the effect of endogenous oestrogens be overwhelmed by these exogenous sources?

Behl: There are reports of certain diets and their association with Alzheimer's disease and atherosclerosis. Indeed, many of the phytooestrogens have phenolic structures. A lot is known about the metabolism of these compounds. After intake they may be metabolized in the liver before they enter the blood flow in the brain.

Gustafsson: It is clear that we are dealing with multiple mechanisms, both genomic and non-genomic. This is true not just for oestradiol but also for vitamin E. Everyone in the nuclear receptor field is looking for a vitamin E receptor, because there are indications that vitamin E can also control gene expression. It is possible that some of these phenomena that you see are mediated by nuclear receptors which might not yet have been cloned. Perhaps one way to find out what's going on would be to see whether by adding all these antioxidants to the cells you could see whether you turn on or off genes. This in itself doesn't necessarily imply a specific mechanism, because the redox status of a cell may in itself determine gene expression.

How can we be sure that these various phenomena in terms of oxidation have anything to do with Alzheimer's? Bruce Ames is trying to sell the message that our cells get more and more destroyed over a lifetime owing to these oxidative phenomena. His pet organelle is the mitochondrion: according to him these are hit hard and we don't have many left by the time we are 80. But I understand that this theory is now challenged. Some competing groups saw no effect on mitochondrial numbers or integrity. What happens when you have a brain infarction with hypoxia? I understand that in the aftermath of that process, a lot of oxidative phenomena occur which are quite destructive to nearby brain tissue. Does this lead to some kind of neurodegeneration like Alzheimer's?

Behl: These are acute models you are referring to. You won't get a similar picture as in Alzheimer's disease, which is an age-related pathological process. The trouble with the Alzheimer's field is that it may never be possible to imitate the long-term disease in mouse models completely since we don't get all the oxidative or other toxic influence on the mouse brain that occurs in the human brain over decades.

Neurohormonal signalling pathways and the regulation of Alzheimer β-amyloid metabolism

Sam Gandy and Suzana Petanceska

Department of Psychiatry, New York University, The Nathan S. Kline Institute for Psychiatric Research, Orangeburg, NY 10962, USA

Abstract. Alzheimer's disease (AD) is characterized by the intracranial accumulation of the 4 kDa amyloid β peptide (Aβ), following proteolysis of a ~700 amino acid, integral membrane precursor, the amyloid β precursor protein (APP). The best evidence causally linking APP to AD has been provided by the discovery of mutations within the APP coding sequence that segregate with disease phenotypes in autosomal dominant forms of familial AD (FAD). Though FAD is rare (<10% of all AD), the hallmark features—amyloid plaques, neurofibrillary tangles, synaptic and neuronal loss, neurotransmitter deficits, dementia—are indistinguishable when FAD is compared with typical, common, 'non-familial', or sporadic AD (SAD). Studies of some clinically relevant mutant APP molecules from FAD families have yielded evidence that APP mutations can lead to enhanced generation or aggregability of Aβ, consistent with a pathogenic role in AD. Other genetic loci for FAD have been discovered which are distinct from the immediate regulatory and coding regions of the APP gene, indicating that defects in molecules other than APP can also specify cerebral amyloidogenesis and FAD. To date, all APP and non-APP FAD mutations can be demonstrated to have the common feature of promoting amyloidogenesis of Aβ. Epidemiological studies indicate that postmenopausal women on oestrogen hormone replacement therapy (HRT) have their relative risk of developing SAD diminished by about one-third as compared with age-matched women not receiving HRT. Because of the key role of cerebral Aβ accumulation in initiating AD pathology, it is most attractive that oestradiol might modulate SAD risk or age-at-onset by inhibiting Aβ accumulation. A possible mechanistic basis for such a scenario is reviewed here.

2000 Neuronal and cognitive effects of oestrogens. Wiley, Chichester (Novartis Foundation Symposium 230) p 239–253

Alzheimer's disease is associated with an intracranial amyloidosis

'Amyloid' is a generic description applied to a heterogeneous class of tissue protein precipitates which have the common feature of β-pleated sheet secondary structure, a characteristic which confers affinity for the histochemical dye Congo red. Amyloids may be deposited in a general manner throughout the body (systemic amyloids) or confined to a particular organ (e.g. cerebral amyloid, renal amyloid). Alzheimer's disease (AD) is characterized by clinical evidence of cognitive failure in association with cerebral amyloidosis, cerebral intraneuronal neurofibrillary pathology, neuronal and synaptic loss and neurotransmitter deficits. The cerebral amyloid of AD is deposited around meningeal and cerebral vessels, as well as in grey matter. In grey matter, the deposits are multifocal, coalescing into miliary structures known as plaques. Parenchymal amyloid plaques are distributed in brain in a characteristic fashion, differentially affecting the various cerebral and cerebellar lobes and cortical laminae.

The main constituent of cerebrovascular amyloid was purified and sequenced by Glenner and Wong in 1984. This 40–42 amino acid polypeptide, designated 'β protein' (or, according to Masters and colleagues, 'A4'; now standardized as 'Aβ' by The Husby Commission), is derived from a 695–770 amino acid precursor, termed the amyloid precursor protein (APP), which was discovered by molecular cloning using oligonucleotide probes from the published amyloid peptide sequence. The primary citations for this molecular neuropathology review can be found elsewhere (Gandy & Greengard 1994).

Aβ is a catabolite of an integral precursor

The deduced amino acid sequence of APP predicts a protein with a single transmembrane domain. While many bioactivities for APP have been catalogued in the literature, Aβ might simply be a by-product of APP metabolism which is not necessarily intimately involved with APP function.

The proteolytic processing steps for APP have by now been definitively identified by purification and sequencing. The first to be identified involves cleavage within the Aβ domain. A large N-terminal fragment of the APP extracellular domain ('protease nexin-II', 'PN-II'; or sAPPα or APPsα, for soluble APP$_\alpha$) is released into the medium of cultured cells and into the cerebrospinal fluid, leaving a small non-amyloidogenic C-terminal fragment associated with the cell. This pathway is designated the α-secretory cleavage/release processing pathway for APP, because the enzyme(s) which perform(s) this non-amyloidogenic cleavage/release has been designated α-secretase. Thus, one important processing event in the biology of APP acts to preclude amyloidogenesis by proteolysing APP within the Aβ domain, between residues Aβ16 and Aβ17.

Until mid-1992, the prevailing notion was that Aβ production might be restricted to the brain and perhaps even then only in association with ageing and AD. This concept became obsolete with the discovery by several groups that a soluble Aβ species (presumably a forerunner of the aggregated fibrillar species which is deposited in senile plaque cores) is detectable in body fluids from various species and in the conditioned medium of cultured cells. Still, 'soluble Aβ' (to distinguish it from deposited Aβ amyloid) is not readily detectable in the lysates of most cultured cells. Soluble Aβ is apparently generated in several cellular compartments, including the endoplasmic reticulum, the Golgi apparatus, the *trans*-Golgi network and endosomes. Aβ production might begin by cleavage at the Aβ N-terminus by another enzyme designated β-secretase, probably beginning in the constitutive secretory pathway, i.e. the *trans*-Golgi network (TGN). This model accounts especially well for providing a codistribution opportunity involving APP and β-secretase prior to APP's encounter with the Aβ-domain-destroying α-secretase which appears to act primarily at the plasma membrane. Data from one familial AD (FAD) mutant (the 'Swedish' FAD mutant; discussed in detail below) support this model strongly. Other data indicate that at least some Aβ arises from APP molecules following their residence at the cell surface; presumably these APP molecules have escaped cell-surface α-secretase and have been internalized intact, encountering β-secretase in an endosome. An Aβ variant bearing Glu11 at its N-terminus has been recently noted as a prominent product of APP metabolism in cultured neurons (Gouras et al 1998). The generation and pharmacological regulation of [Glu11] Aβ formation appears to parallel that observed for standard [Asp1] Aβ. The cellular and biochemical origin of the alternative cleavage at Glu11 and its role in amyloidogenesis, if any, are all yet to be elucidated.

It is important to appreciate that one area of particular current interest is the elucidation of the molecular and cellular mechanisms by which the C-terminus of Aβ is generated, since this region of the APP molecule resides within an intramembranous domain. The protease(s) responsible for that cleavage, designated the γ-secretase(s), are particularly interesting not only because of the novelty of the active site-substrate reaction (i.e. within a membranous domain) but also owing to evidence that another class of FAD (known as presenilin-related FAD; Sherrington et al 1995) appears to act via control of γ-secretase function (Scheuner et al 1996, Borchelt et al 1996). Along this line, the Aβ C-terminus is heterogeneous, being composed mostly of peptides terminating at residue Aβ40; however, a small but important minority of peptides terminate at residue Aβ42, and it is these highly aggregatable peptides which are believed to initiate Aβ accumulation in all forms of the disease (Iwatsubo et al 1994, Lemere et al 1996).

The primary citations for the foregoing review of APP processing can be found elsewhere (Gandy & Greengard 1994).

Pathogenic APP mutations occur within or near the Aβ domain, and yield APP molecules which display proamyloidogenic properties *in vitro* and Alzheimer's-like phenotypes in transgenic mouse models *in vivo*

Certain mutations associated with familial cerebral amyloidoses have been identified within or near the Aβ region of the coding sequence of the APP gene.

These mutations segregate with the clinical phenotypes of either hereditary cerebral hemorrhage with amyloidosis, Dutch type (HCHWAD, or FAD-Dutch) or more typical FAD ('British', 'Indiana' FAD), and provide support for the notion that aberrant APP metabolism is a key feature of AD.

In FAD-Dutch, an uncharged glutamine residue is substituted for a charged glutamate residue at position 22 of Aβ. This mutated residue is located in the extracellular region of APP, within the Aβ domain, where it apparently exerts its proamyloidogenic effect by generating Aβ molecules which bear enhanced aggregation properties and perhaps also alter α-secretase cleavage. A 'Flemish' form of FAD involves a mutation at position 21 of Aβ and generates a phenotype which includes features of both the relatively pure vascular amyloidosis of 'Dutch FAD' and the more typical FAD involving parenchymal and vascular amyloidosis.

Mutations in APP which are apparently pathogenic for more typical FAD have also been discovered. In the first discovered FAD mutation, an isoleucine residue is substituted for a valine residue within the transmembrane domain, at a position two residues downstream from the C-terminus of the Aβ domain, a position equivalent to 'Aβ44' (if such a position existed). Although a conservative substitution, the mutation segregates with FAD in pedigrees of American, European ('British') and Asian origins, arguing against the possibility that the mutations represent irrelevant polymorphisms. Other pedigrees have been discovered in which affected members have either phenylalanyl ('Indiana') or glycyl residues at position 'Aβ44'. Neuropathological examination has verified the similarity of these individuals to typical sporadic AD (SAD) neuropathology. These 'Aβ44' mutant APPs are the most common of the FAD-causing APP mutations, and the mechanism by which the mutations exert their effects appears to be by enhancing generation of hyperaggregable, C-terminally extended Aβ peptides (especially Aβ42).

Another FAD pedigree has been established involving a large Swedish kindred. In this instance, tandem missense mutations occur just upstream of the amino terminus of the Aβ domain. Transfection of cultured cells with APP molecules containing the 'Swedish' missense mutations results in the production of six- to eightfold excess soluble Aβ above that generated from wild-type APP. This is the first (and, to date, only) example of Alzheimer's disease apparently caused by excessive Aβ production. An important issue for clarification in SAD will be to establish whether hyperaggregation or hyperproduction of Aβ (or neither) is/are important predisposing factor(s) to this much more commonly encountered clinical entity. The primary citations for the foregoing review of the genetic–neuropathological correlations of APP mutant FAD can be found elsewhere (Gandy & Greengard 1994).

Some of the most exciting and newest evidence in support of the pathogenicity of these mutants has arisen from the recent demonstration that certain mutant APP transgenic mice exhibit not only cerebral amyloidosis but also neurofibrillary tangle epitopes and neuronal loss (Calhoun et al 1998). If these animals also meet key behavioural criteria, Koch's postulates will have been met for the re-creation of all key features of Alzheimer's disease in a species not otherwise susceptible to the illness.

Signal transduction regulates the relative utilization of APP processing pathways in cultured cell lines, in primary neurons and in rodents *in vivo*

As work continues toward molecular discovery of α-, β- and γ-secretase enzymes, substantial progress has been made towards understanding the regulation of APP cleavage. For example, the relative utilization of the various alternative APP processing pathways appears to be at least partially cell-type determined, with transfected AtT20 cells secreting virtually all APP molecules whereas glia release few or none. In neuronal-like cells, the state of differentiation plays a role in determining the relative utilization of the pathways, with the differentiated neuronal phenotype being associated with relatively diminished basal utilization of the non-amyloidogenic α-secretase cleavage/release pathway.

It is also well-accepted that certain signal transduction systems involving protein phosphorylation are important regulators of APP cleavage, acting in many cases by stimulating the relative activity of non-amyloidogenic cleavage by α-secretase. The role of protein kinase C (PKC) and PKC-linked first messengers has received the most attention. In many types of cultured cells, activation of PKC by phorbol esters dramatically stimulates APP proteolysis (Buxbaum et al 1990) and sAPP$_\alpha$ cleavage/release (Caporaso et al 1992, Gillespie et al 1992, Sinha & Lieberburg 1992). PKC-stimulated α-secretory cleavage of APP may also be induced by the application of neurotransmitters and other first messenger compounds whose receptors are linked to PKC (Buxbaum et al 1992, Nitsch et al 1992). Okadaic acid, an inhibitor of protein phosphatases 1 and 2A, also increases APP proteolysis and release via the α-secretase pathway (Buxbaum et al 1990). Indeed, it has recently been shown in primary neuronal cultures that protein phosphatase inhibition by okadaic acid is much more effective in lowering Aβ production than is PKC activation (Gouras et al 1998). Thus, either stimulation of PKC or inhibition of protein phosphatases 1 and 2A is sufficient to produce a dramatic acceleration of non-amyloidogenic APP degradation. PKC-activated APP processing can be demonstrated to diminish generation of Aβ in virtually every situation tested to date, including primary neuronal cultures and rodent brain *in vivo* (Buxbaum et al 1993, Hung et al 1993, Savage et al 1998). This phenomenon is sometimes referred to as 'regulated cleavage of APP' or 'reciprocal regulation of APP metabolism'; i.e., when certain signals are activated, α-secretase cleavage increases and β- and Glu11-secretase cleavages decrease (Gouras et al 1998).

Hormones such as 17β-oestradiol (Jaffe et al 1994, Xu et al 1998, Chang et al 1997) and dihydroepiandrostenedione (DHEA; Danenboerg et al 1996) are other signal transduction compounds which can apparently regulate APP metabolism in cultured cells. Several of these investigators (Xu et al 1998, Chang et al 1997) have demonstrated that oestradiol diminishes Aβ generation, while others (Jaffe et al 1994, Danenboerg et al 1996, Xu et al 1998) have documented accumulation of sAPP in the conditioned media of cultured cells treated with oestradiol (Jaffe et al 1994, Xu et al 1998) or DHEA (Danenboerg et al 1996). One attractive explanation for the accumulation of sAPP is that steroids might stimulate either the amount or the activity of α-secretase. This explanation is consistent with the finding that the cell-associated product of this activity (the 9 kDa non-amyloidogenic C-terminal fragment) is increased by oestrogen treatment (Jaffe et al 1994). For oestradiol in particular, one means by which the activity of α-secretase might increase is by an increase in the level or activity of PKC, leading to increased 'regulated APP cleavage' along the α-secretase pathway. Oestrogen has been shown to regulate PKC in both normal and neoplastic tissue (Maeda & Lloyd 1993, Maizels et al 1992, Kamel & Kubajak 1988) and to potentiate the phorbol ester-stimulated release of luteinizing hormone from cultured pituitary cells (Drouva et al 1990). The effects of oestrogen on PKC might be mediated by polypeptide growth factors (Dickson et al 1986a,b,c, Bates et al 1988, Lippman et al 1988), many of which activate PKC and a number of which have been demonstrated to enhance the accumulation of sAPP in the conditioned media of cultured cells (Refolo et al 1989, Fukuyama et al 1993). The colocalization of oestrogen receptors in the basal forebrain, cerebral cortex and hippocampus with the receptors for nerve growth factor (NGF), as well as with NGF itself, is consistent with the idea that growth factors might play a role in mediating the effects of oestrogen in the CNS (Miranda et al 1993).

Extending this line of reasoning, a more complex relationship involving intercellular signals and APP metabolism is currently evolving: both steroid hormones and neurotrophins have been demonstrated to potentiate the ability of incoming neurotransmitters to activate the 'regulated cleavage of APP', as measured by sAPP release (Danenberg et al 1995, Fisher et al 1996, Haring et al 1998, Roßner et al 1998a,b). The 'regulated cleavage' model makes it likely that when compounds from either or both of these classes of substances (i.e. steroid hormones, neurotrophins) are present at the time of neurotransmitter application, the fold-diminution of Aβ and [Glu11] Aβ will be greatly enhanced. Assessment of this model is currently underway.

Insights into mechanism(s) of regulated APP processing

In general terms, the possible mechanisms for activated processing of integral molecules can be conceptualized as involving either activation or redistribution of either the substrate (i.e. APP) or the enzyme (i.e. α-secretase). Based on APP cytoplasmic tail mutational analyses (da Cruz e Silva et al 1993, Sahasrabudhe et al 1992), the 'substrate activation' model (Gandy et al 1988) is inadequate to explain activated processing of APP, since abrogation of APP cytoplasmic tail phosphorylation fails to abolish PKC-regulated α-cleavage. In order to test the model of PKC-regulated APP redistribution, Xu et al demonstrated that the PKC-stimulated biogenesis of nascent transport vesicles from the TGN includes vesicles which bear [$^{35}SO_4$]APP (Xu et al 1995). These vesicles would be predicted to fuse with the plasma membrane where codistribution of APP and α-secretase would take place, leading to generation of sAPP$_\alpha$ at the plasma membrane and release of that sAPP$_\alpha$ into the culture medium. Given the preponderant localization of APP within the TGN (Caporaso et al 1994), it is likely that PKC-regulated TGN vesicle biogenesis plays a role in regulating APP metabolism, although its importance cannot be completely assessed until the identification of all relevant PKC targets of this apparatus has been completed.

Along a related line of investigation, Bosenberg et al (1993) succeeded in reconstituting faithful activated processing of the transforming growth factor (TGF)-α precursor in porated cells in the virtual absence of cytosol, and in the presence of N-ethylmaleimide or 2.5 M NaCl. The preservation of activated processing under such conditions suggests that extensive vesicular trafficking is unlikely to be required for all forms of activated processing, and supports a model of direct or indirect enzyme activation which involves an intrinsic plasma membrane mechanism (e.g. direct phosphorylation of an α-secretase or its regulatory subunit). Cell biological and molecular cloning data indicate that APP α-secretases include, or are related to, the metalloproteinase family of adamalysins, and TGF-α and pro-tumour necrosis factor convertases. These enzymes appear to have overlapping specificities for these substrates. Several candidate secretases have recently been successfully cloned, and these data should facilitate resolution of the issue of which enzymes are most important for cleaving which substrates. (Black et al 1997, Peschon et al 1998, Buxbaum et al 1998). Aspartyl proteinases are the most important α-secretases in yeast, and their potential roles, if any, in mammalian APP metabolism, remain to be elucidated (Komano et al 1998).

Therapeutic manipulation of Aβ generation
via ligand or hormonal manipulation

From a therapeutic standpoint, then, the modulation of APP metabolism via signal transduction pharmacology might be beneficial in individuals with, or at risk for,

AD (Sinha & Lieberburg 1992, Gandy & Greengard 1992, 1994), and it is along this line that 17β-oestradiol—via its possible ability to elevate PKC activity—came to be identified as a potential modulator of Aβ metabolism. The ability of oestrogen to modulate Aβ metabolism *in vivo* was recently tested in two animal systems: guinea pigs and plaque-forming mice (Duff et al 2000, Petanceska et al 2000). Prolonged ovariectomy of guinea pigs was associated with a pronounced increase in brain Aβ levels as compared to intact animals. This ovariectomy-induced increase was significantly reduced in guinea pigs receiving 17β-oestradiol. Similarly, ovariectomy of transgenic, plaque-forming mice at an age when Aβ deposition is incipient resulted in enhanced brain amyloidosis. The contributions of altered Aβ generation and altered Aβ clearance to the observed changes in brain Aβ levels remain to be determined. Further studies are also needed to elucidate whether the observed effects on Aβ metabolism occur in response to activation of brain oestrogen receptors or whether they are mediated by oestrogen receptor-independent mechanisms. However, these two studies strongly infer that cessation of ovarian oestrogen production in postmenopausal women might facilitate Aβ deposition by increasing the local concentrations of Aβ peptides in brain and that the negative correlation between hormone replacement therapy of postmenopausal women and AD is at least in part due to the Aβ-lowering effect of oestrogen.

Oestradiol might also modify other factors contributing to Aβ deposition and fibril formation, including the processing of soluble Aβ into an aggregated form (Burdick et al 1992, Pike et al 1993) and/or the association of Aβ with other molecules, such as α1-antichymotrypsin (Abraham et al 1988), heparan sulfate proteoglycan (Snow et al 1994) and apolipoprotein E (Wisniewski & Frangione 1992). In particular, it has recently been recognized that apolipoprotein E is an obligatory participant in Aβ accumulation (Bales et al 1997), and that apolipoprotein E isoforms play important roles as differential determinants of risk for AD (Saunders et al 1993, Strittmatter et al 1993). Human post mortem (Rebeck et al 1993, Polvikoski et al 1995) and cell culture (Yang et al 1999) data are also beginning to indicate that apolipoprotein E isoforms exert at least some of their effects via controlling Aβ accumulation, perhaps at the stage of clearance of Aβ peptides. Apolipoprotein E-isoform-specific synaptic remodelling also appears to be a consistent and potentially relevant observation (Arendt et al 1997, Sun et al 1998).

Events beyond Aβ deposition may also be crucial in determining the eventual toxicity of Aβ plaques. While aggregation of Aβ is important for *in vitro* models of neurotoxicity (Pike et al 1993), the relevance of these phenomena for the pathogenesis of AD is unclear, since Aβ deposits may occur in normal ageing in the absence of any evident proximate neuronal injury (Crystal et al 1988, Masliah et al 1990, Berg et al 1993, Delaere et al 1993). This suggests that other events must distinguish 'diffuse' cerebral amyloidosis with wispy deposits of Aβ alone (Masliah et al 1990) from 'full-blown' AD with neuritic plaques, and neuron and synapse loss. One intriguing possible contributing factor is the association of complement components with Aβ. In cerebellum, where Aβ deposits appear to cause no injury, plaques are apparently free of associated complement, while in the forebrain, complement associates with plaques, perhaps becoming activated and injuring the surrounding cells (Lue & Rogers 1992). Other, as-yet undiscovered plaque-associated molecules may also play important roles.

It should also be possible to apply animal models of Aβ deposition and neurodegeneration (Calhoun et al 1998) to begin to elucidate whether the effect of oestrogen to delay or prevent AD in postmenopausal women is due to one or a combination of its various documented activities relevant to the pathogenesis of AD, i.e. lowering Aβ load, as reviewed above, sustaining the basal forebrain cholinergic system (Luine 1985), modulating interactions with growth factors and/or their receptors (Miranda et al 1994, Sohrabji et al 1994), supporting neuritic plasticity (Woolley 1998, Stone et al 1998), or serving as an antioxidant (Gridley et al 1998). One cannot yet exclude the possibility that oestradiol plays several of these roles in its apparent (Yaffe et al 1998, Mayeux & Gandy 1999) ability to delay and/or prevent AD.

Acknowledgements

This work was supported by the USPHS, the New York State Office of Mental Health, and the Research Foundation for Mental Hygiene. Dr Gandy receives or has received extramural support from the Women's Health Research Institute of Wyeth-Ayerst Pharmaceuticals, a division of American Home Corporation, and honoraria from consulting Parke-Davis Pharmaceuticals, and Hoffman La Roche. Parts of this work have appeared elsewhere (*Trends in Endocrinolgy and Metabolism* vol 10 No 9 pp 273–279) and are included here with permission from Elsevier Sciences.

References

Abraham CR, Selkoe DJ, Potter H 1988 Immunochemical identification of the serine protease inhibitor α₁-antichymotrypsin in the brain amyloid deposits of Alzheimer's disease. Cell 52:487–501

Arendt T, Schindler C, Bruckner MK et al 1997 Plastic neuronal remodeling is impaired in patients with Alzheimer's disease carrying apolipoprotein epsilon 4 allele. J Neurosci 17:516–529

Bales KR, Verina T, Dodel RC et al 1997 Lack of apolipoprotein E dramatically reduces amyloid beta-peptide deposition. Nat Genet 17:263–264

Bates SE, Davidson NE, Valverius EM et al 1988 Expression of transforming growth factor alpha and its messenger ribonucleic acid in human breast cancer: its regulation by estrogen and its possible functional significance. Mol Endocrinol 2:543–555

Berg L, McKeel DW Jr, Miller JP, Baty J, Morris JC 1993 Neuropathological indexes of Alzheimer's disease in demented and nondemented persons aged 80 years and older. Arch Neurol 50:349–358

Black RA, Rauch CT, Kozlosky CJ et al 1997 A metalloproteinase disintegrin that releases tumour-necrosis factor-alpha from cells. Nature 385:729–733

Borchelt DR, Thinakaran G, Eckman CB et al 1996 Familial Alzheimer's disease-linked presenilin 1 variants elevate Aβ1–42/1–40 ratio *in vitro* and *in vivo*. Neuron 17:1005–1013

Bosenberg MW, Pandiella A, Massagué J 1993 Activated release of membrane-anchored TGF-α in the absence of cytosol. J Cell Biol 122:95–101

Burdick D, Soreghan B, Kwon M et al 1992 Assembly and aggregation properties of synthetic Alzheimer's A4/β amyloid peptide analogs. J Biol Chem 267:546–554

Buxbaum JD, Gandy SE, Cicchetti P et al 1990 Processing of Alzheimer β/A4 amyloid precursor protein: modulation by agents that regulate protein phosphorylation. Proc Natl Acad Sci USA 87:6003–6006

Buxbaum JD, Oishi M, Chen HI et al 1992 Cholinergic agonists and interleukin 1 regulate processing and secretion of the Alzheimer β/A4 amyloid protein precursor. Proc Natl Acad Sci USA 89:10075–10078

Buxbaum JD, Koo EH, Greengard P 1993 Protein phosphorylation inhibits production of Alzheimer amyloid β/A4 peptide. Proc Natl Acad Sci USA 90:9195–9198

Buxbaum JD, Liu KN, Luo Y et al 1998 Evidence that tumor necrosis factor alpha converting enzyme is involved in regulated alpha-secretase cleavage of the Alzheimer amyloid protein precursor. J Biol Chem 273:27765–27767

Calhoun ME, Wiederhold KH, Abramowski D et al 1998 Neuron loss in APP transgenic mice. Nature 395:755–756

Caporaso GL, Gandy SE, Buxbaum JD, Ramabhadran TV, Greengard P 1992 Protein phosphorylation regulates secretion of Alzheimer β/A4 amyloid precursor protein. Proc Natl Acad Sci USA 89:3055–3059

Caporaso GL, Takei K, Gandy S et al 1994 Morphologic and biochemical analysis of the intracellular trafficking of the Alzheimer β/A4 amyloid precursor protein. J Neurosci 14:3122–3138

Chang D, Kwan J, Timiras PS 1997 Estrogens influence growth, maturation and amyloid beta-peptide production in neuroblastoma cells and in a beta-APP transfected kidney 293 cell line. Adv Exp Med Biol 429:261–271

Crystal H, Dickson D, Fuld P et al 1988 Clinico-pathologic studies in dementia: nondemented subjects with pathologically confirmed Alzheimer's disease. Neurology 38:1682–1687

da Cruz e Silva OA, Iverfeldt K, Oltersdorf T et al 1993 Regulated cleavage of Alzheimer β-amyloid precursor protein in the absence of the cytoplasmic tail. Neuroscience 57:873–877

Danenberg HD, Haring R, Heldman E et al 1995 Dehydroepiandrosterone augments M1-muscarinic receptor-stimulated amyloid precursor protein secretion in desensitized PC12M1 cells. Ann NY Acad Sci 774:300–303

Danenboerg HD, Haring R, Fisher A, Pittel Z, Gurwitz D, Heldman E 1996 Dehydroepiandrosterone (DHEA) increases production and release of Alzheimer's amyloid precursor protein. Life Sci 59:1651–1657

Delaere P, He Y, Fayet G, Duyckaerts C, Hauw JJ 1993 βA4 deposits are constant in the brain of the oldest old: an immunocytochemical study of 20 French centenarians. Neurobiol Aging 14:191–194

Dickson RB, Huff KK, Spencer EM, Lippman ME 1986a Induction of epidermal growth factor-related polypeptides by 17 beta-estradiol in MCF-7 human breast cancer cells. Endocrinology 118:138–142

Dickson RB, Bates SE, McManaway ME, Lippman ME 1986b Characterization of estrogen responsive transforming activity in human breast cancer cell lines. Cancer Res 46:1707–1713

Dickson RB, McManaway ME, Lippman ME 1986c Estrogen-induced factors of breast cancer cells partially replace estrogen to promote tumor growth. Science 232:1540–1543

Drouva SV, Geronne I, Laplante E, Rérat E, Enjalbert A, Kordon C 1990 Estradiol modulates protein kinase C activity in the rat pituitary *in vivo* and *in vitro*. Endocrinology 126:536–544 (erratum: 1990 Endocrinology 126:1995)

Duff K, Petanceska S, Yu X et al 2000 Enhanced amyloidosis in response to ovariectomy in a transgenic model of Alzheimer's disease. submitted

Fisher A, Heldman E, Gurwitz D et al 1996 M1 agonists for the treatment of Alzheimer's disease. Novel therapeutic properties and clinical update. Ann NY Acad Sci 777:189–196

Fukuyama R, Chandrasekaran K, Rapoport SI 1993 Nerve growth factor-induced neuronal differentiation is accompanied by differential induction and localization of the amyloid precursor protein (APP) in PC12 cells and variant PC12S cells. Brain Res Mol Brain Res 17:17–22

Gandy S, Greengard P 1992 Amyloidogenesis in Alzheimer's disease. Some possible therapeutic opportunities. Trends Pharmacol Sci 13:108–113

Gandy S, Greengard P 1994 Processing of Alzheimer Aβ-amyloid precursor protein: cell biology, regulation, and role in Alzheimer disease. Intl Rev Neurobiol 36:29–50

Gandy S, Czernik AJ, Greengard P 1988 Phosphorylation of Alzheimer disease amyloid precursor peptide by protein kinase C and Ca^{2+}/calmodulin-dependent protein kinase II. Proc Natl Acad Sci USA 85:6218–6221

Gillespie SL, Golde TE, Younkin SG 1992 Secretory processing of the Alzheimer amyloid β/A4 protein precursor is increased by protein phosphorylation. Biochem Biophys Res Commun 187:1285–1290

Gouras GK, Xu H, Jovanovic JN et al 1998 Generation and regulation of beta-amyloid peptide variants by neurons. J Neurochem 71:1920–1925

Gridley KE, Green PS, Simpkins JW 1998 A novel, synergistic interaction between 17β-estradiol and glutathione in the protection of neurons against Aβ 25–35-induced toxicity *in vitro*. Mol Pharmacol 4:874–880

Haring R, Fisher A, Marciano D et al 1998 Mitogen-activated protein kinase-dependent and protein kinase C-dependent pathways link the m1 muscarinic receptor to β-amyloid precursor protein secretion. J Neurochem 71:2094–2103

Hung AY, Haass C, Nitsch RM et al 1993 Activation of protein kinase C inhibits cellular production of the amyloid β-protein. J Biol Chem 268:22959–22962

Iwatsubo T, Odaka A, Suzuki N, Mizusawa H, Nukina N, Ihara Y 1994 Visualization of A beta 42 (43) and A beta 40 in senile plaques with end-specific A beta monoclonals: evidence that an initially deposited species is A beta 42(43). Neuron 13:45–53

Jaffe AB, Toran-Allerand CD, Greengard P, Gandy SE 1994 Estrogen regulates metabolism of Alzheimer amyloid beta precursor protein. J Biol Chem 269:13065–13068

Kamel F, Kubajak CL 1988 Gonadal steroid effects on LH response to arachidonic acid and protein kinase C. Am J Physiol 255:E314–E321

Komano HM, Seeger M, Gandy S, Wang GT, Krafft GA, Fuller RS 1998 Involvement of cell surface glycosyl-phosphatidylinositol-linked aspartyl proteases in alpha-secretase-type cleavage and ectodomain solubilization of human Alzheimer beta-amyloid precursor protein in yeast. J Biol Chem 273:31648–31651

Lemere CA, Blusztajn JK, Yamaguchi H, Wisniewski T, Saido TC, Selkoe DJ 1996 Sequence of deposition of heterogeneous amyloid beta-peptides and *APOE* in Down syndrome, implications for initial events in amyloid plaque formation. Neurobiol Dis 3:16–32

Lippman ME, Dickson RB, Gelmann EP et al 1988 Growth regulatory peptide production by human breast carcinoma cells, J Steroid Biochem 30:53–61

Lue L-F, Rogers J 1992 Full complement activation fails in diffuse plaques of the Alzheimer's disease cerebellum. Dementia 3:308–313

Luine VN 1985 Estradiol increases choline acetyltransferase activity in specific basal forebrain nuclei and projection areas of female rats. Exp Neurol 89:484–490

Maeda T, Lloyd RV 1993 Protein kinase C activity and messenger RNA modulation by estrogen in normal and neoplastic rat pituitary tissue. Lab Invest 68:472–480

Maizels ET, Miller JB, Cutler RJ et al 1992 Estrogen modulates Ca^{2+}-independent lipid-stimulated kinase in the rabbit corpus luteum of pseudopregnancy. Identification of luteal estrogen-modulated lipid-stimulated kinase as protein kinase C delta. J Biol Chem 267:17061–17068

Masliah E, Terry RD, Mallory M, Alford M, Hansen LA 1990 Diffuse plaques do not accentuate synapse loss in Alzheimer's disease. Am J Pathol 137:1293–1297

Mayeux R, Gandy S 1999 Cognitive impairment, Alzheimer's disease and other dementias. In: Goldman MB, Hatch MC (eds) Women and health. Academic Press, San Diego, CA, p 1227–1238

Miranda RC, Sohrabji F, Toran-Allerand CD 1993 Neuronal colocalization of mRNAs for neurotrophins and their receptors in the developing central nervous system suggests a potential for autocrine interactions. Proc Natl Acad Sci USA 90:6439–6443

Miranda RC, Sohrabji F, Toran-Allerand CD 1994 Interactions of estrogen with the neurotrophins and their receptors during neural development. Horm Behav 28:367–375

Nitsch RM, Slack BE, Wurtman RJ, Growdon JH 1992 Release of Alzheimer amyloid precursor derivatives stimulated by activation of muscarinic acetylcholine receptors. Science 258: 304–307

Petanceska S, Nagy V, Frail DE, Gandy S 2000 Ovariectomy and 17β-estradiol modulate the levels of amyloid β peptides in brain. Neurology, submitted

Peschon JJ, Slack JL, Reddy P et al 1998 An essential role for ectodomain shedding in mammalian development. Science 282:1281–1284

Pike CJ, Burdick D, Walencewicz AJ, Glabe CG, Cotman CW 1993 Neurodegeneration induced by β-amyloid peptides in vitro: the role of peptide assembly state. J Neurosci 13:1676–1687

Polvikoski T, Sulkava R, Haltia M et al 1995 Apolipoprotein E, dementia, and cortical deposition of beta-amyloid protein. N Engl J Med 333:1242–1247

Rebeck GW, Reiter JS, Strickland DK, Hyman BT 1993 Apolipoprotein E in sporadic Alzheimer's disease: allelic variation and receptor interactions. Neuron 11:575–580

Refolo LM, Salton SR, Anderson JP, Mehta P, Robakis NK 1989 Nerve and epidermal growth factors induce the release of the Alzheimer amyloid precursor from PC 12 cell cultures. Biochem Biophys Res Commun 164:664–670

Roßner S, Ueberham U, Schliebs R, Perez-Polo JR, Bigl V 1998a The regulation of amyloid precursor protein metabolism by cholinergic mechanisms and neurotrophic receptor signalling. Prog Neurobiol 56:541–569

Roßner S, Ueberham U, Schliebs R, Perez-Polo JR, Bigl V 1998b p75 and trkA receptor signaling independently regulate amyloid precursor protein mRNA expression, isoform composition and protein secretion in PC12 cells. J Neurochem 71:757–766

Sahasrabudhe SR, Spruyt MA, Muenkel HA, Blume AJ, Vitek MP, Jacobsen JS 1992 Release of amino-terminal fragments from amyloid precursor protein reporter and mutated derivatives in cultured cells. J Biol Chem 267:25602–25608

Saunders AM, Strittmatter WJ, Schmechel D et al 1993 Association of apolipoprotein E allele ε4 with late-onset familial and sporadic Alzheimer's disease. Neurology 43:1467–1472

Savage MJ, Trusko SP, Howland DS et al 1998 Turnover of amyloid beta-protein in mouse brain and acute reduction of its level by phorbol ester. J Neurosci 18:1743–1752

Scheuner D, Eckman C, Jensen M et al 1996 Secreted amyloid beta-protein similar to that in the senile plaques of Alzheimer's disease is increased in vivo by the presenilin 1 and 2 and APP mutations linked to familial Alzheimer's disease. Nat Med 2:864–870

Sherrington R, Rogaev EI, Liang Y et al 1995 Cloning of a gene bearing missense mutations in early-onset familial Alzheimer's disease. Nature 375:754–760

Sinha S, Lieberburg I 1992 Normal metabolism of the amyloid precursor protein (APP). Neurodegeneration 1:169–175

Snow AD, Sekiguchi R, Nochlin D et al 1994 An important role of heparan sulfate proteoglycan (Perlecan) in a model system for the deposition and persistence of fibrillar Aβ amyloid in rat brain. Neuron 12:219–234

Sohrabji F, Greene LA, Miranda RC, Toran-Allerand CD 1994 Reciprocal regulation of estrogen and NGF receptors by their ligands in PC12 cells. J Neurobiol 25:974–988

Stone DJ, Rozovsky I, Morgan TE, Anderson CP, Finch CE 1998 Increased synaptic sprouting in response to estrogen via an apolipoprotein E-dependent mechanism: implications for Alzheimer's disease. J Neurosci 18:3180–3185

Strittmatter WJ, Saunders AM, Schmechel D et al 1993 Apolipoprotein E: high-avidity binding to β-amyloid and increased frequency of type 4 allele in late-onset familial Alzheimer disease. Proc Natl Acad Sci USA 90:1977–1981

Sun Y, Wu S, Bu G et al 1998 Glial fibrillary acidic protein-apolipoprotein E (apoE) transgenic mice: astrocyte-specific expression and differing biological effects of astrocyte-secreted apoE3 and apoE4 lipoproteins. J Neurosci 18:3261–3272

Wisniewski T, Frangione B 1992 Apolipoprotein E: a pathological chaperone in patients with cerebral and systemic amyloid. Neurosci Lett 135:235–238

Woolley CS 1998 Estrogen-mediated structural and functional synaptic plasticity in the female rat hippocampus. Horm Behav 34:140–148

Xu H, Greengard P, Gandy S 1995 Regulated formation of Golgi secretory vesicles containing Alzheimer beta-amyloid precursor protein. J Biol Chem 270:23243–23245

Xu H, Gouras GK, Greenfield JP et al 1998 Estrogen reduces neuronal generation of Alzheimer β-amyloid peptides. Nat Med 4:447–451

Yaffe K, Sawaya G, Lieberburg I, Grady D 1998 Estrogen therapy in postmenopausal women: effects on cognitive function and dementia. JAMA 279:688–695

Yang DS, Small DH, Seydel U et al 1999 Apolipoprotein E promotes the association of amyloid-β with Chinese Hamster Ovary cells in an isoform-specific manner. Neuroscience 90:1217–1226

DISCUSSION

Levin: Your conclusion is that it takes several days for oestrogen to decrease Aβ formation in your system. If you are going to implicate any of the signal transduction systems, they will have to stay up for a long period. Do you know the temporal nature of oestradiol-induced PKC, for example?

Gandy: The best way to address this would be to do the primary culture experiment (or use an oestrogen-sensitive cell line), give treatments, and do kinase assays and amyloid metabolism assays in parallel. I don't know whether the literature on oestrogen elevation of protein kinase activity is particularly clear on that issue.

Levin: Does the ApoE polymorphism phenotype that's connected to generation give us any other insights as to the mechanisms of oestrogen protection? Is that also a more highly susceptible oxidative product that could be analogous to apolipoprotein (Apo)D and low density lipoprotein oxidation in the vasculature in terms of the deleterious effects on the cell?

Gandy: There is an oestrogen response element in ApoE, and it has been demonstrated to be oestrogen responsive. It is therefore possible that oestrogen is acting in AD exclusively or partially by regulating transcription of ApoE.

However, one wouldn't expect that to be isoform dependent, and the major linkage risk factor story for ApoE in Alzheimer's is the ApoE E4 variant versus others. In terms of the antioxidant capacity, it is true that the rank order for ApoE risk factors in terms of AD exactly parallels their activity as antioxidants: E4 increases risk, E3 is intermediate and E2 may actually be protective. For example, the ApoE genotype can modulate FAD caused by APP mutations. That is, there are two individuals with APP mutations who have the E2/E3 genotype, and they're far out from their expected age of onset: it's not clear whether they're going to die free of disease or whether they're just going to get disease very late.

Murphy: You know that I'm interested in the possibility that APP is a synaptic structural protein. When you treat the primary neurons with oestrogen for several days, this is quite enough time to see an increase in synaptogenesis. Did you see an increase in overall levels of APP between control and oestrogen-treated cultures? One of the mechanisms through which oestrogen is working may have to do with cleavage of APPs because it is a synaptic protein and oestrogen can cause synaptogenesis.

Gandy: I couldn't exclude redistribution, but total levels of APP holoprotein don't change. The APP mRNA levels don't change.

Gibbs: My understanding of current thinking about amyloid plaques in Alzheimer's is that there are two schools of thought. The first is that amyloid plaques are interesting but unimportant with respect to the pathophysiology of Alzheimer's; that they are like gravestones showing where the damage has previously been done, and that the neurofibrillary tangles are thought to be the real problem. In the second school of thought, people think that the amyloid is important. Where do you stand on this issue?

Gandy: It is possible to explain all genetic forms of Alzheimer's by placing amyloidosis as an early step. All APP mutations, presenilin mutations and ApoE polymorphisms can be demonstrated to facilitate amyloidosis. It is possible that amyloid plaques are just gravestones, but there is nothing about the clinical or pathological phenotype to suggest this is the case. Starting with amyloidosis is the best lead we have now.

Gibbs: Since Alzheimer's occurs much more frequently in women than in men because women live longer, isn't there tissue available from women who have been on oestrogen versus women who have not, where one could then quantify numbers of amyloid plaques and see whether there is a difference?

Gandy: I have seen grant proposals on this issue, but no data or published papers. The 'neuropathology of menopause' is terrifically exciting, but I am unaware of substantial data which address it.

Henderson: At our Alzheimer's centre, my colleagues have retrospectively collected reproductive histories on women with Alzheimer's disease. For women who use oestrogen after the menopause, the duration of therapy is variable, and very few women use hormones continuously. An observational study therefore must deal with variable durations of oestrogen exposure and variable intervals between last usage and date of death. In our exploratory analyses — perhaps not surprisingly — we were unable to discern a relationship between prior oestrogen use and semi-quantifiable neuropathological measures of neurofibrillary tangles or neuritic plaques.

Toran-Allerand: You gave your animals oral oestrogen replacement. This means that it goes through the liver and there will be endproducts that may not be the same as what the cultures were originally exposed to. Why don't you give them oestrogen by patch?

Gandy: Everything is a compromise. We are basing this largely on the fact that human females take oestrogen predominantly by mouth. I'm not sure how well animals will wear patches for weeks on end.

Toran-Allerand: A number of groups have shown that activation of the MAP kinase cascade (Mills et al 1997, Desdouits-Magnen et al 1998) and exposure to oestradiol (Jaffe et al 1994, Xu et al 1998) significantly increase α-secretase activity and may or may not decrease Aβ generation. APP has been described as a component of caveolae. It is conceivable that when there is an initial activation of APP via a rapid phenomenon such as the MAP kinase cascade, this could then trigger subsequent genomic effects which could explain why it might take longer for one to see the increase in APP.

Gandy: There are examples of first messenger signals where there's not a perfect reciprocal relationship between enhanced sAPPα release and Aβ-diminished Aβ release. This is particularly important as shown by the Cephalon group *in vivo*, where they treated living mice with intracerebral phorbol ester and drove down Aβ generation but didn't change sAPPα levels (Savage et al 1998).

References

Desdouits-Magnen J, Desdouits F, Takeda S et al 1998 Regulation of secretion of Alzheimer amyloid precursor protein by the mitogen-activated protein kinase cascade. J Neurochem 70:524–530

Jaffe A, Toran-Allerand CD, Greengard P, Gandy S 1994 Estrogen regulates metabolism of Alzheimer amyloid beta precursor protein. J Biol Chem 269:13065–13068

Mills J, Laurent Charest D, Lam F et al 1997 Regulation of amyloid precursor protein catabolism involves the mitogen-activated protein kinase signal transduction pathway. J Neurosci 17:9415–9422

Savage M J, Trusko SP, Howland DS et al 1998 Turnover of amyloid beta-protein in mouse brain and acute reduction of its level by phorbol ester. J Neurosci 18:1743–1752

Xu H, Gouras GK, Greenfield JP et al 1998 Estrogen reduces neuronal generation of Alzheimer β-amyloid peptides. Nat Med 4:447–451

Oestrogens and dementia

Victor W. Henderson

Department of Neurology, University of Southern California School of Medicine, 1420 San Pablo Street (PMB-B105), Los Angeles, CA 90089, USA

Abstract. The decline in circulating oestrogen concentrations that occurs after the menopause has the potential to impact Alzheimer's disease and other forms of dementia. Relevant actions include neurotrophic and neuroprotective effects; effects on acetylcholine and other neurotransmitters; and effects on proteins implicated in Alzheimer's disease pathogenesis. Since 1990, 14 case-control and cohort studies have considered the relation between postmenopausal oestrogen therapy and Alzheimer's disease. Most, but not all, report that oestrogen therapy is associated with a reduced Alzheimer risk of approximately one-half. Almost no epidemiological data address the potential link between oestrogen and other forms of dementia. Several small interventional trials have considered whether oestrogen might improve cognitive function of women with Alzheimer's disease. Data, however, are limited, and there is no compelling evidence that the short-term use of oestrogen monotherapy has a substantial impact on dementia symptoms. In summary, the use of oestrogen to reduce Alzheimer risk is biologically credible, and the preponderance of epidemiological evidence suggests that oestrogen therapy is indeed protective. This potentially important role of oestrogens for the primary prevention of Alzheimer's disease remains to be verified through well-designed randomized controlled trials.

2000 Neuronal and cognitive effects of oestrogens. Wiley, Chichester (Novartis Foundation Symposium 230) p 254–273

Introduction: ageing, oestrogen and dementia

Of the various causes of cognitive loss associated with ageing, Alzheimer's disease is by far the most important, representing two-thirds of dementia cases in western countries. It is approximately five times more common than the second most prevalent form of dementia, that attributed to multiple cerebral infarcts and related forms of ischaemic vascular disease. In Alzheimer's disease, the loss of memory and other cognitive skills is insidious and progressive. Characteristic pathological features include neurofibrillary tangles in neuronal cell bodies and neuritic plaques within the neuropil of the brain. Tangles consist largely of a modified form of tau, a microtubule-associated protein. An abnormal protein fragment known as β-amyloid is prominent in plaque cores, and various proteins associated with the acute phase inflammatory response are also associated with plaques.

Both genetic and non-genetic factors are believed to be important in Alzheimer pathogenesis. Early-onset disease, in which symptoms of dementia first appear in the fourth, fifth, or sixth decades of life, is often associated with dominantly inherited mutations of genes that encode the amyloid precursor protein (chromosome 21) and so-called presenilin proteins (chromosomes 14 and 1). Alzheimer's disease is very much a disorder of ageing, however, and early-onset forms are uncommon in comparison to Alzheimer's disease that first manifests itself in old age.

Late-onset Alzheimer's disease is phenotypically similar to early-onset forms, but dominantly inherited mutations are not believed to play an important role in this age group. Rather, it is likely that a number of genes modify disease susceptibility. The best recognized susceptibility gene is on chromosome 19 and encodes a lipid transport protein known as apolipoprotein E. There are three common isoforms of apolipoprotein E (ε2, ε3, and ε4), and increased susceptibility to Alzheimer's disease is conferred by possession of the ε4 allele (Strittmatter et al 1993). Gender appears to modify risk, with possession of the ε4 allele acting to increase risk more for women than men (Bretsky et al 1999).

Other factors, including oestrogen status, may also affect Alzheimer susceptibility. Menopause, associated with the loss of ovarian oestrogen production, occurs on average at age 51 years. In the absence of hormone therapy, a woman therefore spends about 40% of her adult life in a state of relative oestrogen deprivation. The brain is an important target organ for oestrogen, as well as for other steroid hormones, and consequences of oestrogen loss on central nervous system function have only recently been considered.

Oestrogen and brain function: potential relevance to dementia

Like other steroid hormones, oestrogen interacts with specific intranuclear receptors to regulate protein synthesis. Neurons express two receptor types, the alpha oestrogen receptor (ERα) and the recently described beta receptor (ERβ) (Shughrue et al 1997). Regional selectivity is implied by the observation that individual neurons can express ERα, ERβ, neither receptor, or occasionally both receptor types. Some oestrogen actions occur within a matter of seconds or minutes, too rapidly to invoke genomic activation; these are postulated to involve membrane receptors (Wong et al 1996). Other effects, such as antioxidant effects, are independent of oestrogenic properties of the steroid compound and may not necessarily require interactions with specific receptors. Finally, oestrogens can influence brain function indirectly through actions on non-neural tissues, such as the cerebral vasculature and the immune system.

Oestrogen has clear effects on a number of neurotransmitter systems of the brain, including cholinergic, noradrenergic and serotonergic neurons known to be

TABLE 1   Oestrogen actions that may be germane to Alzheimer's diease

---

Neurotrophic actions
   Interactions with neurotrophins
   Neurite extension
   Synapse formation
Neuroprotective actions
   Protection against apoptosis
   Antioxidant properties
   Anti-inflammatory properties
   Augmentation of cerebral blood flow
   Enhancement of glucose transport into the brain
   Blunting of corticosteroid response to behavioural stress
Effects on neurotransmitters
   Acetylcholine
   Noradrenaline
   Serotonin
   Others (dopamine, glutamate, γ-aminobutyric acid, neuropeptides)
Effects on proteins involved in Alzheimer's disease
   Apolipoprotein E
   Tau protein
   Amyloid precursor protein

Adapted from Henderson (1997).

vulnerable to pathological changes of Alzheimer's disease. For example, widely-projecting cholinergic neurons in the basal forebrain possess receptors for oestrogen (Toran-Allerand et al 1992), and oestrogen boosts cholinergic markers in the basal forebrain and in projection target areas (Gibbs & Pfaff 1992, Luine 1985).

Neurotrophic and neuroprotective effects of oestrogen appear especially germane to Alzheimer pathogenesis and progression (Table 1). Oestrogens modulate neurotrophic factors and growth-associated proteins (Miranda et al 1993, Shughrue & Dorsa 1993), promote the growth of nerve processes (Brinton et al 1997, Toran-Allerand 1991) and enhance synaptic plasticity (Foy et al 1999, Woolley & McEwen 1992). Oestrogens may also protect against neuronal death due to apoptosis (Pike 1999), reduce oxidative damage (Behl et al 1997) and inflammation (Grossman 1985), and protect against neuronal damage linked to the stress-induced secretion of cortisol (Lindheim et al 1992).

Finally, oestrogen affects proteins associated with Alzheimer's disease. Oestrogen increases the expression of apolipoprotein E within select brain regions (Stone et al 1997), induces tau production (Ferreira & Caceres 1991), and inhibits the formation of β-amyloid from the amyloid precursor protein (Xu et al 1998).

Very few data directly address the issue of whether oestrogen affects dementia due to disorders other than Alzheimer's disease. It is reasonable to hypothesize that neuroprotective effects, as well as effects on cerebral blood flow (Ohkura et al 1995) and glucose uptake (Shi & Simpkins 1997), would have a favourable impact on multi-infarct dementia. Indeed, healthy older women are more apt to use oestrogen than women diagnosed with vascular forms of dementia (Mortel & Meyer 1995). Although oestrogen protects the brain in experimental models of acute cerebral ischaemia (Simpkins et al 1997), stroke incidence is not reduced by oestrogen therapy (Grodstein et al 1996).

Oestrogen therapy and the primary prevention of Alzheimer's disease

Early case-control studies, in which a number of potential risk factors for Alzheimer's disease were examined, reported no significant association between the use of hormone therapy after the menopause and Alzheimer's disease. However, relatively few women in these studies used oestrogen, and the statistical power to detect a significant oestrogen effect was low. The issue was reconsidered in a 1994 analysis of women enrolled in a longitudinal study of ageing and dementia (Henderson et al 1994). Comparisons between 143 women with Alzheimer's disease and 92 healthy women of the same mean age showed oestrogen use to be more common in the latter group. Expressed as an estimate of relative risk (RR), oestrogen use was associated with about a two-thirds risk reduction for Alzheimer's disease (RR=0.33, 95% confidence interval [CI]=0.15–0.74). Although similar analyses by others yielded comparable results (Birge 1994, Mortel & Meyer 1995), estimates of risk derived from patterns of current hormone therapy are particularly subject to bias. For example, if physicians are more apt to prescribe oestrogens for women with memory loss, then the degree of risk reduction would be underestimated by studies that focus on current patterns of oestrogen use. Of greater concern, if physicians are less likely to prescribe oestrogens for women with cognitive deficits, then the apparently beneficial association could be entirely spurious.

Since 1990, 14 case-control and cohort studies have considered the relation between postmenopausal oestrogen therapy and Alzheimer's disease (Baldereschi et al 1998, Birge 1994, Brenner et al 1994, Broe et al 1990, Graves et al 1990, Henderson et al 1994, Kawas et al 1997, Lerner et al 1997, Mortel & Meyer 1995, Paganini-Hill & Henderson 1994, 1996, Tang et al 1996, van Duijn et al 1996, Waring et al 1999). In several recent studies, analyses have been based on data for oestrogen use that had been collected earlier in the postmenopausal period, before the onset of dementia symptoms (Henderson 2000). Among the first and largest of these was a nested case-control study conducted in Leisure World (Paganini-Hill & Henderson 1994, 1996), an upper-middle class retirement community in southern California. Detailed information on hormone use was collected from each woman at the time of her enrolment into the cohort. Subsequently, death certificate records were obtained, from which were identified 248 women thought to have had Alzheimer's disease; these cases were matched to 1198 control subjects. When patterns of hormone therapy were evaluated, oestrogen use was associated with a one-third lower risk of Alzheimer's disease (RR=0.65, 95% CI=0.49–0.88). Although case ascertainment in Leisure World was undoubtedly incomplete and some cases were therefore misclassified as controls, the likely effect of this bias would be to diminish the magnitude of any association between oestrogen and Alzheimer's disease rather than to imply a false association. Most, but not all, recent epidemiological studies have similarly concluded that the use of oestrogens after the menopause appears to protect against Alzheimer's disease (Fig. 1).

If oestrogen therapy does in fact reduce Alzheimer risk, then it might be predicted that greater oestrogen exposure would be associated with greater risk reductions. Oestrogen exposure can be assessed in terms of dose or the duration of oestrogen use. In Leisure World, risk estimates for Alzheimer's disease did decrease significantly with increasing dose of the longest used oral oestrogen preparation (Paganini-Hill & Henderson 1996) (Fig. 2). Similarly, significant associations between the duration of oestrogen use and the degree of risk reduction were noted in analyses from Leisure World (Paganini-Hill & Henderson 1996), New York City (Tang et al 1996) and Rochester, Minnesota (Waring et al 1999), but not from Baltimore (Kawas et al 1997) (Fig. 3).

Several investigators have examined surrogate, or indirect, markers of endogenous oestrogen exposure. Markers include earlier age at menarche, later age at menopause and greater body weight. Results of these analyses have been mixed, with some showing a significant relation between inferred oestrogen exposures and reduced Alzheimer risk, and others showing no association. In the Italian ageing study (Baldereschi et al 1998), early menarche was linked to significant reductions in Alzheimer risk, and a similar but weaker trend in the same direction was reported from Leisure World (Paganini-Hill & Henderson 1996). A Dutch study of early-onset Alzheimer's disease found later menopause was linked to reduced Alzheimer risk (van Duijn et al 1996), but other investigators have found no association between age at menopause and Alzheimer's disease (Baldereschi et al 1998, Paganini-Hill & Henderson 1996, Tang et al 1996). Finally, because oestrogen production occurs to some extent in adipose tissue, body weight after the menopause correlates with circulating levels of oestrogens. In Leisure World, greater body weight at the time of cohort

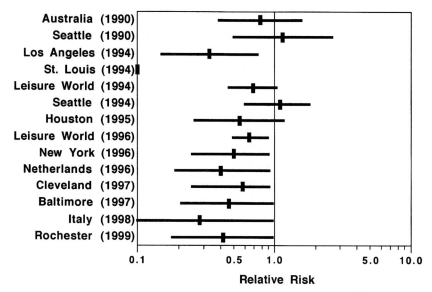

FIG. 1. Oestrogen therapy and the relative risk of Alzheimer's disease. 95% confidence intervals, where available, and point estimates of risk are shown for case-control and cohort studies reported between 1990 and 1999. Relative risks of less than one imply a protective effect of oestrogen therapy, and 95% confidence intervals of less than one imply a statistically significant effect. Information of oestrogen use from Leisure World (Paganini-Hill & Henderson 1994, 1996), Seattle (Brenner et al 1994), New York (Tang et al 1996), Baltimore (Kawas et al 1997) and Rochester (Waring et al 1999) were collected prospectively, before the onset of dementia symptoms. Other data are from Australia (Broe et al 1990), Seattle (Graves et al 1990), Los Angeles (Henderson et al 1994), St. Louis (Birge 1994), Houston (Mortel & Meyer 1995), the Netherlands (van Duijn et al 1996), Cleveland (Lerner et al 1997) and Italy (Baldereschi et al 1998). The two Leisure World analyses were not independent, in that a subset of subjects were included in both. Figure modified with permission from Henderson (2000).

enrolment was predictive of subsequent Alzheimer's diagnoses, but retrospective estimates of weight at age 55 years were not associated with Alzheimer's disease in the Italian study.

Oestrogen therapy and the treatment of Alzheimer's disease symptoms

The possibility the oestrogen may have a role in improving dementia symptoms receives indirect support from results of randomized controlled trials of postmenopausal women without dementia. Subjects treated with oestrogen are reported to perform better than control subjects on measures of choice reaction time, attention and concentration, distractibility, verbal memory, or abstract reasoning (Fedor-Freybergh 1977, Phillips & Sherwin 1992, Sherwin 1988). However, in contrast to these positive studies, a randomized controlled trial

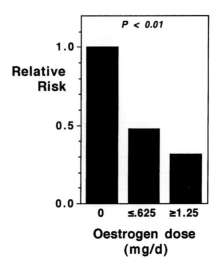

FIG. 2.   Oestrogen exposure and Alzheimer risk: dosage. In Leisure World, the longest-used dosage of the most commonly prescribed oral oestrogen (conjugated equine oestrogens) was assessed for 198 cases and 912 controls. Greater doses were associated with lower estimates of Alzheimer risk (Paganini-Hill & Henderson 1996). mg/d = milligrams per day. Figure modified with permission from Henderson (2000).

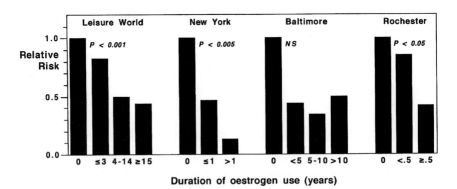

FIG. 3.   Oestrogen exposure and Alzheimer risk: duration. Increasing durations of oestrogen therapy were associated with lower estimates of Alzheimer risk in Leisure World (Paganini-Hill & Henderson 1996), New York (Tang et al 1996) and Rochester (Waring et al 1999), but not in Baltimore (Kawas et al 1997). NS = not significant. Figure modified with permission from Henderson (2000).

involving 62 subjects detected no benefit of oestrogen on measures of psychomotor speed, attention, working memory, or visual memory (Polo-Kantola et al 1998).

Other indirect evidence that oestrogen could be useful for the treatment of Alzheimer's disease comes from an analysis of cognitive performance as a function of body weight, where weight is a surrogate marker for oestrogen exposure. In a study of 347 women with clinically diagnosed Alzheimer's disease — after adjustments were made for age, education, symptom duration and height — greater body weight was significantly linked to better test scores on each of the two cognitive tasks examined (Buckwalter et al 1997). Although endogenous levels of circulating oestrogens were presumably greater among heavier women, findings are linked to oestrogen only inferentially.

Among women with Alzheimer's disease, those who take oestrogen are described as scoring substantially better on a variety of cognitive tasks than women who do not (Henderson et al 1996). Although encouraging, the results of this comparison are open to several interpretations. First, and most optimistically, results could imply that oestrogen prescriptions for women with Alzheimer's disease would lead to clinically meaningful improvement. Second, since most women in this study who took oestrogen had done so for many years, findings could suggest that oestrogen users who do develop Alzheimer symptoms will progress more slowly but that demented women given a new prescription for oestrogen would show no short-term improvement. Finally, this was not a randomized study; and although groups were carefully matched for age, education and duration of dementia symptoms, group differences could have been due to chance or to some unrecognized factor other than oestrogen therapy.

The use of oestrogen for women with symptomatic Alzheimer's disease has been formally evaluated in only a few placebo-controlled, double-blind trials. A Japanese study of 14 women with Alzheimer's disease compared oral conjugated equine oestrogens (1.25 mg/d) to placebo (Honjo et al 1993). Three weeks after treatment was begun, women in the oestrogen group showed significant improvement on three psychometric tests, whereas women in the placebo group were unchanged. A Seattle study (Asthana et al 1999) compared transdermal oestradiol (0.05 mg/d) to placebo among 12 demented women. Subjects were assessed over an eight week period on a well-constructed neuropsychological battery that included a large number of tasks. On most measures, performances were similar in the two treatment groups, but women given oestrogen outperformed other women on two tasks. A three month study involving eight women with Alzheimer's disease detected no effect of low-dose transdermal oestradiol (0.05 mg, twice per week) (Fillit 1994), whereas a nine month study of 20 women yielded positive results on a clinician's global impression of change for those treated with oral conjugated oestrogens (0.625 mg/d) (Birge 1997). Finally, a recent study of 42 women detected no substantial effect of conjugated equine oestrogens (1.25 mg/d) on cognitive, functional, or global outcome measures after 16 weeks of treatment (Henderson et al 2000).

Strategies to increase brain levels of acetylcholine are a current mainstay in the treatment of Alzheimer's disease. Efficacy is limited, but cholinesterase inhibitors (e.g. tacrine, donepezil, rivastigmine) are approved for Alzheimer patients in several countries. The possibility that oestrogen might enhance effects of cholinergic therapy was considered in *post hoc* analyses of a 30 week, multicentre trial of tacrine (Schneider et al 1996). A small number of women using oral oestrogen at the time of study enrolment had been randomized to the active treatment arm. In *post hoc* analyses, these demented subjects performed significantly better on primary outcome measures than women in the placebo arm. In the active treatment arm, women not taking oestrogens performed more like those in the placebo group. Although oestrogen was not a randomized intervention in this study, results imply that oestrogen 'priming' may enhance effects of cholinergic intervention.

Summary and conclusions

A number of oestrogen actions affect brain functions in ways that might favourably influence Alzheimer's disease and other forms of dementia. Analytical studies during the past decade imply that the risk of Alzheimer's disease among women who have taken oestrogen is reduced by about a half, the equivalent of postponing the onset of Alzheimer's disease by five years. If this assessment proves accurate, this reduction in risk would indeed be of enormous public health importance. However, in the absence of randomized controlled trials, there remains the possibility that unrecognized bias or confounding could account for the reported associations in epidemiological analyses. Adequately powered randomized, placebo-controlled intervention trials would provide still stronger support for the contention that oestrogen does in fact reduce Alzheimer risk. Such trials are now underway, but results are not anticipated for several years.

In contrast to generally positive results of studies examining the association between oestrogen and Alzheimer prevention, results of clinical trials are generally discouraging with respect to a role for oestrogen in ameliorating Alzheimer symptoms. At present, there are few convincing data that short-term oestrogen therapy is effective as a single therapeutic agent.

If oestrogen does help prevent Alzheimer's disease, then additional research will be required to determine relevant mechanisms of action; relative efficacy of different oestrogen compounds, including selective oestrogen receptor modulators; optimal timing of oestrogen use in the postmenopausal period;

preferred dose and dosing regimen; and, as a progestogen is typically prescribed for women with a uterus who take oestrogen, the influence of progesterone.

References

Asthana S, Craft S, Baker LD et al 1999 Cognitive and neuroendocrine response to transdermal estrogen in postmenopausal women with Alzheimer's disease: results of a placebo-controlled, double-blind, pilot study. Psychoneuroendocrinology 24:657–677

Baldereschi M, Di Carlo A, Lepore V et al 1998 Estrogen-replacement therapy and Alzheimer's disease in the Italian Longitudinal Study on Aging. Neurology 50:996–1002

Behl C, Skutella T, Lezoualch F et al 1997 Neuroprotection against oxidative stress by estrogens: structure–activity relationship. Mol Pharmacol 51:535–541

Birge S J 1994 The role of estrogen deficiency in the aging central nervous system. In: Lobo R A (ed) Treatment of the postmenopausal woman: basic and clinical aspects. Raven Press, New York, p 153–157

Birge S J 1997 The role of estrogen in the treatment of Alzheimer's disease. Neurology (suppl 7) 48:S36–S41

Brenner DE, Kukull WA, Stergachis A et al 1994 Postmenopausal estrogen replacement therapy and the risk of Alzheimer's disease: a population-based case-control study. Am J Epidemiol 140:262–267

Bretsky PM, Buckwalter JG, Seeman TE et al 1999 Evidence for an interaction between apolipoprotein E genotype, gender, and Alzheimer's disease. Alzheimer Dis Assoc Disord 13:216–221

Brinton RD, Tran J, Proffitt P, Montoya M 1997 17β-estradiol enhances the outgrowth and survival of neocortical neurons in culture. Neurochem Res 22:1339–1351

Broe GA, Henderson AS, Creasey H et al 1990 A case-control study of Alzheimer's disease in Australia. Neurology 40:1698–1707

Buckwalter JG, Schneider LS, Wilshire TW, Dunn ME, Henderson VW 1997 Body weight, estrogen and cognitive functioning in Alzheimer's disease: an analysis of the tacrine study group. Arch Gerontol Geriatr 24:261–267

Fedor-Freybergh P 1977 The influence of oestrogens on the wellbeing and mental performance in climacteric and postmenopausal women. Acta Obstet Gynecol Scand (suppl) 64:1–99

Ferreira A, Caceres A 1991 Estrogen-enhanced neurite growth: evidence for a selective induction of tau and stable microtubules. J Neurosci 11:393–400

Fillit H 1994 Estrogens in the pathogenesis and treatment of Alzheimer's disease in postmenopausal women. Ann NY Acad Sci 743:233–238

Foy MR, Xu J, Xie X, Brinton RD, Thompson RF, Berger TW 1999 17β-estradiol enhances NMDA receptor-mediated EPSPs and long-term potentiation. J Neurophysiol 81:925–929

Gibbs RB, Pfaff DW 1992 Effects of estrogen and fimbria/fornix transection on p75[NGFR] and ChAT expression in the medial septum and diagonal band of Broca. Exp Neurol 116:23–39

Graves AB, White E, Koepsell TD et al 1990 A case-control study of Alzheimer's disease. Ann Neurol 28:766–774

Grodstein F, Stampfer M J, Manson JE et al 1996 Postmenopausal estrogen and progestin use and the risk of cardiovascular disease. N Engl J Med 335:453–461 (erratum: 1996 N Engl J Med 335:1406)

Grossman CJ 1985 Interactions between the gonadal steroids and the immune system. Science 227:257–261

Henderson VW 1997 Estrogen, cognition, and a woman's risk of Alzheimer's disease. Am J Med (suppl 3A) 103:S11–S18

Henderson VW 2000 Hormone therapy and the brain: a clinical perspective on the role of estrogen. Parthenon Publishing, New York

Henderson VW, Paganini-Hill A, Emanuel CK, Dunn ME, Buckwalter JG 1994 Estrogen replacement therapy in older women. Comparisons between Alzheimer's disease cases and nondemented control subjects. Arch Neurol 51:896–900

Henderson VW, Watt L, Buckwalter JG 1996 Cognitive skills associated with estrogen replacement in women with Alzheimer's disease. Psychoneuroendocrinology 21:421–430

Henderson VW, Paganini-Hill A, Miller BL et al 2000 Estrogen for Alzheimer's disease in women: randomized, double-blind, placebo-controlled trial. Neurology 54:295–301

Honjo H, Ogino Y, Tanaka K et al 1993 An effect of conjugated estrogen to cognitive impairment in women with senile dementia — Alzheimer's type: a placebo–controlled double blind study. J Jpn Menopause Soc 1:167–171

Kawas C, Resnick S, Morrison A et al 1997 A prospective study of estrogen replacement therapy and the risk of developing Alzheimer's disease: the Baltimore Longitudinal Study of Aging. Neurology 48:1517–1521 (erratum: 1998 Neurology 51:654)

Lerner A, Koss E, Debanne S, Rowland D, Smyth K, Friedland R 1997 Smoking and oestrogen-replacement therapy as protective factors for Alzheimer's disease. Lancet 349:403–404

Lindheim SR, Legro RS, Bernstein L et al 1992 Behavioral stress responses in premenopausal and postmenopausal women and the effects of estrogen. Am J Obstet Gynecol 167:1831–1836

Luine V 1985 Estradiol increases choline acetyltransferase activity in specific basal forebrain nuclei and projection areas of female rats. Exp Neurol 89:484–490

Miranda RC, Sohrabji F, Toran-Allerand CD 1993 Presumptive estrogen target neurons express mRNAs for both the neurotrophins and neurotrophin receptors: a basis for potential developmental interactions of estrogen with neurotrophins. Mol Cell Neurosci 4:510–525

Mortel KF, Meyer JS 1995 Lack of postmenopausal estrogen replacement therapy and the risk of dementia. J Neuropsychiatr Clin Neurosci 7:334–337

Ohkura T, Teshima Y, Isse K et al 1995 Estrogen increases cerebral and cerebellar blood flow in postmenopausal women. Menopause 2:13–18

Paganini-Hill A, Henderson VW 1994 Estrogen deficiency and risk of Alzheimer's disease in women. Am J Epidemiol 140:256–261

Paganini-Hill A, Henderson VW 1996 Estrogen replacement therapy and risk of Alzheimer's disease. Arch Intern Med 156:2213–2217

Phillips SM, Sherwin BB 1992 Effects of estrogen on memory function in surgically menopausal women. Psychoneuroendocrinology 17:485–495

Pike CJ 1999 Estrogen modulates neuronal Bcl-x_L expression and β-amyloid-induced apoptosis: relevance to Alzheimer's disease. J Neurochem 72:1552–1563

Polo-Kantola P, Portin R, Polo O, Helenius H, Irjala K, Erkkola R 1998 The effect of short-term estrogen replacement therapy on cognition: a randomized, double-blind, cross-over trial in postmenopausal women. Obstet Gynecol 91:459–466

Schneider LS, Farlow MR, Henderson VW, Pogoda JM 1996 Effects of estrogen replacement therapy on response to tacrine in patients with Alzheimer's disease. Neurology 46:1580–1584

Sherwin BB 1988 Estrogen and/or androgen replacement therapy and cognitive functioning in surgically menopausal women. Psychoneuroendocrinology 13:345–357

Shi J, Simpkins JW 1997 17β-estradiol modulation of glucose transporter 1 expression in blood–brain barrier. Am J Physiol 272:E1016–E1022

Shughrue PJ, Dorsa DM 1993 Estrogen modulates the growth-associated protein GAP-43 (neuromodulin) mRNA in the rat preoptic area and basal hypothalamus. Neuroendocrinology 57:439–447

Shughrue PJ, Lane MV, Merchenthaler I 1997 Comparative distribution of estrogen receptor-α and -β mRNA in the rat central nervous system. J Comp Neurol 388:507–525

Simpkins JW, Rajakumar G, Zhang Y-Q et al 1997 Estrogens may reduce mortality and ischemic damage caused by middle cerebral artery occlusion in the female rat. J Neurosurg 87:724–730

Stone DJ, Rozovsky I, Morgan TE, Anderson CP, Hajian H, Finch CE 1997 Astrocytes and microglia respond to estrogen with increased apoE mRNA *in vivo* and *in vitro*. Exp Neurol 143:313–318

Strittmatter WJ, Saunders AM, Schmechel D et al 1993 Apolipoprotein E: high-avidity binding to β-amyloid and increased frequency of type 4 allele in late-onset familial Alzheimer disease. Proc Natl Acad Sci USA 90:1977–1981

Tang MX, Jacobs D, Stern Y et al 1996 Effect of oestrogen during menopause on risk and age at onset of Alzheimer's disease. Lancet 348:429–432

Toran-Allerand CD 1991 Organotypic culture of the developing cerebral cortex and hypothalamus: relevance to sexual differentiation. Psychoneuroendocrinology 16:7–24

Toran-Allerand CD, Miranda RC, Bentham WDL et al 1992 Estrogen receptors colocalize with low-affinity nerve growth factor receptors in cholinergic neurons of the basal forebrain. Proc Natl Acad Sci USA 89:4668–4672

van Duijn C, Meijer H, Witteman JCM et al 1996 Estrogen, apolipoprotein E and the risk of Alzheimer's disease. Neurobiol Aging (suppl) 17:S79–S80 (abstract)

Waring SC, Rocca WA, Petersen RC, O'Brien PC, Tangalos EG, Kokmen E 1999 Postmenopausal estrogen replacement therapy and risk of AD: a population-based study. Neurology 52:965–970

Wong M, Thompson TL, Moss RL 1996 Nongenomic actions of estrogen in the brain: physiological significance and cellular mechanisms. Crit Rev Neurobiol 10:189–203

Woolley CS, McEwen BS 1992 Estradiol mediates fluctuation in hippocampal synapse density during the estrous cycle in the adult rat. J Neurosci 12:2549–2554

Xu H, Gouras GK, Greenfield JP et al 1998 Estrogen reduces neuronal generation of Alzheimer β-amyloid peptides. Nat Med 4:447–451

DISCUSSION

Levin: Your observation that oestrogen can prevent Alzheimer's but that it may have little effect on established disease is absolutely consistent with what has now been shown in the cardiovascular system. While oestrogen is clearly an excellent means of preventing the development of cardiovascular disease, a recent large study published in *JAMA* on secondary prevention showed that it has no effect on established disease (Hulley et al 1998).

It looks like oestrogen can be protective under a situation of ever exposed. It may be that we don't have to tell women if they are menopausal that they need to take oestrogen for the rest of life. Instead there may be a 10 or 12 year window. Considering the data from your Leisure World study, the 4–14 years of oestrogen replacement was no different than the greater than 15 years group in terms of prevention. This is also consistent with other findings in the cardiovascular system where women who were 'ever exposed' did almost as well as women who had been taking oestrogen for a long time.

With regard to the age of menopause issue, you showed that disappointingly the age of menopause didn't seem to be a particular parameter. In other words you would expect the women who underwent menopause at a later age and hence had longer exposure to oestrogen to be more protected than a woman who underwent menopause at a relatively premature age. There are similar data with the bone studies. The explanation given there is that there is a massively accelerated loss of bone in the peri-menopausal period. You have a woman who is at risk with low bone density, and then she undergoes menopause with a tremendous loss of bone in that early period. Might a similar thing be happening with the induction of Alzheimer's, where there's an accelerated phase of stress that results in neuronal jeopardy and that is a critical parameter?

Sherwin: What are the neurobiological explanations for why neuroprotection would be enduring if there was exposure for a period of time to oestrogen, but then the hormone was withdrawn?

Henderson: There are many potential mechanisms by which oestrogen could be protective for Alzheimer's disease. Some effects may be enduring, for example changes in synaptic plasticity or neurite growth. As another speculative example, a short-term reduction in oxidative stress might have long-term benefits. This is an important question, and the epidemiological data don't answer very well whether there's a critical window along the lines alluded to by Dr Levin where oestrogen exposure may be particularly important.

Sherwin: I ask because many of these epidemiological studies break their population down into 'never-users' and 'ever-users'. So, an 80 year old woman who took oestrogen between the ages of 52 and 55 would fall into the category of an ever-user. I always wondered whether there was any biological justification for doing this.

Henderson: There is a methodological, but not biological, justification for comparing never-users to ever-users when these data are collected retrospectively. This dichotomy is more reliable than continuous estimates of exposure duration. Even when one obtains exposure data prospectively, one must still deal with different hormone regimens, given for different durations and at different ages.

Sherwin: In your last study, where you showed us your analyses (Henderson et al 2000), I was thinking that four months of oestrogen treatment may not be sufficient. Oestrogen might not ameliorate cognitive functioning in someone who has Alzheimer's disease, but what it might conceivably do is retard deterioration. Four months is probably not long enough for you to be able to see that.

Henderson: You are right. Where feasible, it is clearly advantageous in a progressive neurodegenerative disorder to extend treatment trials for longer periods of time.

Fillit: I think in this room, where we have more biologists than clinical people, it seems there's a dissociation among those of us who are interested in Alzheimer's disease and late life cognitive decline, between much of the experimental work which is short term and the human model which is long term. I don't think we have seen the pivotal experiments which place an emphasis on the neurobiology of the human condition, which involves the long-term effects of oestrogen deficiency. I would encourage us to move the neurobiology into looking at the long-term effects. Is there a critical period after which this enters into an irreversible state, and how much of it is reversible after long-term deficiency?

Along these lines, there was a multicentre treatment trial that went on for a year to see whether oestrogen replacement therapy slowed progression in postmenopausal women with Alzheimer's disease. It's unofficial at this time, but apparently that study did look at long-term outcomes. The primary outcome measure was slowing the rate of progression using oestrogen, and there was no effect. One thing I've learned here from Bob Gibbs and others is that this is a drug discovery process: if there's an effect, we are trying to find the right drug delivery process with the right doses. In the field we have put a lot of hope on the idea that oestrogen therapy has good rationale, and ultimately in the clinic will have an effect in preventing cognitive decline and Alzheimer's disease, and maybe treating it. With some of these negative studies coming out, we need to defend or at least explain the neurobiological basis for why they didn't work.

Gandy: In response to Barbara Sherwin's question about the idea of 'windows' of treatment, in Alzheimer's disease there are several examples where you can identify critical periods. One could imagine temporally superimposing a critical insulting event with a transient exposure to oestrogen, and this might lead to a good outcome. That is, transient oestrogen might protect against an insult which was received at the same time. There are three examples I can think of straight away. The first is head trauma. If you analyse the evidence associating head trauma as a risk for Alzheimer's disease, this works in a sort of 'never-ever' kind of way. Second, pathology is clearly non-linear over the lifetime. No one at age 40 has pathology. Most people who are going to develop disease do so over the 25-year period between 40 and 65. In both these cases, it is possible that transient oestrogen exposure (during head injury or during the period when pathology is initiated) could be protective. The third example is genetics. With ApoE ε4, it clearly exerts its effect prior to age 80. If you have an ε4 allele but live to age 80 without dementing, you've escaped the ε4 risk.

Resnick: This issue of critical periods is important. We know from the early organizational effects of hormones that there are critical periods for hormone exposure early in development. In some species, early hormone exposure has been linked to sexually dimorphic behavioural and brain development. Perhaps at other periods in the lifespan hormone exposure is important to cognitive and brain changes.

McEwen: The brain is a plastic organ, and as things change it does tend to reorganize itself. In the normal period of oestrogen withdrawal it may still be responsive to oestrogens but perhaps in a somewhat different way.

Toran-Allerand: Most of the clinical studies have used the most commonly prescribed oestrogen, which is conjugated equine oestrogens. During her lifetime a woman secretes primarily 17β-oestradiol from the ovaries, and most of the neurobiological studies of oestrogen action of the brain have used some form of oestradiol and not the ill-defined conjugated horse oestrogens, some of which have never really been characterized, and others of which attach to the oestrogen receptor for far longer than does 17β-oestradiol.

Fillit: Actually, our first trial used Estrace (Fillit et al 1986), which is 17β-oestradiol.

Toran-Allerand: Since then, they haven't because they are all funded by one company. It seems to me really important to at least consider whether or not the use of 17β-oestradiol as replacement is in any way different from oral conjugated horse oestrogens.

Henderson: I agree with you. Studies of conjugated oestrogens are justified on the basis that conjugated equine oestrogens represent the most commonly prescribed formulation, at least for North American women. It would be nice if the clinical studies meshed more closely with those of the bench scientist, who for obvious reasons tends to use 17β-oestradiol. The epidemiological data, which are based on what older women in the real world are taking, largely reflect effects of conjugated oestrogens.

Toran-Allerand: But the respect to treating people who already express symptoms of Alzheimer's disease it might be very different, and the use of 17β-oestradiol might have different effects than conjugated horse oestrogen.

Gibbs: Barbara Sherwin commented that four months might not be long enough. I would also like to raise the possibility that four months of continuous oestrogen treatment might be too long. Most of the studies that have seen increases in performance have looked at six weeks.

In our own studies, repeated administration of oestrogen was considerably less effective than oestrogen plus progesterone, or continuous oestrogen in the form of a capsule. In old animals, where we waited 10 months before we started treating, the animals were non-responsive. In contrast, when we only waited three months before we started treating, the animals were responsive. This raises the issue of critical period again. Did you look to see how long the women were oestrogen-deprived before you started treating?

Henderson: Selection criteria in our study required that oestrogens not have been used within the previous three months (Henderson et al 2000).

Gibbs: How do you explain the oestrogen plus tacrine data which you showed first, which look very convincing?

Henderson: The tacrine analysis (Schneider et al 1996) was *post hoc*, and clearer conclusions would come from a placebo-controlled study in which oestrogen was a randomized variable. My understanding is that such trials are ongoing, and we may have a more clear-cut answer soon.

Gustafsson: Is there preliminary information on the possible protective effect of oestrogen on vascular dementia? In view of the vessel protective effect of oestrogens one might foresee some beneficial effect even in that type of disease.

Henderson: There are few direct data on vascular dementia, which by the way is much more difficult to diagnose validly than Alzheimer's disease. We heard from Dr Simpkins and others that oestrogens are protective in an acute stroke model. From the epidemiological literature, there is no evidence that oestrogen has an appreciable effect on the incidence of ischaemic stroke. On the other hand, oestrogen might be of benefit if the development of vascular dementia reflected stroke severity. The one study that looked at vascular dementia (Mortel & Meyer 1995) found that current oestrogen use was less common in women diagnosed with vascular dementia than in healthy control subjects.

Fillit: There was also a treatment trial of elderly women with cerebrovascular disease and cognition. Oestrogen improved cerebral blood flow and cognition.

Gustafsson: What about Parkinson's disease? There you have oxidative phenomena. An antioxidative effect of oestrogen could help here.

Henderson: In terms of Parkinson's disease, there are few clinical data. However, a multicentre clinical trial has been organized and will soon be enrolling Parkinson's women.

Baulieu: Did you include progesterone in your list of preventive drugs? We have seen progesterone effects in myelination in the peripheral nervous system (Koenig et al 1995). We still do not know whether this is the same in the CNS.

Henderson: In terms of disease prevention, the epidemiological data that I referred to do not distinguish between women using unopposed oestrogen and women using oestrogen plus a progestogen.

Baulieu: Do you believe that progesterone would have an influence?

Henderson: There are hints from the clinical literature (Ohkura et al 1995, Schneider et al 1996) that women taking a progestogen may not realize the putative cognitive benefits of oestrogen taken alone.

Bethea: Which progesterone?

Henderson: Medroxyprogesterone acetate (MPA) is the most commonly prescribed progestin in the USA.

Toran-Allerand: MPA antagonizes oestrogen.

Baulieu: Do you think that in addition to the hormone-directed events that you wish to see, the selection of people is a big factor in determining the 'cerebral' success of hormone replacement therapy (HRT)? In France this is very clear: up to now, the women taking HRT do not represent a typical cross-section of women in society. Are the results due to oestrogen replacement, or some other component of personality?

Henderson: This remains an unanswered question. One attempts to control or adjust for potentially confounding variables. Even in the USA, where hormone use is more prevalent than in many other countries, there are important differences between oestrogen users and non-users, including level of education, socioeconomic status, and use of health care resources.

Baulieu: Has anyone tried preventing cognitive decline using hormones in males?

Henderson: Several studies have examined cognition in relation to testosterone in non-demented men, and small testosterone treatment trials in Alzheimer's disease are now underway. There are no studies of testosterone for the primary prevention of cognitive decline or Alzheimer's disease.

Levin: I wouldn't implicate testosterone decline as a causative factor for Alzheimer's. I don't know why you would want to replace it.

Henderson: The rationale for looking at testosterone in men is driven in large part by the allure of the oestrogen hypothesis in women, plus the recognition that there are androgen receptors in the brain and that testosterone probably does have some effect on some cognitive skills.

Levin: My comment is driven by the fact that the androgen levels in most healthy men decline very modestly.

Bethea: Perhaps we should mention the work from Fred Gage's laboratory on cell division that is ongoing in the hippocampus (Eriksson et al 1998). In response to Barbara Sherwin's earlier question, this could be one of neurobiological bases for the 'ever or never' effects of oestrogen. That is, oestrogen may keep a stem cell population in the hippocampus viable. Then, after a certain period of time with no oestrogen you may lose that stem cell population altogether and never be able to bring it back. This cell division is also correlated with improved performance in memory. Gage is finding this in primates, which is really exciting. My second comment is that in the nucleus of the diagonal band of Broca, we see a lot of the serotonin 2C receptor. This is a stimulatory receptor that acts through the phosphatidyl inositol pathway. I think that maintaining the serotonergic input to those cells could be critically important, and certainly oestrogen would facilitate that.

Kushner: In the studies of breast cancer and oestrogen, it's been found that bone density is an excellent surrogate measure of lifelong oestrogen exposure. Have you ever thought of using this?

Resnick: We haven't used this, but we have the data and could easily examine it. It is a good suggestion.

Luine: In Howard Fillit's study, the women that responded the best had more thinning of the bones (Fillit et al 1986).

Sherwin: There were two out of seven patients who did have lower bone density who were also the responders.

I want to return to the crucial issue of critical periods we've been talking about. There is no increase in breast cancer if a woman takes oestrogen for five years. After five years this changes. Will five years of oestrogen treatment provide a woman with neuroprotection down the line? If you look at another organ system, namely bone metabolism, we know that if a woman takes oestrogen for five years, it will protect her bone density, but when she stops taking it in the sixth year she will lose bone. Oestrogen protects bone density if you have a blood level: is it any different for the brain? Is that initial neuroprotective effect induced by oestrogen treatment enduring after withdrawal of the drug for up to 20 years?

Bethea: Perhaps there is an increase in cell death, and oestrogen intervention might decrease the slope, so that you would have fewer cells dying at a critical period. But there's cell death going on all the time in the brain.

Simpkins: One of the major events that occurs that we can detect in the brain is the hot flush at the time of menopause. Hot flushes are transient responses in the brain that are treatable by oestrogens. If you look at the Glut1 transporter in the endothelium of the brain you can demonstrate this, as well as looking at transport of glucose into the brain. One event that is happening in the menopause that could be particularly damaging is neuronal damage that occurs as a result of these brief periods of deprivation of glucose. These glucose deprivation periods also cause overexpression of a form of amyloid precursor protein that has a kunitz protease inhibitor and should be processed to β-amyloid. It is possible that those women who go on to show disease in their lifetime are ones that had very severe episodes of hot flushes during the menopause, and the women who chose to take oestrogens did so to treat these hot flushes and that was the benefit that they received. One of the things I think we as a research community do very poorly is document events that are happening in the menopause. It is unfortunate that even the woman's health initiative is not doing a good job of getting lots of details about women's menopausal experiences. It is possible that if oestrogen is used to treat these flushing events that one may get benefit long after the menopause.

Levin: I want to reiterate that the cardiovasculature literature, which is very clear, shows that ever exposed is protective. If you look at the atherosclerotic process, which is a long-term process, having atherosclerosis for seven years less as a result of taking oestrogen for seven years is clearly better than having atherosclerosis for 15 or 20 years. This is translated into the outcome. This may be part of what is happening in the brain. If you are interested in oestrogen action in the brain, I would urge you to look at other organ systems.

Gandy: There is a series of papers from a group in Israel, which report that when cells are treated with DHEA prior to application of carbachol, there's a dramatic facilitation of the carbachol-stimulated anti-amyloid effect.

Toran-Allerand: I would like to address the issue of males, androgens and Alzheimer's. It is true that the male brain does not always respond to oestrogen the way the female does, even though the male brain may have oestrogen receptors. However, it is also true that regions like the basal forebrain, even in adults, have aromatase activity, and circulating androgens in the male can be converted to oestrogens in the basal forebrain. But what we don't know is whether circulating androgens are not the problem: perhaps there are problems with the aromatase activity in the basal forebrain, or problems with receptors in the basal forebrain. I would not give up looking at the male endocrinological patterns.

Gustafsson: It is important not to think about the oestrogen receptors as being only activatable by oestrogens. In males it is quite conceivable that certain androgenic metabolites could be the physiological ligands for ERβ. It is well known that is a good ligand for ERβ. Maybe more importantly, DHEA can given rise to a similar metabolite, 5-androstene-3β,17β-diol, which is as good an ERβ binder as 5α-androstane-3β,17β-diol.

Levin: That is a good point. Although free testosterone declines modestly in the healthy ageing of male, the DHEA levels decline precipitously.

Sherwin: In the 72 and 74 year old males that I mentioned yesterday, their free testosterone levels were in the lower one-third of the male range, so they do decline significantly beyond the sixth decade of life. Bioavailable testosterone declines even more profoundly.

Wolf: We also see a decline of 20–30%. With respect to DHEA, from the epidemiological studies there's no evidence relating its decline to any cognitive measures in healthy individuals or Alzheimer's patients.

Resnick: In terms of the human clinical studies, it's important to keep in mind that when we look at oestrogen effects on age-associated decline in memory in non-demented women, we may be dealing with different mechanisms than the effects on Alzheimer's disease.

Becker: There is a lot of interest and excitement about understanding how oestrogen acts in humans in Alzheimer's disease. But as you see from this meeting, there's not a lot of work being done on behaviour in any species, let alone humans. We need to apply all this incredible work on the action of oestrogen in the nervous system, to figure out what that means for behaviour of animals and people.

Pfaff: I wanted to mention the considerable degree of conservation of some of the mechanisms we've been talking about, from lower mammals to the human brain. Elsewhere, I made a list of 18 items, which were grouped into basic endocrine mechanisms, neuroanatomical and neurochemical steps, fundamental molecular and biophysical steps, and also gonadotropin-releasing hormone neuronal biology (Pfaff 1999). Sometimes the degree of conservation is really remarkable. I do therefore retain the faith that findings from us mouse people have some relevance to human people!

References

Eriksson PS, Perfilieva E, Bjork-Eriksson T et al 1998 Neurogenesis in adult hippocampus. Nat Med 4:1313–1317

Fillit H, Weinreb H, Cholst I et al 1986 Observations in a preliminary open trial of estradiol therapy for senile dementia-Alzheimer's type. Psychoneuroendocrinology 11:337–345

Henderson VW, Paganini-Hill A, Miller BL et al 2000 Estrogen for Alzheimer's disease in women: randomized, double-blind, placebo-controlled study. Neurology 54:295–301

Hulley S, Grady D, Bush T et al 1998 Randomized trial of estrogen plus progestin for secondary prevention of coronary heart disease in postmenopausal women. Heart and Estrogen/progestin Replacement Study (HERS) research group. JAMA 280:605–613

Koenig HL, Schumacher M, Ferzaz B et al 1995 Progesterone synthesis and myelin formation by Schwann cells. Science 268:1500–1503

Mortel KF, Meyer JS 1995 Lack of postmenopausal oestrogen replacement therapy and the risk of dementia. J Neuropsychiatry Clin Neurosci 7:334–337

Ohkura T, Isse K, Akazawa K, Hamamoto M, Yaoi Y, Hagino N 1995 Long-term estrogen replacement therapy in female patients with dementia of the Alzheimer type: 7 case reports. Dementia 6:99–107

Pfaff DW 1999 Drive: neural and molecular mechanisms for sexual motivation. MIT Press, Cambridge, MA

Schneider LS, Farlow MR, Henderson VW, Pogoda JM 1996 Effects of estrogen replacement therapy on response to tacrine in patients with Alzheimer's disease. Neurology 46:1580–1584

Index of contributors

Non-participating co-authors are indicated by asterisks. Entries in bold indicate papers; other entries refer to discussion contributions.

A

*Agard, D. **20**

B

Baulieu, E.-E. 15, 30, 38, 51, 53, 88, 131, 170, 197, 198, 235, 269, 270
Becker, J. B. 53, 88, 89, **134**, 146, 147, 148, 149, 150, 151, 152, 185, 200, 217, 272
Behl, C. 52, 54, 131, 217, **221**, 234, 235, 236, 237, 238
Bethea, C. L. 17, 18, 30, 31, 37, 39, 72, 91, 92, **112**, 130, 131, 132, 133, 219, 237, 269, 270, 271
*Bicknell, R. J. **74**

F

*Faber, D. S. **155**
*Feng, W.-J. **20**
Fillit, H. 19, 35, 90, 91, 183, 237, 266, 268, 269
*Frye, C. A. **155**

G

Gandy, S. E. 39, 54, 88, **239**, 251, 252, 253, 267, 271
Gibbs, R. B. 18, 35, 39, 54, 87, 89, 91, **94**, 107, 108, 109, 110, 150, 153, 171, 183, 197, 198, 199, 200, 216, 236, 252, 268
*Green, P. S. **202**
*Greene, G. **20**
*Gundlah, C. **112**
Gustafsson, J.-Å. **7**, 15, 16, 17, 18, 19, 110, 130, 132, 146, 148, 149, 154, 169, 184, 186, 196, 214, 218, 234, 238, 269, 272

H

*Heck, S. **221**
Henderson, V. W. 30, 39, 87, 92, 149, 199, 200, 216, 237, 252, **254**, 266, 268, 269, 270
Herbison, A. E. 16, 29, **74**, 85, 86, 87, 88, 89, 90, 91, 149, 153, 169, 170, 185
*Hsu, F.-C. **155**
Hurd, Y. 86, 93, 132, 150, 196, 218

K

Kirschbaum, C. 17, 87
Kushner, P. J. **20**, 27, 28, 29, 30, 31, 33, 34, 35, 72, 132, 153, 218, 220, 270

L

Levin, E. R. 28, 35, 36, 38, **41**, 51, 52, 53, 54, 55, 70, 85, 90, 133, 146, 147, 186, 196, 214, 217, 220, 251, 265, 270, 271, 272
*Li, X. **155**
*Lopez, G. **20**
Luine, V. N. 27, 28, 29, 37, 39, 85, 88, 108, 109, 153, 154, 199, 270

M

McEwen, B. **1**, 17, 18, 27, 33, 35, 37, 38, 50, 55, 69, 70, 72, 85, 87, 107, 130, 131, 132, 145, 147, 152, 154, 196, 200, 201, 214, 268
*Manthey, D. **221**
*Markowitz, R. S. **155**
*Mirkes, S. J. **112**
*Moosmann, B. **221**
Murphy, D. D. 38, 52, 169, 171, 172, 181, 182, 186, 200, 252

274

Subject index